GARDENS AND THE PASSION FOR THE INFINITE

ANALECTA HUSSERLIANA

THE YEARBOOK OF PHENOMENOLOGICAL RESEARCH

VOLUME LXXVIII

For sequel volumes see the end of this volume.

GARDENS AND THE PASSION FOR THE INFINITE

Edited by

ANNA-TERESA TYMIENIECKA

The World Phenomenology Institute

Published under the auspices of
The World Institute for Advanced Phenomenological Research and Learning
A.-T. Tymieniecka, President

KLUWER ACADEMIC PUBLISHERS
DORDRECHT / BOSTON / LONDON

Library of Congress Cataloging-in-Publication Data

International Society for Phenomenology, Fine Arts, and Aesthetics. Conference (6th: 2002? : Harvard Divinity School)
Garden and the passion for the infinite / edited by Anna-Teresa Tymieniecka.
p. cm. – (Analecta Husserliana; v. 78)
Includes index.
ISBN 1-4020-0858-9 (alk. paper)
1. Aesthetics – Congresses. 2. Phenomenology – Congresses. 3. Gardens – Congresses. I. Tymieniecka, Anna-Teresa. II. Title. III. Series.

2002030115

ISBN 1-4020-0858-9

Published by Kluwer Academic Publishers,
P.O. Box 17, 3300 AA Dordrecht, The Netherlands.

Sold and distributed in North, Central and South America
by Kluwer Academic Publishers,
101 Philip Drive, Norwell, MA 02061, U.S.A.

In all other countries, sold and distributed
by Kluwer Academic Publishers,
P.O. Box 322, 3300 AH Dordrecht, The Netherlands.

Printed on acid-free paper

TABLE OF CONTENTS

SECTION III

SECTION IV

SECTION V

ACKNOWLEDGEMENTS

The present collection gathers the studies presented at the Sixth Annual Conference of our International Society for Phenomenology, Fine Arts and Aesthetics which took place at the Harvard Divinity School in Cambridge, Massachusetts on May 12–14, 2000.

I am bringing it to the world forum of our readers with a particular satisfaction because the treatment of the theme shows the wide breadth of our work: indeed, by going into the depth of the human significance of the garden, we see how in all major cultures it comes prominently from the innermost springs of the Human Condition within the ontopoiesis of life.

Our thanks go first of all to the authors, then to our devoted copy-editor and proofreader Robert Wise.

A-T.T.

THEME

GARDENS AND THE PASSION FOR THE INFINITE

A beautiful field with straight furrows methodically seeded, well watered, all green with tender leaf – what a lovely sight that is, one with the promise of a good, needed harvest. A lovely picturesque meadow, a pasture graced with grazing sheep, but also the habitat of rabbits, woodchucks, moles, birds of the air, etc. A thick forest, whose stout and tall trees provide wood for building homes, fences, bridges, tools, whose thick foliage is the abode of innumerable larger and smaller living beings. The human owner of these parcels, proud of his/her dominion over this corner of earth, dreams of extending the bounds of that dominion to further ranges. This is a desire to expand frontiers.

One may venture to say that while out in the fields and meadows, one rides the surface of our human engagement with Mother Earth. In gardening, on the contrary, one enters, through personal care, into the growth of plants, into deepest communion with Mother Earth, into participation in the networks of nature-life.

When we cultivate a garden, our innermost personal engagement enters into play. The interest of the gardener goes beyond the produce that the plot yields, beyond increase in quantity and size, in soil improvement, and better technique, all that leads to a better harvest.

And what about gardens? Be they vegetable, herb, or green gardens, pleasure gardens, ornamental formal gardens, or romantic informal gardens, all play symbolic roles. Gardens figure centrally in some of the oldest myths of humanity. Paradise expresses the deepest yearning of the human heart for the harmony of everything alive, for a perfect human condition. An ideal state lost in immemorial time and never retrieved is symbolized by the garden. In contrast, as a symbol of human accomplishment, of the glory of the consumate monarch and of the highest cultural acheivement there stand the gardens of Versailles, an enduring reflection of *le Roi-Soleil.* Chinese gardens are meant to be miniatures of the world. Getty's park is a mysterious labyrinth of meanings. And the garden that I cultivate around my house expresses the layout of my soul with its temperament, aesthetic taste, ingrownness with nature-life. Common and unassuming, it the "essential garden."

These larger or smaller or altogether small spaces, perhaps encircled by a fence meant to protect from herbivores, dogs, and trespassers, are devoted to

A.-T. Tymieniecka (ed.), Analecta Husserliana LXXVIII, 1–4.

the cultivation of plants. A garden assumes the tiniest dimensions on a windowsill. A garden is a space for the cultivation of specially selected plants meant for personal use – vegetables, greens, herbs, flowers, ornamental shrubs, all cultivated for the pleasure of their color, shape, fragrance, grace, taste, for direct enjoyment.

As our human cultures manifest, there is a particular significance in the cultivation of a garden that exceeds by far any practical life-sustaining benefit. The sensibilities, emotions active in the gardener's involvement with his or her piece of land run through the entire range of the human soul. Emerging from its subliminal, specifically human sphere, these sensibilities soar on the wings of Imaginatio Creatrix above attachment to possessions, enterprises, ambitions, power ... and express the full gamut of aesthetic, moral, and intellectual enjoyment of our beingness, extending out toward the furthest frontiers of human longing.

To understand this privileged life significance of the garden, we have to grasp it as being inscribed within the web of human life, as having the phenomenology of life as its context.

In short, the gardener's and viewer's immersion in a garden extends through all the rays of logical expansion in the ontopoietic unfolding of the individual human person. A garden stirs one's entire beingness. Our whole vital and creative system, as it were, feeds on the garden and draws out from it the entire spectrum of rays of significance. A garden may be seen as a mirror of the affectivity of the cultivator and of that of the viewer as well. Gardens appear to be a quintessential locus for our full experience of nature-life. A garden serves as a mirror of our inner existence; a garden is seen as a miniature of the world; it functions as a symbolic expression of our highest longings for order in beauty. We see in a garden an "Eden," an ideal of harmony, beauty, sublimity, innocence, peace, and fulfillment, a prototype of "Paradise."

A garden partakes in all the "passions of the earth" (see "Passions of the Earth in Human Existence, Creativity, and Literature," *Analecta Husserliana* LXXI). As living beings, we cannot refrain from at least tacitly partaking in these passions. They measure our very beingness. Through the cultivation of a garden the human being digs deeper and deeper *ad infinitum* into the life sedimentations of this participation, descending toward the vital-organic womb of life itself. In doing so, our living beingness reenacts this participation through our entire imaginative system, assuming it intuitively. Yet simultaneously it draws on all the significant rays of the specifically human preoccupations of Imaginatio Creatrix, which suffuses the human sphere of life with the highest ideals of mankind. In short, the range of significant human sensibilities provides the inspiration to descend from the highest elevation into the entire logoic spread.

On the one hand, the experience of the gardener cultivating the life of plants from germination through unfolding and growth is already preceded by an aesthetic project and a vision of the garden's flourishing, and the harvest encompasses all the human passional strivings: vital, subliminal, communicative. This follows in the imaginative reworking and appropriation through experience of all the ontopoietic phases of life's unfolding. Yet in following their temporal lines, this aesthetically transformed passion is not bound to their temporal succession; it moves freely, albeit in a synchronizing aesthetic that oscillates between the never graspable promise of the origins at one end and the fulgurating rays of fulfillments at the other.

By provoking these two passional strivings of the human being, the garden brings together the originary, the elementary, and that which recedes from sight into the infinite, elevating the human spirit.

Does not this passional oscillation between the termini of the Human Condition, which embrace the entire spread of the ontopoietic unfolding of human beingness – in particular, in the sphere of aesthetic experiences – fall in line with the innermost device of the Logos of Life, which moves so constantly between these poles?

On the other hand, and on reverse course, do we not in our immersion in a garden's life *nolens volens* follow the flow and duration of its progress? Do we not merely postpone, and only for short intervals, the wild onward course of changing life? For a period we bring nature-life to a stanstill. Indeed, the garden that is planned in our imagination, that is shaped at our own hands, and which reposes in a seemingly finished state is from instant to instant undergoing an imperceptible transformation, one that merges in its duration with our own inner flux. We immerse ourselves in this duration, expanding into an infinite duration.

Dwelling in that duration we achieve an infinite repose. Does not the midsummer blossoming seem to last forever?

This compressing of the successive stages of becoming into an imaginative illusion of endurance is particularly exemplified in symbolic commemorative forms of the garden. In some gardens we seek to retain past glories of humankind, great achievements and ideas, longings and dreams, our human passion for perdurance. We attempt through them to stop the flow of change, to retain elusive being, the shapes that this ceaseless flux once took, by immortalizing them in a faint image rescued from oblivion and brought into the palpitating present course of life in defiance of the inexorable passage of time.

The pulls and tensions of spontaneities, forces, dynamisms struggling onwards in the growth-and-decay progress of the living being that is human, on being worked upon in imaginative interpretation, are reappropriated in the human being's most intimate participation in nature-life and lose their present urgency, so that we may repose in nature, have confidence in its logoic rules, and hope for continuing fulfillment in the harvest to come.

Could we not say then that the aesthetic passion carrying the project of the garden lies at the crossing of two innermost passional drives of human life: the drive toward infinite renewal and a drive for infinite repose?

The Garden – a unique aesthetic synthesis of our passions.

Anna-Teresa Tymieniecka

SECTION I

MARIE-ODILE SWEETSER

MIRRORS OF AFFECTIVITY AND AESTHETICS: GARDENS, PARKS, AND LANDSCAPES AS SEEN BY THEOPHILE DE VIAU AND LA FONTAINE

Thanks to the pioneering scholarly work done in the past decades, the baroque flowering in French literature and arts has been recognized as a major event and an important contribution to the aesthetic landscape of early modern culture. In the recently published issue of *Littératures classiques*, "Le Baroque en question(s)," the editor Didier Souiller states with good reason that the relationship between the baroque and classicism has changed . . . baroque aesthetics were still in full bloom during the reign of the Sun King. He adds that today we can better appreciate the works of Molière and La Fontaine, "le plus grand poète du siècle, qui n'est plus enfermé dans l'image réductrice d'auteur des seules *Fables*."[1]

The purpose of this essay is to explore this continuity seen in selected works of two great poets, one productive in the early seventeenth century, the other in the second half of that century, and to show what they had in common in their basic tastes and values, beyond the obvious differences in their personalities. It appears that a shared taste for freedom of expression, refusing total submission to rules and strict generic distinction, finding ways to allow their inner self and emotions to come through, as well as their aesthetic views, were common goals for these poets belonging to different generations, in spite of the conventional labels of baroque applied to the former and classical to the latter.

These important traits: the passionate interest for natural beauty perceived in settings such as country landscapes, parks and gardens, and the power of emotions and passions felt in such privileged locations, will become apparent in the works considered.

Two of Théophile de Viau's early works illustrate his taste and stated preference for the beauty of natural settings and freedom of expression, asserted in no uncertain terms:

> Malherbe a très bien fait, mais il a fait pour lui
> . . . . .
> La règle me déplaît; j'écris confusément:
> Jamais un bon esprit ne fait rien qu'aisément.[2]

The early ode *Le Matin*, in spite of some conventional traits such as allusions to mythological deities, to a lion leaving his den, an incongruous sight in a

A.-T. Tymieniecka (ed.), Analecta Husserliana LXXVIII, 7–24.

French landscape,[3] is a sumptuous hymn to the awakening of life at dawn. It is a painterly rendition of a country landscape at sunrise, artfully composed, starting with the brilliant colors of the sky and progressing downwards:

L'Aurore sur le front du jour
Sème l'azur, l'or et l'ivoire
Et le Soleil, lassé de boire
Commence son oblique tour.

(*Le Matin, O. C.* I, XI, 158, 1–4)

The poet then evokes the hills, alive with the activities of bees gathering nectar from aromatic plants that will be transformed into honey, and of lambs frisking about on the green hills; this is obviously a spring landscape, animated by the sounds of birdsongs, a hymn of praise to the morning light. Then the poet turns to human activities: the farmer driving his oxen in his field, the young woman working at her spinning wheel. The whole scene vibrates with manifestations and tensions of life:

Une confuse violence
Trouble le calme de la nuit;
Et la lumière avec le bruit,
Dissipe l'ombre et le silence.

(*O. C.* I, 159, 41–44)

Life triumphs over repose, light over darkness, sounds over silence: the world is awake, the cycle of daily activities resumes in full. A pantheistic view of the world is present.

The sudden change taking place in the last stanza has puzzled commentators.[4] From the macrocosm of life, the poet turns to the microcosm of his personal life:[5] he addresses his lady-love and invites her to enjoy with him the beauty and the freshness of the morning in the intimate setting of their garden, ending with a conventional homage to her beauty. Her complexion he compares to the color of the garden flowers, lilies and roses:

Il est jour, levons-nous, Phyllis!
Allons à notre jardinage,
Voir s'il est, comme ton visage,
Semé de roses et de lis.

(*O. C.* I, 160, 59–64)

One may wonder whether the poet's shift indicates a preference for the quiet moment of a morning stroll in the microcosm of a garden, shared with the woman he loves, instead of the world view he has so powerfully evoked in the major portion of the poem. The personal is obliterating the cosmic, the ending stressing the togetherness, with happiness found in the intimacy of a quiet moment opposed to the bustle of activities in an open and populated landscape. The morning stroll in the flower garden will prolong the delights of the night and seal the harmonious complicity of the lovers in their enjoyment of the morning beauty.

It is again as an ode celebrating his lady-love, addressed under the conventional poetic name of "Corinne," that *La Solitude* can be read. The framework is again a natural setting, the traditional *locus amoenus* inherited from Greek and Latin bucolic, pastoral tradition, rediscovered by Medieval and Renaissance poets and writers, authors of romances and pastoral plays.[6] Again the presence of traditional mythological creatures, divinities of the forests and waters – nymphs, dryads, naiades – is commonplace. Visual and aural sensations are presented as pleasurable and contributing to the enjoyment of the lovers, secluded from the rest of the world, in a state of perfect intimacy:

Dans ce val solitaire et sombre,
Le cerf, qui brame au bruit de l'eau
Penchant ses yeux dans un ruisseau,
S'amuse à regarder son ombre.

(*La Solitude, O. C.* I, xiii, 160, 1–4)

The stag's voice and the pleasant sound of running water are suggested by the alliterations that Théophile uses. The stag is personified; like Narcissus, he enjoys looking at his reflexion. Recent commentators have stressed, in the first part of the poem, allusions to tragic love stories: Philomèle, Hyacinthe, victims of violent acts of passion.[7] However the poet also presents the forest as a place of safety, an asylum, a haven where Cupid and Venus are present:

Ici l'Amour fait ses études;
Vénus y dresse des autels,
Et les visites des mortels
Ne troublent point ces solitudes.

(*O. C.* I, 161, 29–32)

The first fifteen stanzas are dedicated to the description of the forest, the *locus amoenus*, shady, cool and quiet, where victims of violent passion have found a refuge, but also a desirable retreat for a couple to consummate their love. The poet, in the second part, addresses his beloved, Corinne, inviting her to lie down with him on a patch of moss, or better, to enter an even more remote rocky hollow:

Corinne, je te prie, approche;
Couchons-nous sur ce tapis vert,
Et pour être mieux à couvert,
Entrons au creux de cette roche.

(*O. C.* I, 162, 61–64)

The lover is obviously attempting to convince the young woman of his admiration for her charms and of his love, and to entice her to respond to it. She appears to be in the preliminary stage before yielding to the entreaties of her lover for a tender embrace in this intimate setting, a natural expression of love in a natural beautiful spot. Corinne's initial shyness is quite attractive and allows her wooer to reassure her:

Ah! je vois que tu m'aimes bien:
Tu rougis quand je te regarde.
Dieux! que cette façon timide
Est puissante sur mes esprits!

Ma Corinne, que je t'embrasse!
Personne ne nous voit qu'Amour.

(*O. C.* I, 165, 151–158)

The expected outcome of such an amorous encounter in a discreet and safe setting is alluded to, without expressing any doubt on the role taken by the lady:

Les vents qui ne se peuvent taire,
Ne peuvent écouter aussi,
Et ce que nous ferons ici
Leur est un inconnu mystère.

(*O. C.* I, 165, 161–164)

The couple will celebrate love's mysteries in a communion together. Théophile had presented in *Le Matin* the postlude of a night of love, communion in a garden. Here we have the prelude to an afternoon of love in a protected natural setting. The poem can be seen as an epithalamion[8] expressing in fervent terms the expectation of a lover ready to share with an attractive partner the joys of natural pleasures. As Guido Saba has pointed out in his recent book, two poles of inspiration are central in Théophile's poetic work: sensual love and nature. The aesthetic form of the poem is consonant with this vision. The subtle fusion of painting and music, a series of refined impressionist touches, coexists in the perfection of each stanza.[9]

Although *La Maison de Sylvie* belongs to a later period in his career and represents the more mature person and artist who has known many trials and tribulations, the same basic tenets and qualities of expression and style can be observed. In this extraordinary suite of ten odes, Théophile celebrates his patrons, the Duc and Duchesse de Montmorency and their splendid country estate, Chantilly,[10] which was a haven and a temporary asylum for the persecuted poet, hounded by the forces of Justice. Thus Chantilly, the château and especially the park are for him

Les lieux les plus révérés
Où la vertu se réfugie,
Et dont le port me fut ouvert
Pour mettre ma tête à couvert
Quand on brûla mon effigie.
(*La Maison de Sylvie,* Ode I, *O. C.* II, 201, 6–10)

The poet's admiration and devotion are especially directed to the Duchesse, Marie-Félice des Ursins, from the famous Italian Orsini family, celebrated under the poetic name of Sylvie, well chosen since the lady is represented as thoroughly attuned to the natural beauty of the park and fully deserving the homage of the poet. Her kindness and virtue are an image of God:

Après lui je m'en vais louer
Une image de Dieu si belle
Que le Ciel me doit avouer
Du travail que je fais pour elle.
(Ode I, *O. C.* II, 202, 51–54)

If Théophile's poems do not carry her name to immortality,

Ces eaux, ces rochers et ces bois
Prendront des âmes et des voix
Pour en conserver la mémoire. . . .
(Ode I, *O. C.* II, 203, 68–70)

Sylvie has supernatural powers, she can bestow life and speech on inanimate objects. She is therefore given the rank of a divinity, able to tame the waters:

Je sais que ces miroirs flottants
Où l'objet change tant de place,
Pour elle devenus constants
Auront une fidèle glace. . . .
(Ode I, *O. C.* II, 204, 101–104)

She participates in the beauty of the cosmos and embodies heavenly virtues:

Mais sa gloire n'a pas besoin
Que mon seul ouvrage en réponde:
Le ciel a déjà pris le soin
De la peindre par tout le monde.

(204, 111–114)

Ode II paints a lovely portrait of the Duchesse in a pastoral scene; she is fishing at sunset in the park of Chantilly. This corresponds to a "scène de genre" by an impressionist. Again, she is endowed with supernatural gifts: the fish are eager to be caught by her. The sun, the stars, the breeze and the grass stop their natural course in order not to disturb her sport and let her enjoy the last moments of daylight. The mythological water creatures, the Tritons, are metamorphosed into white deer. There is a painterly contrast, a chiaroscuro between the white complexion of Sylvie, symbol of her pure, angelic essence, and her dark eyes, as well as a natural contrast between the white fur of the deer, the snow-covered ground and the darkened alleys of the park:

Et dans ses vêtements modestes
Où le front de Sylvie est peint
Fait briller l'éclat de son teint
A l'envi des neiges célestes. . . .

(Ode II, *O. C.*, 207, 95–100)

The kindness of Chantilly's lord and lady extends to the care and protection of the wild creatures roaming in the park during the winter, as they provide them with food and shelter and protect them from "des outrages de la Nature."

Again, in Ode III, the park is described as a *locus amoenus*, combining the freshness and cooling shade provided by trees and running brooks,[11] so that even unhappy lovers find some relief there. Théophile, as he did in *La Solitude*, evokes doomed legendary characters. Thus the swans of Chantilly recall the legend of Cygnos mourning the death of Phaeton, his relative and friend metamorphosed by Jupiter into a swan. This brings the poet to his own sad memories in Ode IV, of a close relationship with his friend Tircis who was impugned by his enemies. Thus the view of the swan brings up a painful memory and reveals an important aspect of Théophile's affectivity, as several critics have pointed out.[12]

It is again the natural beauty of the park and the fecundity of Nature that Théophile celebrates in Ode VII, reflected in the harmonious union of the Duc and Duchesse:

Elle a mis toute sa bonté
Et son savoir et sa richesse
Et les trésors de sa beauté
Sur le Duc et sur la Duchesse;
Elle a fait les heureux accords
Qui joignent leur âme et leur corps.

(Ode VII, *O. C.* II, 224, 31–36)

In Ode VIII, the poet evokes another tragic legend, recorded in Ovid's *Metamorphoses*, linked to the song of the nightingale, heard in Chantilly's park: Philomela, ravished by her brother-in-law Tereus, was changed into a nightingale. The story of her misfortune brings to the poet's mind a comparison with his own:[13] only the knowledge of his patron's kindness can soothe the horror of the prison, and in the darkness, he sees Chantilly:

Et mon oeil, qui suit mon désir,
Voit Chantilly dans ces ténèbres.

(Ode VIII, *O. C.* II, 229, 99–100)

In contrast with the dispiriting darkness of his cell, memories of the edenic park and its charming lady come alive and provide pleasurable visions and sensations to the unfortunate prisoner:

Mes sens en ont tout le tableau:
Je sens les fleurs au bord de l'eau,
Je prends le frais qui les humecte,
La Princesse s'y vient asseoir,
Je vois, comme elle y va le soir,
Que le jour fuit et la respecte.

(Ode VIII, *O. C.* II, 230, 105–110)

In the following ode, the gorgeous picture of Dawn and the evocation of the goddesses Juno, Minerva, Venus, and Diana who reigns over the forests, provide Théophile with the opportunity for a special homage to Sylvie who inspired him. He places her above all those deities, since by looking at her poet in a kind way, she could stir him to write his best poetry:

Un de ses regards seulement,
Qui partent si nonchalamment,
Donne à mes chansons tant d'amorce
Et de si douces vanités,
Que les autres divinités
N'en jouissent plus que de force.

(Ode IX, *O. C.* II, 231, 35–40)

All the bucolic charms of Chantilly appear again: the soft breezes, the fountains and ponds, the birds' songs.

The poet precisely addresses the nightingale in the last ode, but every aspect of Chantilly deserves his praise because they are all linked to Sylvie:

> Une goutte d'eau, une fleur,
> Chaque feuille et chaque couleur
> Dont Nature a marqué ces marbres,
> Mérite tout un livre à part,
> Aussi bien que chaque regard
> Dont Sylvie a touché ces arbres.
>
> (Ode X, *O. C.* II, 233–234, 15–20)

It appears clearly from the text that Théophile's feelings toward his patrons, his gratitude for their protection and gracious hospitality, and the poetic inspiration received from Sylvie are linked to the many natural beauties of Chantilly which he remembers so keenly and are part and parcel of his own affectivity. They are also an expression of his impressionistic aesthetics, with constant shifts from one aspect of the picture to another, with juxtaposition of painterly touches, with, however, the recurrence of the central character, the portrait of Sylvie, in various guises,[14] as a nymph presiding over the waters, a fairy endowed with magical powers, above all a Muse.

Are we so far from La Fontaine, who was also claiming freedom of expression and the superiority of a "natural" style?[15] These traits will appear especially in the works dealing with park settings. La Fontaine, like Théophile, intends to celebrate generous and hospitable patrons in *Le Songe de Vaux*, with a special regard for Madame Fouquet who appears in his work under the poetic name of Sylvie. Both women are depicted in the beautiful surroundings of a château and its park. Nevertheless, the similarity of the name chosen by both poets is striking. J-P. Collinet remarks about it in his perceptive study, where he shows the affinities between the two poets.[16] Although La Fontaine did not mention Théophile by name, the author assumes that he knew the poetic works of his predecessor, still often printed at the time he wrote the fragments that remain of *Le Songe de Vaux*. The implicit attitude of both poets towards their patron's wife and the lady of his domain is similar, befitting their status in the Chantilly or Vaux circle: admiration, gratitude, tender if respectful feelings. Vaux, like Chantilly, embodies the beauty and charm of its mistress who is also the poet's Muse, Sylvie.

The temporal setting of *Le Songe de Vaux* is a spring night when Acante, the narrator, falls asleep and asks the God of Sleep to show him the palace and gardens of Vaux:

> Il me fit voir en songe un palais magnifique,
> Des grottes, des canaux, un superbe portique,
> Des lieux que pour leurs beautés
> J'aurais pu croire enchantés.[17]

Acante in his dream sees a wealth of beautiful objects belonging to the various elements used in the creation of the domain of Vaux. However, this most pleasurable dream is interrupted by the apparition of Aminte, the woman he pursues in vain. She will become the central figure in an episode of the work:

> Et je dormirais encore,
> Sans Aminte et ses appas.
> Cette fière beauté, qui s'érige un trophée
> Du cruel souvenir de mes voeux impuissants,
> Souffrit que cette nuit les charmes de Morphée
> Aussi bien que les siens régnassent sur mes sens.
>
> (*O. D.*, p. 82)

Two major feminine figures intimately connected to the narrator's inner feelings and emotions as well as to the setting of the park, are therefore present in the text. Because of these relationships, *Le Songe de Vaux* is far from simply being a descriptive and encomiastic piece of writing, commissioned by a generous patron; the affectivity and aesthetics of its creator are embodied in it.

The following fragment deals with the contest between four fairies, representing four arts: Architecture, Painting, Landscape Architecture and Poetry. Thus, we become acquainted with the narrator's aesthetic reflexions.[18] Is there hierarchy among the arts, what are the major virtues of each of them? These questions have preoccupied critics in a long debate over the centuries. The first fairy to speak, Palatiane, acknowledges the pleasures provided by the other contestants. As for gardens, she does admit their charm, but does not consider them essential since they do not afford protection and comfort to human beings:

> Errer dans un jardin, s'égarer dans un bois,
> Se coucher sur des fleurs, respirer leur haleine,
> Ecouter en rêvant le bruit d'une fontaine,
> Ou celui d'un ruisseau roulant sur des cailloux,

Tout cela, je l'avoue, a des charmes bien doux;
Mais enfin on s'en passe, et je suis nécessaire. . . .

(Le Songe de Vaux II, *O. D.*, p. 86*)*

In reply to this utilitarian statement, Apellanire, the Fairy of Painting, takes up the ongoing debate between *utile* and *dulce*:

Car il ne s'agit pas d'être la plus utile;
C'est assez de causer le plaisir seulement. . . .

(O. D., p. 87*)*

Hortésie can certainly pursue this point and bring fitting arguments to make a vibrant apology for gardens and parks.[19] She celebrates the shade given by trees, the beauty of flowers, the peaceful, quiet atmosphere of parks offering a retreat from urban and even state cares:

L'air est plus pur, le jour plus beau;
Les nuits sont douces et tranquilles;
Et ces agréables séjours
Chassent le soin, hôte des villes,
Et la crainte, hôtesse des Cours.

(O. D., p. 89*)*

Les vergers, les parcs, les jardins
De mon savoir et de mes mains
Tiennent leurs grâces non pareilles. . . .

The art of horticulture provides not only pleasure but useful production in orchards and vegetable gardens. Both elaborate, decorative parks and useful orchards and country gardens have their charms. One remembers that La Fontaine in one of his fables celebrates the charm of a humble garden where the farmer-owner of the property cultivates both vegetables and flowers, "Le Jardinier et son Seigneur" (*Fables*, IV, 4). From an aesthetic point of view, the poet had declared "la grâce plus belle encore que la beauté" (*Adonis, O. D.*, p. 6): so for La Fontaine-Acante, parks and gardens combine all three qualities: beauty, usefulness, and supreme above all others, gracefulness. This is reasserted about the presence of waters: their ornamental, pictorial and musical value in formally designed parks, such as Vaux and Versailles, later conceived by the same master gardener, André Le Nôtre.

Je donne au liquide cristal
Plus de cent formes différentes,
Et le mets tantôt en canal,
Tantôt en beautés jaillissantes;
. . . . .

Parfois il dort, parfois il coule,
Et toujours il charme les yeux. . . .

(*O. D.*, p. 91)

The water displays of Vaux were singled out not only for their aesthetic value, but probably also to celebrate the master of the domain, Fouquet, who had a definite interest in maritime projects. La Fontaine wrote a fanciful episode, "Aventure d'un saumon et d'un esturgeon," the fishes being presented as ambassadors to the Finance Minister, sent by Neptune himself,[20] but ending with a "compliment galant" to Sylvie: the ambassadors from the Sea God do not wish to leave Vaux, because they are happy near its lovely lady, "l'adorable Sylvie."

It is she who is the central figure in the next episode. "Comme Sylvie honora de sa présence les dernières chansons d'un cygne qui se mourait," she is shown as an intellectually curious person who wants to test the validity of an old legend, that of the dying swan said to sing his most beautiful song before he dies. Sylvie-Madame Fouquet has sent for the famous musician Lambert. With her guests, she is going to compare the song of the dying swan to the musician's. The swan's song is judged far inferior to Lambert's (*O. D.*, p. 100), thus showing that art improves on nature, a view quite consonant with the aesthetics of classicism, which, according to Malherbe, "Aux miracles de l'Art fait céder la Nature."

This episode also constitutes a delicate and humorous homage to Sylvie's intelligence and wit, and shows the admiration of the poet for the gracious hostess of Vaux who shares the new scientific spirit of the age and tries her hand at experimentation.

Contrasting with the elaborate design of the park, the adjacent woods of Mainsy offer a spot of simple, natural bucolic beauty. In his dream, Acante finds himself in a lovely *locus amoenus* at dawn, depicted in brilliant colors. There he discovers his lady-love, Aminte, asleep on a carpet of violets in a relaxed pose, with her clothes in suggestive disarray, the breeze having partly unveiled her bosom. Acante's eroticism is clearly the focus of the picture; sensual love and natural beauty are united, as they were in Théophile's poetry. The contemplation of the charms of the sleeping beauty bring up passionate emotions:[21]

Je les [ses charmes] considérai longtemps avec des transports qui ne peuvent s'imaginer que par ceux qui aiment. Encore est-ce peu de dire transports car, si ce n'était véritable enchantement, c'était au moins quelque chose qui en avait l'apparence: il semblait que mon âme fût accourue toute entière dans mes yeux.

(*O. D.*, p. 110)

The sense of sight and its power on the emotions is stressed. It brings to Acante the desire to have another sense participate in his delight, and to kiss Aminte:

> Ces lèvres où les Cieux ont mis tant de merveilles,
> Auraient pu m'excuser;
> Et tout autre que moi, les voyant si vermeilles,
> Eût voulu les baiser.
>
> (*O. D.*, p. 110)

The lady is cast in the role of the "belle insensible" and the lover is thwarted in his attempt to woo her. He asks what brings her to Vaux: that allows him to contrast art and nature in reverse to the previous position implied in the superiority of the musician Lambert to the dying swan: The park of Vaux, to be sure, is a work of art:

> L'art y brille partout; cependant la nature
> Est plus belle en vos yeux.
>
> (*O. D.*, p. 111)

Aminte refuses to use art and therefore will not compete with the fairies representing various arts.

The *Songe de Vaux* does reflect the affectivity of the narrator, and fragment vii depicts his erotic self in his encounter with Aminte to whom the printed version of *Adonis* was dedicated. Marlies Kronegger and Jean-Pierre Collinet have perceptively shown the shift from the surroundings, the frame of the park, garden and woods, to the dream landscape and the dreamer, Acante, transforming the work into a subjective, poetic and oniric vision.[22] The importance of Aminte is quite apparent in the second, short preface to *Le Songe de Vaux* where he discusses the possible interpretation, whether allegorical or personal, of the characters. The author would prefer that Aminte be seen as a specific individual, to make the episode more passionately felt (*O. D.*, p. 81). It also reflects his aesthetic views that art can and does in some cases surpass nature. Vaux celebrates the triumph of art over raw nature, while the musician Lambert, with his artistic training, produces more beautiful sounds than the dying swan. On the other hand, the unspoiled, natural beauty of Aminte is more pleasing, more touching to Acante than if it were contrived, manipulated by artifice.

Les Amours de Psyché et de Cupidon brings the reader back to dream gardens, those built by the God of Love, as well as those of Versailles, built by a young and spirited king. In his stimulating "Introduction" to his edition of the text, Michel Jeanneret stresses the propitious circumstances in 1669, the baroque energy, vitality and creative force that prevailed during the first half of the century and were still in full swing.[23]

La Fontaine obviously hoped to become part of the magic circle of artists catering to "la jeune Cour" with *Psyché*, but was disappointed by the lack of success of this original work. Graceful park structure, ornamental pools, grottoes, and statues of mythological characters were placed on Versailles' walkways, representing variations on the theme of metamorphosis. They are present in the text of *Psyché*, where Louis XIV plays the part of the Sun God, Apollo. From the actual park to a dream world, through the mediation of a poetic prose mixed with verse, the transition is hardly noticeable.

The gardens of Versailles constitute the framework in which the reading of the tale takes place, a *locus amoenus* in which not only the reading but lively and friendly literary discussions happen in the most attractive surroundings, replicating in a clever "mise en abyme," the beauty of Cupidon's palace and gardens.[24] One of the four friends, Acante, "who loved above all gardens, flowers, the shade of trees," suggests an outing in the country, for the reading of the tale composed by Poliphile, far from the city, where they would not be disturbed. The site of Versailles is proposed by Ariste: recent improvements have made it a splendid location and the fall season is very pleasant, with fairly long days (*O. D.*, pp. 127–128). The gardens of Versailles will provide a suitable spot for an exchange of aesthetic views, in the time-honored tradition of Platonic dialogue. The first part of the reading takes place in the grotto of Thetys. This gives the poet the opportunity to link the myth of the Sun spending the night and resting from its daily course in the realm of the Sea Goddess to Louis XIV the Sun King, devoting most of his time and energy to affairs of State, and needing the well-deserved relaxation that Versailles provides for a hardworking monarch:

> Quand le Soleil est las, et qu'il a fait sa tâche,
> Il descend chez Thétys, et prend quelque relâche,
> C'est ainsi que Louis s'en va se délasser
> D'un soin que tous les jours il faut recommencer.
>
> (*O. D.*, p. 131)

The palace and gardens of Cupidon provide a fitting parallel to those of Versailles. The narrator celebrates the latter, most notable for their splendid features: fountains, cascades, rare and exotic trees and shrubs. Psyché thoroughly enjoys the beauty of the park, where the art of landscape gardening adds to Nature. But she finally wanders into the woods, and enters a grotto, driven by curiosity and the pleasures of discovery. In the darkness of the cave, she meets her husband who reproaches her for not having come inspired by love:

> Si vous aimiez, vous chercheriez le silence et la solitude. . . . Vous chercheriez les antres sauvages, et auriez bientôt appris que de tous les lieux où on sacrifie au dieu des amants, ceux qui lui plaisent le plus sont ceux où on peut lui sacrifier en secret.
>
> *(O. D.*, p. 150*)*

We find here the same optimal conditions for a love encounter in a natural, hidden place as those evoked in Théophile's *Solitude*: solitude and discretion, a secret shared by the lovers. But Cupidon does not want to share the secret of his identity with his wife who, like a romantic heroine, expresses her suffering to the natural elements (*O. D.*, p. 155).

After having enjoyed the lavish, gloriously allegorical setting of the grotto, Acante proposes a walk in the gardens and a less formal spot to listen to the continuation of the story, a lawn where they will sit on the fine, soft grass. In the course of their walk, they are able to admire the marvellous features of the gardens created by Le Nôtre, especially the water displays. The description of fountains becomes a poem according to J-P. Collinet,[25] celebrating the mystic union of two major natural elements: water and fire, the Sun God Phébus and the Sea Goddess Thétys:

> Phébus sortant de l'onde
> A quitté de Thétys la demeure profonde:
> En rayons infinis l'eau sort de son flambeau;
> On voit presque en vapeur se résoudre cette eau.
>
> *(O. D.*, p. 186*)*

The novelty of such creations is praised. The new aesthetics are in evidence in the gardens of Versailles:

> On ne connaissait point autrefois ces beautés.
> Tous parcs étaient vergers du temps de nos ancêtres,
> Tous vergers sont faits parcs: le savoir de ces maîtres
> Change en jardins royaux ceux de simples bourgeois,
> Comme en jardins de dieux il change ceux des rois.
>
> *(O. D.*, p. 187*)*

The knowledge and expertise of master landscape gardeners means that their art can improve on nature. The creations of Le Nôtre made possible the extraordinary display of the 1668 festivities, "Les Plaisirs de l'Ile enchantée":

> Tout le monde a ouï parler des merveilles de cette fête, des palais devenus jardins, et des jardins devenus palais, de la soudaineté avec laquelle on a créé, s'il faut dire ainsi, ces choses, et qui rendra les enchantements croyables à l'avenir.
>
> *(O. D.*, p. 187*)*.

When Psyché loses everything because of her transgression, she finds some solace in the woods near the dwelling of the kind old man who has given her asylum and assistance. The natural setting of the wilderness inspires her to express her sorrow in elegiac verse (*O. D.*, pp. 204–205). She decides to seek forgiveness by going to Venus. The goddess's temple is located in a spacious, open valley, surrounded by pleasant hills: it is a domain of peace where even animals do not prey on each other, and of love. Birds are numerous, and a wide canal with clear, transparent water, bordered by green meadows shaded by trees, makes it an edenic spot. However, those lovely surroundings are deceptive and a series of trials and tribulations start for the heroine. It is in the private, natural setting of a cave that the reunion and reconciliation take place, recalling their previous meeting described in the first part. The triumph of love, source of all beauty in nature and in artistic creation, is celebrated in the "Hymne à la Volupté":

> Pourquoi sont faits les dons de Flore,
> Le Soleil couchant et l'Aurore,
> Pomone et ses mets délicats,
> Bacchus, l'âme des bons repas,
> Les forêts, les eaux, les prairies,
> Mères des douces rêveries?
> Pourquoi tant de beaux arts, qui tous sont tes enfants?
>
> (*O. D.*, p. 258)

All the pleasures of the senses, those provided by the contemplation of beauty in Nature and in the creations of the fine arts, are invoked as they all contribute to human happiness, including the poet's, who does acknowledge the catholicity of his tastes:

> J'aime le jeu, l'amour, les livres, la musique,
> La ville et la campagne, enfin tout; il n'est rien
> Qui ne me soit souverain bien,
> Jusqu'au sombre plaisir d'un coeur mélancolique.
>
> (*O. D.*, p. 258)

It appears again that the poet entertains two different aesthetic tendencies which may appear contradictory but do not constitute for him a conflict: a taste for unadorned natural beauty as well as for works of art, both of which are admired and cherished. J-P. Collinet has noted those two postulations and sees their coexistence as the essence of his aesthetic.[26]

After some comments made by Poliphile's friends on the emotions generated by the tale in the listeners, Acante invites them to contemplate the magnificent colors of the sunset and the brilliant light of the full moon. There is beauty everywhere, in the framing of the story, the gardens of Versailles, and the various settings of the tale, the palace and gardens of Cupidon, the woods, the cave, and the contemplation of the sky at the conclusion of their journey of discovery, paralleling Psyché's initiation.

From the examination of texts dealing with parks, gardens, and landscapes by two poets writing at different periods in the seventeenth century, the "baroque" and the "classical," common features do emerge. There are individual differences of temperament and style, to be sure, but no break in the continuity of a major tradition. In their evocation, both poets express their affectivity and their aesthetic leanings. Both express a longing for a bucolic, pastoral retreat protected from the outside world where love or friendship can be realized in full intimacy. They also express their fervent admiration for the creations of art, combining various natural elements inspired by the desire to dazzle the human eye. Excellent and thoughtful commentators, especially in recent years, have stressed this continuum in literature and in a wider context, in culture. We are grateful for the results of their worthwhile research.

University of Illinois, Chicago

NOTES

1 Didier Souiller, "Le Baroque en question(s)," *Littératures clasiques*, No. 36, Printemps 1999, 5–13. Quotations, 7. Jean-Pierre Chauveau, "La France et le Baroque," *Ibid.*, 149, note 22: "Les fêtes de Cour, auxquelles contribuent la plupart des écrivains et des artistes, Molière en tête, sont, au temps de Louis XIV, le refuge le plus éclatant d'une certaine esthétique baroque."

2 Théophile de Viau, *Oeuvres complètes*, éd. critique par Guido Saba (Paris: Honoré Champion, 1999), 3 vols. I, *Première partie, Poésies*. All quotations and notes of the editor refer to this edition and will be given in the text like this: *Elégie à une Dame*, 203, 72; 205, 119–120 under *O. C.*, volume, page and line numbers.

3 Ernst Robert Curtius, *European Literature and the Latin Middle Ages*, transl. Willard R. Trask (N. Y.: Pantheon Books, 1953), p. 184, notes the presence of wild animals and plants in works written about northern settings. The poets found them in the tradition coming from Antiquity.

See also *O. C.* I, 319, note 3.

4 See G. Saba's notice listing the various explanations offered about this shift, *O. C.* I, 316–317.

5 Marlies Kronegger entitles chapter II of her book *The Life Significance of French Baroque Poetry* (New York: Peter Lang, 1988), "From Macrocosm to Microcosm," and quotes, "Gardens of Vaux-le-Vicomte and Versailles" that "can be a place for feasts, or meditation, and at all times they are restorative for both body and soul," p. 29.

6 E. R. Curtius, *European Literature, op. cit.*, ch. 10, "The Ideal Landscape," pp. 183 & sq., especially p. 195: "The *locus amoenus* . . . forms the principal motif of all nature descriptions. It is . . . a beautiful, shaded natural site," with trees, a meadow, a spring or brook, birdsong, flowers, a light breeze.

7 James R. Leva, "Latency and Metamorphosis in Théophile de Viau's *La Solitude*," *Actes de Las Vegas*, ed. Marie-France Hilgar, Biblio 17, vol. 60 (Paris, Seattle, Tübingen: PFSCL, 1991), pp. 115–119, followed by Robert T. Corum, Jr., "Response," pp. 121–122.

Robert T. Corum, Jr., "La Quête de l'autre dans la poésie amoureuse de Théophile de Viau," in *Théophile de Viau. Hommage à Guido Saba*, ed. Roger Duchêne, Biblio 17, vol. 65 (Paris, Seattle, Tübingen: PFSCL, 1991), pp. 63–73.

David Michael Roberts, "Théophile's Cygnus and the Vulnerable *Locus Amoenus*," *Actes de Las Vegas, op. cit.*, pp. 123–127.

Robert E. Hill, "In Context: Théophile de Viau's *La Solitude*," *Bibliothèque d'humanisme et Renaissance*, vol. 30, 1968, 499–536. This very detailed and well researched study is of interest to the specialists. However, it seems that, replaced in the framework of Théophile's thought, stressing the freedom to fulfill natural drives, the moral question does not bother the speaker or his partner.

8 John D. Lyons, "De temps à autre: mélancolie et solitude," *XVIIe Siècle*, n. 180, vol. 45, n. 3, Juillet-Sept. 1993, 435–447. See p. 441. The critic interprets this part of the poem as a scene of seduction, the outcome of which is not described because the suggestion is far more alluring than any description. The mention of Valéry's *Les Pas* seems very pertinent.

9 Guido Saba, *Théophile de Viau: Un poète rebelle* (Paris: P. U. F., 1999), p. 77.

10 *Ibid.*, pp. 97–106, on "La Maison de Sylvie." Marlies Kronegger, *The Life Significance of French Baroque Poetry, op. cit.*, pp. 107–110. "The gardens of Chantilly, in Théophile de Viau's 'La Maison de Sylvie,' breathe the life of the cosmos, in which mythological figures emerge," p. 107.

11 The importance of water, fountains, and waterworks in seventeenth-century gardens and parks has been studied by historians of art and architecture and literary critics. See William Howard Adams, *The French Garden 1500–1800* II, "Grottos, Water and Other Garden Fantasies" (New York: George Braziller, 1979), pp. 21–36. Jean Rousset, *L'Intérieur et l'extérieur*, "XVIIe Siècle: Miroirs et eaux dormantes" (Paris: José Corti, 1968), pp. 199–211.

12 David Michael Roberts, "Théophile's Cygnus and the Vulnerable *Locus Amoenus*," *loc. cit.*, pp. 123–127.

Guido Saba, *O. C.* II, 379, notes to pp. 211–212.

13 Jean Marmier, "La Poésie de Théophile de Viau, théâtre du moi," *PFSCL*, N. 9, 1978, 56: "Dans *la Maison de Sylvie*, un savant entrelacement de thèmes mène du parallélisme à l'identification entre le personnage du rossignol et celui du prisonnier qui 'sait chanter.'"

14 Gisèle Mathieu-Castellani, "Théophile de Viau: une poétique du discontinu," in *Théophile de Viau. Hommage à Guido Saba, op. cit.*, pp. 89–100. See especially pp. 90–91.

15 Marie-Odile Sweetser, "La Modernité de La Fontaine," in *Refiguring La Fontaine. Tercentenary Essays*, ed. Anne L. Birberick (Charlottesville: Rookwood Press, 1996), pp. 1–21.

16 Jean-Pierre Collinet, "La Fontaine et Théophile," in *Théophile de Viau. Hommage à Guido Saba, op. cit.*, pp. 151–170, see especially p. 168, showing the interest of La Fontaine in *La Maison de Sylvie* at the time when he was working on *Le Songe de Vaux*.

17 La Fontaine, *Le Songe de Vaux*, in *Oeuvres diverses*, ed. Pierre Clarac, Pléiade (Paris: Gallimard, 1958), p. 82. All references to *Le Songe de Vaux* and *Les Amours de Psyché et de Cupidon* are to this edition and will be indicated in the text, under *O. D.* and page number.

18 Robert N. Nicolich, "The Triumph of Language: The Sister Arts and Creative Activity in La Fontaine's *Songe de Vaux*," *L'Esprit créateur*, Jean de La Fontaine issue, ed. David Lee Rubin, vol. XXI, N. 4, Winter 1981, 10–21.

19 Jean-Marie Pérouse de Montclos, *Vaux-le-Vicomte*, photos by Georges Fessy, preface by Marc Fumaroli (Paris: Editions Scala, 1997). "Les Jardins," p. 97 ff. Jean-Christian Petitfils,

Fouquet (Paris: Perrin, 1998) ch. X, "Les Jardins de Vaux," about the Le Nôtre project, p. 187.

William Roberts, "The Claims of Hortésie in *Le Songe de Vaux*," in *Voyages. Récits et Imaginaire.* Actes de Montréal, éd. Bernard Beugnot, Biblio 17, vol. 11 (Paris, Seattle, Tübingen: PFSCL, 1984), 316–336.

20 Jean-Christian Petitfils, *Fouquet, op. cit.*, ch. XVII, "Le Rêve colonial et les canons de Belle-Ile."

Marc Fumaroli, *Le Poète et le Roi. Jean de La Fontaine en son siècle* (Paris: Ed. de Fallois, 1997), p. 180.

21 M. Fumaroli, *Le Poète et le Roi, op. cit.*, p. 179: "Symbolisé par une jeune Ariane française endormie et révélée aux sens du poète par le Sommeil, Vaux-le-Vicomte n'est pas seulement une "villa" destinée à délasser un ministre, c'est la capitale de la contemplation, du loisir et du repos voluptueux."

22 Marlies Kronegger, "Poetic Inspiration and the Renewal of Life: *Le Songe de Vaux*," *Analecta Husserliana* LX, 1999, 169–200: "The poet La Fontaine had faith in the human potential to create and reach the invisible and unknown, in the imaginative expression of deep-seated human impulses revealed in dreams, myths and allegories," 170. Jean-Pierre Collinet, "Réflexions sur *Le Songe de Vaux*," *Le Fablier*, n. 6, 11–16, see especially p. 12 and p. 15.

23 Michel Jeanneret, "Introduction" to his edition, with Stefan Schoettke, de La Fontaine, *Les Amours de Psyché et de Cupidon*, Le Livre de Poche classique (Paris: Librairie Générale Française, 1991), pp. 5–6.

24 Jean-Pierre Collinet, "La Fontaine et ses châteaux," *Travaux de littérature* XII, 1999, 377, "Architectes et architecture dans la littérature française," Actes du Colloque international de 1997, published by Madeleine Bertaud.

25 *Ibid.*, 370–371.

26 *Ibid.*, p. 366, "Bipolarisation toute lafontainienne . . . le dispense . . . de choisir entre les deux postulations contraires du grandiose et du gracieux . . . utopies symétriques où se résumeront . . . à la fois l'esthétique du poète et son éthique."

MATTHEW LANDRUS

LEONARDO'S *ANNUNCIATION HORTUS CONCLUSUS* AND ITS REFLEXIVE INTENT

I would like to preface my discussion by recalling the meeting's introduction by President Tymieniecka and the divertimenti of strings that followed.[1] It would seem that her discussion of gardens and the "three vectors of our human being," in addition to the performance of Mozart's composition for string quartet, refer in my presentation to Leonardo da Vinci's concepts of *diligenza, imaginativa* and *memoria* in painting. These concepts were core principles in Leonardo's painting of the *Annunciation* garden, or *hortus conclusus*. One may consider that diligence, *diligenza*, refers to the first vector, or to that of organizing the garden into a set of rationally configured plots. Imagination, *imaginativa*, would be a second vector of poetic impulse, the expression of the life of the garden's varied flora and aesthetically arranged objects. These vectors contribute to the third vector, that of creating a *memoria*, or memory system, which contributes to nostalgia. Syncopations of the divertimenti express mathematical diligence and poetic improvisation, which refer to nostalgic feelings of similar order and poetry. Gardens and divertimenti, first divided by our imposed dimensions, and secondly as poetic expressions of their and our being, also engage us along the third vector of our nostalgia. Certain combinations of mathematical diligence and poetic intuition define such creations as ineluctable proofs of the marriage between measure and imagination, science and art, institution and culture, mind and body, rule and desire, Psyche and Eros. As typical binary referents, these terms are loaded with nostalgic connotations. Previous emotional interests tend to govern directions that these binary vectors will take. Artist/scientists, such as landscape architects, musicians, painters, sculptors, ballet dancers, public speakers, and the like, reaffirm these innate vectors of our human being. These individuals interpret and express the order of the first vector and the creativity of the second vector in order to form the third vector, the nostalgia for exceptional examples that combine the two vectors.

To be more specific, the present paper proposes an interpretation of these vectors as Leonardo da Vinci's requirements for the composition of his *Annunciation hortus conclusus*. I propose that the referents of the first and second vectors, in this case, were chosen for their abilities to refer, in the

A.-T. Tymieniecka (ed.), Analecta Husserliana LXXVIII, 25–46.

painting, to the individual looking at the painting. These vectors therefore create a memory system that appeals to the nostalgic sense of having been capable of being part of the painting. These references, or bracketed aspects of the painting, include the perspective grid upon which the *hortus conclusus* is set, and the traditional placements of Gabriel, the Madonna, and flora. In this way the compositional elements, or referents, set up a one-to-one relationship between the viewer and painting. The perspective illusion of space and the expressive realism of its composition are meant to offer a scale model of the actual life-size event of the *Annunciation*. Leonardo intended that the viewer have the feeling of once being in the painting's garden. Hence, the vectors of diligent technique and imaginative naturalism would serve the third vector: the feeling of a divine presence in the form of a painting. Today the term "nostalgia" identifies this kind of feeling for having been in that garden. Husserl referred to this kind of finding as a reference to the *Lebenswelt* (lived world). He recognized the three parts or vectors as *unselbständig* (non-independent) Hegelian "moments."[2] Leonardo became famous for his unique ability to link such moments in a single painting.

To begin discussing these *moments*, I would like to start with his concept of creation, re-making or invention. In his surviving notebooks, he rarely mentions the word *creare* – to create. To express a similar idea, he preferred the more highly regarded term, *invenzione*, to denote a completely original conception. I will argue that he intended to make his paintings into completely original inventions. While this interpretation would not surprise most anyone familiar with Leonardo's works, there has been no discussion of a specific proof of Leonardo's actual intention to make paintings his inventions. The reason for this approach to his inventions is to point out the extent to which such works were meant to refer back to themselves more than to anything else. The more unique the work of art, the more reflexive its relation to an individual's projected vision of him/herself in the painting. In this manner, painting can be an invention of the self, a re-creation our human being. This could be what Leonardo intended, even if he recognized that the painted illusion would naturally be limited by the scope of possible interpretations.

The inventive example to be discussed here is his first known completely autograph painting, the *Annunciation* of 1472. It is believed that he painted this work without assistance from anyone, as it may have been the "masterpiece" required for him, at twenty years of age, to graduate from apprentice to master.[3] Of particular interest is the composition's *hortus conclusus*, the sacred enclosure in which Gabriel delivers his message to the

Madonna. For Leonardo this could be no ordinary garden, but an environment of vibrant, rich and highly detailed scenery. Most textures and colors are represented, from over thirty distinct species of plants, to various kinds of garments, to carved marble and several other kinds of stone masonry, architectural features, the endless landscape in the distance, various kinds of trees, etc. In this sense, the *Annunciation* not only re-presents a narrative, it also shows off the complete repertoire of the artist's skills. It is also the diagrammatic visualization of a highly sophisticated memory system.[4] Tendencies in Leonardo's writings suggest that he had developed methods for making paintings such as the *Annunciation* into more than mere paintings, but the inventions capable of being sophisticated memory systems. It will not be possible to use such writings as proof of his intentions for the *Annunciation*, since the majority of them were written over twenty years after the painting. Instead, these writings could refer to Leonardo's intentions for paintings in general. But the fact that these writings are in agreement with techniques used in the *Annunciation* suggests a consistent pattern of approach in both early painting and later writing.

In what follows, four short case studies about this inventive approach will be discussed with regard to the reflexive quality of a painting. These are: I) the *Annunciation's hortus conclusus* and perspective field; II) cause and effect; III) the non-paradox of human vision; and IV) the painter's creation of what nature has yet to create. These approaches begin with Leonardo's garden and end with his support of the individual's ability to make something new in nature. From the general viewpoint of the garden, to the viewpoint of the reflexive value of a painted illusion, there appear to be tendencies with respect to his intent to invent an illusion containing the self.

For example, such an inventive methodology may be outlined in the diagram of Figure 1. This illustrates portions of Leonardo's discussions on how to portray the nose. The proportional use of all four approaches to the nose will make the invention noted at the illustration's center. Along one axis, he tends to refer to an artist's unique, unteachable talent, as opposed to his/her taught, repeatable skills. On the opposite axis, he often considered *imaginativa* (imagination) opposite to *ragione* (reason). Four kinds of effects inside the circle are subject to the various causes around the circle's periphery. The main point of this diagram is to compare the naturally linked abilities with the abilities that require extra training. Concentration especially on these latter abilities would produce the central aim of an invention. By developing such techniques in proper proportion on the picture plane, one's invention could work as a kind of memory system.

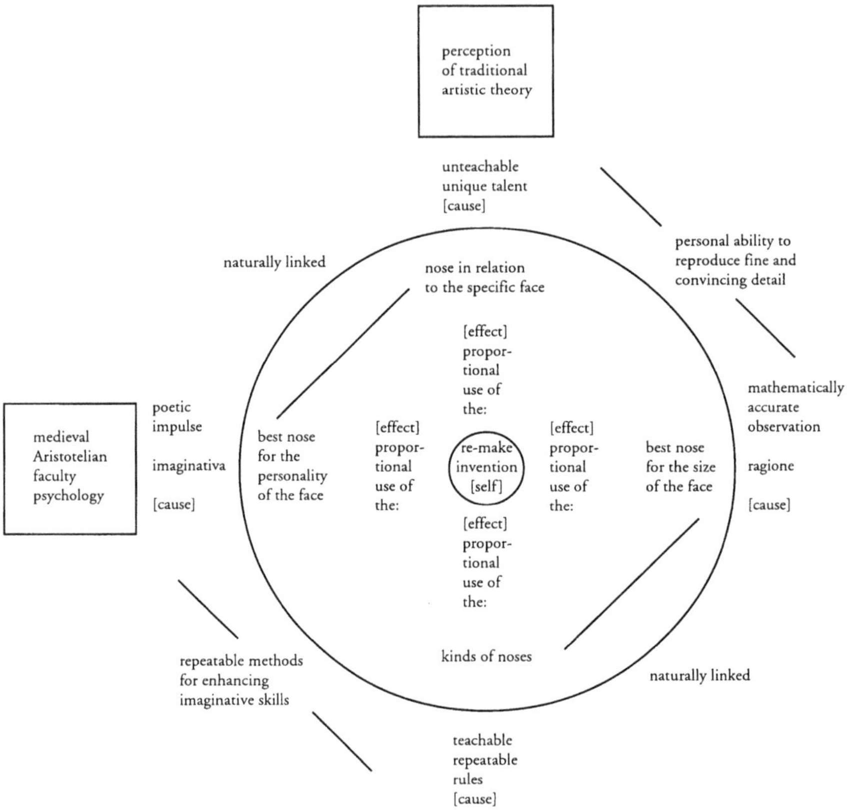

Fig. 1. Tendencies in Leonardo's writings and drawings about proportional method.

I

Though Leonardo's first activities in Verrocchio's studio remain a mystery, we have the earliest evidence of him as an artist in 1472: a record of his dues paid to the Compagnia di San Luca in Florence. This date also approximates the one given the *Annunciation*, the first painting attributed to Leonardo as the primary artist. Remarkably, several highly finished drawings relative to this work survive, suggesting the possible life-long importance of such studies, and the painting, to Leonardo. One written piece of evidence suggests, according to David Alan Brown, that this could have been Leonardo's first demonstration of a "masterpiece" after apprenticeship and close to his twentieth birthday. The Florentine confraternity ledger lists him, like

Verrocchio and Pollaiuolo, as a *dipintore*, as opposed to his fellow studio assistants, Botticelli, Ghirlandaio, and Perugino, who are listed as *dipintori* (cf. note 3).

For the late *quattrocento* Florentine painter, *relievo* was more important than *disegno* in the conversion of three-dimensional reality to two-dimensional illusion. This kind of illusion would therefore have its own kind of three-dimensionality. The optical effects of *relievo* were meant to cause a reflexive relationship between the viewer and the work of art. Thus the painted illusion became more than a medieval icon, it was physical proof of the world depicted therein. By expanding the illusion field, perspective tools fragmented and complicated the painted forms. Rather than simplify the viewer's approach to the two-dimensional object, perspective method gave the viewer an opportunity to look inside, to its multiple aspects of view. The two-dimensional *relievo* described, directed, framed, built and diagrammed the plane field. Adding three-dimensional illusions to this pictorial surface during the Renaissance involved tools of perspective developed by the architect, sculptor and painter Filippo Brunelleschi. Thereby the viewer could see into the picture plane. Such perspective was intended to keep the viewer, and the most important objects in the picture, normally within a few meters of the picture plane which divided internal and external worlds. These internal and external worlds therefore could mirror one another at a specific moment, in a specific place. The internal world of – say – the Madonna and Child with Saints could equal an external world of the same group at a specific time and place. Renaissance viewers could identify with these "vanishing points" of view.

The perspective field's landscape contains interactions between these diagrams and the less structural content. For Leonardo, these interactions between contents at similar and differing scale levels, normally refer to the artist and spectator, as these two are at opposite ends of the perspective double pyramid. In this referential manner, a specific artist communicates directly through the surface rhetoric to a specific spectator. Looking at the surface of Leonardo's work in this way, as a rhetorical landscape of technical and creative material, one explores the structure underlying its composition. Such an examination of the surface explores his artistic methodology.

Part of this methodology consists of a unification of two concepts essential to his artwork: the creative instinct and diagrammatic reasoning. These concepts remain central to an understanding of the work of "artist-engineers" such as Leonardo, Brunelleschi, Ghiberti, Taccola, and Francesco di Giorgio. For those individuals, inventive instinct requires the use of one's *immaginativa*, whereas diagrammatic reasoning requires the use of one's *ingegno*, or unique talent. The former interest concerns *grazia* and *fantasia*, whereas the latter deals with *intrinsica forma*. As Leonardo states in his *Paragone, buono disegno* and *maniera* depend equally on these creative and technical abilities. The present study addresses these modes of visual reasoning by examining some of Leonardo's early work in terms of its perspective landscape. Rather than approach the aesthetic material of a pictorial window, the following examples will discuss the architectonic diagrams which construct this visual field of the perspective landscape. At issue, as Martin Kemp notes, is "the balance between optical description and intuitive expression" [which] "is miraculously struck with such subtlety that at one moment the system seems to dominate; at another an ineffable magic appears to rule."[5] In this sense, Leonardo sought not just to illustrate Nature's effects, but to remake its visual structure. He felt that this structure was based on, as Kemp notes, "visual principles – what he called 'causes' or *ragioni*. The rules of tone, colour, reflection and shine . . . ."[6]

II

Leonardo's creations and studies express his interest in almost every kind of causal relationship in nature. Numerous examples of his interest in *causa* and *effetto* exist from his earliest artwork and writings. In what follows about his work prior to 1494, it will be shown that causal explanations assisted his development of mechanical approaches to proportions in art and *invenzione*. This is due to proportion being more about the organization of observable effects than about theoretical concepts. Leonardo worked with *proporzione* in *natura misura* (measure), *disegno* and *colore* in a manner fully consistent with his contemporary language, logic and world-view.[7] I argue that his studies of mechanics and proportion served the purposes of causal explanation more so than theoretical rhetoric. This kind of causal explanation of proportional effect is an *explanandum* (gerund) more so than an *explanans* (active present participle).

A study of developments leading up to Leonardo's *Trattato della pittura* expresses his early use of proportional tools to manipulate and create nature's effects. These tools enabled the spatial conceptualization of causes and effects in art, engineering and biology. Such devices formed the most effective practical invention or they formed the most eidetic and reflexive reality in an artwork. For these reasons, *quattrocento* proportional tools may seem to be the same as the theological tools of the medieval "proportionists." Except for the view of proportion as Logos, God's order of the cosmos, Leonardo's concept of proportion differs from the medieval definition to the degree that it is rooted in visible causes of the world of lived experience. Though to early Italians the Aristotelian God was not considered physically on Earth, his *forza* or *causa* was not physically exterior, but proportionately transcendent in the macrocosm and microcosm. It was this medieval concept of proportion in Nature which Leonardo tried to master through a combination of observational, experimental and scholarly methods.

If Leonardo may be said to have inherited a theological point of view, it was similar to a statement by the Apostle Paul: "For what man knoweth the things of a man, save the spirit of man which is in him."[8] Only "actual" causes in the outside world were known, whereas "potential" causes of the soul were unknown. A comparison of this belief to medieval sociolingual traditions indicates an Italian departure from Latin tradition. The medieval Italian word for "object," *cosa*, meant the same as *res* in Latin. Similarly, the Italian *effetto* stood for a combined "truth of" – *vero* – and "object" – *res*, for the Latin *re vero*. The Medieval Latin and Italian "cause" – *causa*, retained the same general meaning as in Roman Latin. There was no major difference in the *trecento* between *cosa* and *causa*, in the sense that an object could be its own cause and visa versa; for this agreed with the medieval adaptation of Aristotelianism. Though Leonardo never fully denied this view that the cause of an object's effects and motions were proportional to its particular elemental nature, he questioned the approach especially in his writings after 1492. Perhaps the closest he came to the attribution of an external force to an object, and thus his near discovery of inertia, was his comment that "no element is able to move itself without percussion. . . every weight. . . moves in the direction of its own weight" (*Codex Atlanticus* 109va (303v) *"Lo elemento non pesa nel suo elemento, se non per percussione. Ogni peso, il quale sai sospeso per traverso al suo firmamento, quando fia lasciato in libertà, passerà tanto di là da . . ."*). He is not known to have studied the "*quantita discontinua*" (the discontinuous quantity) or mathematics of *cosa, causa, effetto* and *forza*. He linked these effects and powers "continuously" to one another when studying "*quantita continua*" – geometrical, proportional, and mechanical quantities. His concerns about cause and effect, more so than theoretical rhetoric and mathematical theorems, grew out of his practical interests. This kind of work on the physical mechanics of nature's effects, rather than on the scholastic, theological, and theoretical aspect of natural philosophy, locates him within a tradition reaching back to ontological discourses on cause (αιτιο, aka. logos) and effect (ενερλεια) by Hero of Alexandria, Erisistratus, Strato of Lampsacus, and Aristotle.[9]

Leonardo restricted his interest in the soul specifically to its effects on the body. "If anyone wishes to see how the soul dwells in its body, let him observe how this body uses its daily habitation; that is to say, if this is devoid of order and confused, the body will be kept in disorder and confusion by the soul."[10] "The rest of the definition of the soul," he states, "I leave to the consideration of the friars . . . who by inspiration know all the mysteries."[11] His view that the soul is capable of being "*senza ordine e confusa*" was written shortly after 1500. The part of the soul's definition that he leaves to the friars was noted around 1510, when mentioning the Aristotelian/Galenic view of its formation of the fetus. The corruptibility and "mystery" of the soul was, for Leonardo, considered later in life. His earlier writings nearly oppose these points of view, suggesting his early access to, and partial agreement with, the ideas of Socrates, Aristotle and Galen. Throughout his notebooks, however, his approach was no less mechanistic. In the late 1480s, he noted a Galenic approach to the body and its soul – "the spirit of man" – and the soul's incorruptibility: "The soul can never be corrupted with the corruption of the body, but is in the body as it were the air which causes the sound of the organ, where when a pipe bursts, the wind would cease to have any good effect."[12] Here he compares the "*causa*" of the soul, the *spiritus animalis*, to a specific type of *spiritus*, that being "*vento*" (wind). He interprets or copies a belief similar to Galen's that *anima* form therefore consists of the material and motive *pneuma* (i.e. breath, *spiritus*, wind). This compares to the Socratic tradition that form, in terms of measure and proportion, enables matter to survive.[13] Aristotle agrees that "the soul must be a substance in the sense of the form of a natural body having life potentially within it."[14] In these contexts, *pneuma*, measure, proportion and form all represent good qualities, by virtue of their causes and effects, of the natural elements. Leonardo's studies prior to 1494 indicate his general agreement with these concepts. Proportion and ratio (αναλογια), cause and reason (αιτιο), account and explanation (λογοδοσια), were themselves the kind of logos (λσγοζ) governing his language, logic and world-view as artist and scientist.

Another belief of the medieval proportionists was that the experience of *proporzione* and *effetto* was a revelation of the soul's "ideal" form. The modern association of the word "revelation" with the kind of serious prophecy in the *Book of Revelations* refers appropriately in this case to the strength of revelation possible in a painting's *proporzione* and *effetto*. The effect of something became more noticeable with the increase of its proportional qualities. To give an artwork a heightened *effetto*, one had to reveal the work's *proporzione* to the viewer's soul.

Music commonly provided the metaphor for the relationship between the microcosmic soul and the macrocosm as either resonant or dissonant.

Socrates referred to this in the famous comment: "Give me beauty in the inward soul; and may the outward and inward man be at one." A performance of dissonant, disproportionate music could ban a musician from the court, as in the case of Martino Guastalino of Cesena, expelled from the Sforza Court in 1483 because he "peressere le aere rue dissonante et discorde."[15]

For an Italian Renaissance painter, the medieval concern for a reflexive relationship between the external and internal proportional logos could make more sense in terms of perspective. As Leonardo states: "Each thing transmits the image of [itself] to all those places where the thing itself is visible, and so conversely this object is able to receive into itself all the images of the things which are in front of it."[16] This repeats John Peckham's claim in his *Perspective communis* that objects and the eye itself transmit their images. For, "if you look at someone's eye, you will see your image reflected," states Leonardo around 1492 in Manuscript A.[17] On the sheet with that statement he adds the drawing of a visual pyramid. At one end of the pyramid, lines radiate diagonally out from an eye, meeting at a flat plane perpendicular to the lateral direction of the eye's center. At the other end of the pyramid a similar cone-shaped set of lines fans out from the distance point. The flat plane between these cones is the picture plane.

From around 1492 to 1494, especially in manuscripts A and H, and on pages later included in Codexes Urbinas and Atlanticus, Leonardo wrote much of what exists of his views on perspective. He consistently noted that one finds the best perspective construction as two pyramids connected at their bases, with one tip at the eye, the other tip at the vanishing point. On the first sheet of his Manuscript A, Leonardo confidently writes that "perspective is nothing else than the seeing of an object behind a sheet of glass . . . these things approach the point of the eye in pyramids, and these pyramids are cut by the said glass."[18] A convincing perspective construction must enable the same vision at both ends of the double pyramid. The vanishing point sends its image to the eye; the eye sends its image back to the vanishing point; this point therefore refers back to the eye; this eye therefore refers to the vanishing point. Thus perspective constructed a rational, reflexive relationship between the specific "viewer" and "his reflection in what's viewed."

III

Around 1490, Leonardo began writing a *paragone*, an argument for the superiority of painting over the other arts. Later editors considered this to be the introduction to his *Treatise on Painting*. One of the reasons Leonardo considered painting to be greater than any other art form was its ability to invent a lasting impression of beauty. Whereas music and gardens change like the flowing stream, which is never the same as it was when one last stepped into it, music and the garden are two of the least mimetic experiences because they both involve unpredictably infinite variations of changes as they develop. Leonardo was well aware of this problem when he wrote extensively about how to reproduce the infinite variety of nature's effects. He said, "Among the praiseworthy and marvelous things which are apparent in the works of nature, it happens that none of the proportions of any species in themselves precisely resemble in any of their details those of any other."[19] Thus all things in nature have different details.

Leonardo also consistently noted the irrefragable unity of everything in Nature. He writes about a "pyramidal law" which keeps everything in proportional order to the whole in such a way that the infinite variety of nature's effects extends out in pyramid form from the same origin.[20] This origin of an effect's power and direction is the eye, the vanishing point, stem of a plant, beginning of a stream, or the emanation of sound from a plucked string. As is the case with pyramidal perspective, the study of a lived experience begins with a review of the artist's physical senses. The garden is an ideal venue in which to begin this review, since one's experience in it can involve all five senses: olfactory, auditory, tactile, visual and (depending on the garden) taste. When the experience of any one of those senses triggers the memory of a garden, cognitive patterns can trigger other senses relative to the previous garden experience. If, for example, the graceful melodies of a Mozart divertimento, or the beautiful plants in a painting, inspire nostalgic feelings of a previous experience such as that of a garden, then one could recall the scent of roses, the site of many colors, or the feel and taste of a fresh peach. This is not to suggest that music and painting necessarily trigger one's nostalgia for gardens, or vice versa. Relationships between stimuli and one's nostalgia are too complicated to comprehend in exact terms. What experiences mean cannot be reduced to specific ideas and mental concepts any more than one can precisely copy the garden's flowers or the nuances of an instrumentalist's chord changes.

At issue when studying the artist's sense experience is one's method for examining the available materials. Merleau-Ponty offers a direct approach to such problems in his essay, "Eye and Mind." He addresses the "paradox of human vision" which can lead to an ambiguous vision if the world that one sees is also a vision of the self. To Leonardo, this sort of Neoplatonic problem indicates the error of auto-mimesis, for which he clearly gives the following warning:

Measure on yourself the proportion of the composition of your limbs, and if you find any discordant part, take note of it and make very sure that you do not adopt it in the figures that are composed by you, because it is a common vice of painters to take delight in making things similar to themselves. This happens because our judgement is that which directs the hand in the creation of the delineations of figures in various configurations until it is satisfied. Because the judgement is one of the powers of the soul, through which it composes the form of the body in which it lives according to its will, when it has to reproduce with the hands a human body, it willingly reproduces that body of which it was the original inventor, and from it arises that he who loves will eagerly fall in love with things similar to himself.[21]

Leonardo did not see an unsolvable paradox of human vision. He associates the auto-mimesis "fault in the extreme" – as he calls it – with the soul's automatic movements which form one's own body, thereby limiting one's ability to imitate the actual variety of other forms of Nature's effects.[22] Beyond defining the soul as the "formative agent of the body," he says, "the rest of the definition of the soul I leave to the consideration of the friars, the fathers of the people, who by inspiration know all the mysteries."[23] He avoids mysterious difficulties with his use of an Aristotelian inductive method, for which he has been commonly appreciated as a kind of modern scientist.

Referring to a luminous image's rays imprinting themselves onto the eye, he explains that "we learn in the first place by the inductive method which teaches us that the eye when wide open does not show us such rays round luminous bodies."[24] But in this statement there is more of the Aristotelian natural philosophy of Nicholas Oresme, than the empirical natural science of Francis Bacon, Galileo, Newton, Mill and Hume. The novelty of Leonardo's exhaustive writing about the importance of observational method coincides with his careful reading of Oresme's experimental methods. In this case, induction is an adduction of Oresme and personal experiment. Leonardo uses adduction to explain causation. He expects that the consequence of an effect of nature can be explained by combining evidence from previous and personal experiences. To prove the above statement about there being no "rays round luminous bodies," he refers first to an experiment: "Only the image of a star or other light should pass to the eye through the smallest perforation made in the paper placed before the eye, such luminous [images] will always be without rays."[25] The second proof, he says, "is shown by the ninth of perspective where it is stated: – the angle of the incidence is always equal to the angle of reflection."[26] The third proof adds that two of the three images, when impressed upon the almost closed eye, are reflected away by the eyelids, leaving only the central image to reach the pupil. These three proofs range in methodology from the inductive experiment, to the adductive reference, to the causal explanation of images forced into or away from the eye. A fourth proof combines methods of previous proofs by adding a proof that lifting or lowering the head causes one to lose sight of lower or upper rays of a luminous body, since one of the three images bounces off the lower or upper eyelid. If, for example, the face is lowered, then "the thick part of the upper covering of the eye neither sees nor is seen by that luminous body."[27]

Herein lies the most compelling part of Leonardo's argument. Both the observed object and the observing subject are at opposite ends of a perspective continuum.[28] It must be possible for the object at the other end of that continuum to see the observer in the same way that the observer sees the object. A physical relationship exists between the luminous body and the viewer due to the extent of perspective connection between them. Leonardo writes: "The rays, which seem as though they extend from the luminous body to make contact with the eye that beholds it, start when the eye, being almost closed up, looks through the narrow crack that intervenes between the eyelids. . . ."[29] This proves his ability to perceive a kind of paradox of human vision, since the world being viewed can look back with the same perspective view. Only there was nothing paradoxical, ambiguous or enigmatic about this kind of vision in the pre-Cartesian world-view.

The post-Cartesian paradox of human vision forms a basis for misunderstandings about perception. This is one reason why Merleau-Ponty wrote "Eye and Mind." He argues in the article that "scientific thinking, a thinking which looks on from above, and thinks of the object-in-general, must return to the 'there is' which precedes it; to the site, the soil of the sensible. . . ."[30] Thus, looking at Leonardo's art and science requires a method which transcends such scientific thinking. A modern scientific misconception, for example, leads one to believe that all objects have precise linear boundaries which can be represented in a three-dimensional graph of x, y, z axes. "But this line has been contested by all modern painting, and probably by all painting," argues Merleau-Ponty.[31] In support of this issue, he refers to comments by Leonardo, Ravisson, Bergson, Rodin and Klee about infinitely variable flexuous lines. With respect to linear boundaries, one could note Leonardo's view: "The boundaries of bodies are the least of all things . . . [they are] the boundary line of the surface, which as a line is indivisible. Therefor, painter, do not surround your bodies with drawn lines."[32]

As for the "there is," this echoes one of Leonardo's first and most sustained beliefs: "All our knowledge has its foundation in our sensations."[33] He is thought to have written this as one of his earliest writings in Milan, around 1485 to 1487. The statement reflects a primary point in Aristotle's *de Anima*:

"No one can learn or understand anything in the absence of sense (432a7)."[34] Medieval and Renaissance debates on Aristotle's theory of the intellect focused on this fundamental precept. Of particular importance to Leonardo were comments that immediately followed this precept, specifically those dealing with the image as a link between sense and intellect. Aristotle wrote, "When the mind is necessarily aware of anything it is necessarily aware of it along with an image: for images are like sensuous contents except in that they contain no matter (432a8–9)."[35] The popular interest in differing approaches to Aristotle commentaries in northern Renaissance Italy are noted in Ermolao Barbaro's dedicatory letters for his Latin edition of Themistius' *de Anima* paraphrase, published at Treviso in 1481.[36] Otherwise, prior to 1487, the best known examples of Aristotle's *de Anima* in Italy were in the form of Averroes's commentary of 1472 and Thienesis's paraphrase of 1486. Leonardo could have heard or seen Themistius' paraphrase of *de Anima* 432a7–9: "From the outset the intellect collects [the concepts of] one, two, or number from the objects of perception. That is why even now when it contemplates such objects, it must do so along with the image, for images are like sense-impressions, except that they are without matter."[37] Themistius excludes the phrase "No one can learn . . . in the absence of sense," as a lengthy and unnecessary double negative, replacing it with the positive paraphrase, "From the outset the intellect collects. . . ." For "understand" he incorporates "intellect collects," whereas "anything . . . of sense" means "number from the objects of perception." In this case, it would be important to delete the modern translator's "[the concepts of]," and explain why the numbers are not *concepts* of Kantian things-in-themselves, but that they are physical elements of nature. Prior to Descartes's analytic geometry, the numerical logos, for the most part, retained its Pythagorean and Euclidian status as a building block for "objects of perception." These objects, whether being of number or being of reality, were in this case believed to be very similar, though it is important to distinguish between the two kinds of being. As Hans-Georg Gadamer states,

> As a matter of fact, the distinction here in modes of being is one which only such a "Pythagorean" is actually capable of making who has already been alerted by Plato or Socrates to the difference between noetic being and being for the senses, and who has abandoned the naive identification of the being of number with the being of reality as a whole, an identification which was typical of Pythagorean thought.[38]

Form – such as being of number, and matter – such as being of reality, were both distinct modes of the self and, by extension, of the illusion of reality in Leonardo's painting.

Nonetheless, no less essential than air, water, fire and earth, the number and the logos dictated, for example, that the heart-shaped proportions of a rock or plant could cure heart disease.[39] For Themistius, this numerical logos had universal appeal due to its link between the ideas of Aristotle and Plato, as well as between Hellenism and Christianity. These numbers, rather than reflect ideas and thoughts in the analytic sense, actually form the logic that *physically* links language and world-view.[40] Before Kant, thoughts could not easily – if at all – be perceived as *known* entities. "For what person knows a man's thoughts, except the spirit of the man which is in him?" (*Corinthians* 2: 11). Thus Plato, Aristotle, the Apostle Paul, Themistius, and Leonardo knew that it was much easier to train the senses to understand *matter* than it was to understand *form*. Since only matter could be easily observed, understanding form required the most ingenious intellect. Plato's hierarchical order of good possessions leads from form to matter, as Socrates indicates to Protarchus in the last paragraphs of *Philebus*: (first) measure, moderation, fitness, (second) proportion, beauty and perfection, (third) mind and wisdom, (fourth) sciences, arts and true opinions, and (fifth) pure pleasure. Aristotle supports this hierarchy in *De Anima*. Leonardo knew of these statements in *De Anima*.

This latter issue of a trajectory from form to matter is the sort of distinction Gadamer makes in his essay on "Idea and Reality in Plato's *Timaeus*."[41] Thought (*noesis*) is a kind of being not mediated by the senses, whereas opinion (*doxa*) is a "being for the senses."[42] This is consistent with one of Leonardo da Vinci's earliest known statements, which could have been inspired by Plato's *Timaeus*: "Four in number are the powers, memory and intellect, licentious-lascivious desire and covetousness (self-preservation). The first two are reasonable, the others sensual. Of the five senses seeing, hearing and smelling cannot easily be prohibited, touching and tasting not. Smelling is connected with tasting in the dogs and other greedy animals."[43]

In this, written around 1485–87, Leonardo expresses the remaining influence of Florentine Platonists from as early as 1472, and especially between 1480 and 1482, at the Piazza San Marco of Lorenzo di Medici. Such a statement refers to comments in *Timaeus* like: "the Divine Power keeps upright our whole body. Whoso, then, indulges in lusts or in contentions and devotes himself overmuch must of necessity be filled with opinions that are wholly mortal. . . . But he who has seriously devoted himself to learning and to true thoughts . . . thinks thoughts that are immortal and divine."[44] Though Leonardo's writings confirm his repeated concern for a more Aristotelian world-view, subject more to sensory experience than to non-empirical opinion, occasionally one can see in his work the profound influence of the Platonic and Christian Florentine environment in which he lived from the ages of ten to thirty.

IV

Eighteen to twenty years after painting the 1473 *Annunciation*, Leonardo wrote the first portions of his plans for a *Trattato della Pittura*. The treatise's *parte prima* presents a *paragone*, which defines painting as a science of much nobler rank than the arts of poetry, music and sculpture. One of the earliest arguments in this *paragone*, dated to around 1490, responds to its heading, "The Reply of King Matthias to a Poet Who Challenged a Painter."[45] Leonardo quotes the king as saying that the proportional components of a painted surface, "give me so much pleasure in their divine proportions that I judge that nothing else on earth made by man could give me greater pleasure."[46]

This is Claire Farago's translation of the last line in the King's polemic against the poet. She states that Leonardo could have copied the statement from a "tongue-in-cheek embroidery based on the fame of the Hungarian ruler's acumen. . . ."[47] The subject of a jest or not, this statement is a rhetorical battle against the poet. The fame of this statement, if recreated contemporaneously on embroidery or elsewhere, owes much to common courtly debate on the *paragone*. The King's words and Leonardo's *paragone* state a rhetorical Truth, similar to, but taken more seriously than mere opinion. Only matters of *veritas* and *virtus* were worthy of the court and *institutum*. With respect to this appropriateness of speech, King Matthias and Leonardo, steeped in the humanist courtly tradition, followed the advice of Cicero and Quintillian. Linguistic decorum, according to Cicero, should be "correct, lucid, ornate and suitably appropriate . . ." ("*quam ut Latine, ut plane, ut ornate, ut ad il quodcumque agetur apte congruenterque dicamus*").[48] Quintillian recounts this rule of decorum at the start of his eleventh book, discussing "what style is most suitable for conciliating, instructing or moving the judge" ("*quid conciliando, docendo, movendo iudici conveniat*").[49] He ends the book with a reminder that oratory "consists in serious pleading, not in mimicry" ("*actione enim constat, non imitatione*").[50] Thus, while King Matthias may seem to jest when saying "do not blame my choice to put your work under my elbow while I hold the work of the painter," his comments indicate an educated speaker's intent to move, as well as inform and instruct an audience.[51] The King's *actione enim*, or truthful action, is an ironic overstatement that the poet's work deserves a place under his elbow while he looks at a good painting. Poetic originality takes an ironically lesser position to that of the painter/craftsman's *imitatione*. Also, to dishonor the poet's gift by making it an armrest, overstates its lower status than that of painting. To disrespect a book or manuscript in the late fifteenth century was to disrespect its knowledge, not simply the material on which it was printed. So Leonardo argued against this profound contemporary respect for poetry and the written word when in praise of the painter's *imitatione* of nature. He quotes King Matthias: "[Poetry] does not content the mind of the listener, or viewer, as does the proportionality of very beautiful members, components of divine beauties in this face in front of me."[52] The King had also stated: "Therefore, I judge your invention to be greatly inferior to the painter's solely for the reason that it does not compose any harmonic proportionality."[53] The first of these statements bases the virtue of the art on its ability to "content the mind" with its proportionality. The second argument notes the greater value of a painter's "invention" according to its immediate display of harmonic proportionality. The painter, hereby, invents and imitates nature's harmonic proportionality. The King insults the poet with the question, "Do you not know that our soul is composed of harmony, and that harmony is only generated in those instants when the proportionality of objects is seen or heard?"[54] Poets were well aware of this fact and tried to create such an instant of proportionality with words. But the painter's results, argue the King and Leonardo, are more instantaneous, therefore inventing and imitating the harmonious works of God for the more immediate viewing pleasure of the harmonious soul. The chapter on King Matthias therefore appropriately concludes: "For painting extends not only to the works of nature but to the infinite [others] which wait for nature ever to create them."[55] Leonardo hereby wraps up a chapter which elevates painting from an art of mere imitation, to that of *inventing* the works of God.

Although Farago's translations of the Matthias chapter were used in this case, differing interpretations of the chapter's concluding statement are worth noting. Farago's transcription reads: "Con ciò sia che questa non ch'elle opere di natura ma ad infinite attende che la natura mai le creò."[56] She identifies the subject, "*ciò*" as "painting," refers to "*sia*" as "extends," adds "[others]" to "*infinite*," and translates "*le creò*" as "to create them." Irma Richter translates the statement as, "Because she deals not only with the works of nature but extends over an infinite number of things which nature never created."[57] Richter's additions include "deals" for "*sia*," "extends over" between "*ma*" and "*ad infinite*," "number of things" attached to "*infinite*," and the combination of "*attende*" and "*mai*" to make the word "never." A. Philip McMahon's interpretation reads: ". . . Since painting gives heed not only to the works of nature but to an infinite number of things that nature never created."[58] This statement mainly copies Richter's syntax, with "gives heed" for "*sia*." Martin Kemp and Margaret Walker translate the statement as follows: ". . . Since she embraces not only the works of nature but also an infinite number that nature never created."[59] Kemp and Walker view "*sia*" as "embraces," they add "number" to "*infinite*," and combine "*attende*" with "*mai*" to form "never." After removing some of the above translators' additions and adopting a combination of their approaches, a literal assessment of Leonardo's statement may also read: "For she embraces not only the works of nature, but an infinite [number] awaiting what nature ever [would have] created." Converting the last phrase to past perfect preserves much of the original syntax. The Kemp/Walker word "embraces" signifies the painter's ability to invent and imitate nature more so than "deals . . . with," "extends to," or "gives heed." Painting "embraces" the works of nature as its "true daughter," as stated in the Matthias chapter, capable of recreating elements of her mother, in addition to creating elements never before seen by her mother. As the daughter of nature, painting would not necessarily *deal* with her mother nor *extend* to her, though she may *give heed.*

Leonardo uses the Latin phrase, "*ad infinite*," to denote an unlimited quantity of works of nature. Richter's translation, "extends over an infinite number of things," can be misleading, depending on the way one defines "things." Leonardo refers to the "works of nature," rather than to its things. Such works, like the flow of water, growth of a tree, and movement of an animal, operate as microcosms of the macrocosm. They operate things – water, wood, muscle. These things cannot represent concepts in the post-Kantian sense. Things and number, to Leonardo, denote physical elements of nature. Physical elements necessary for the operation of the works of nature include water, wood, muscle, and number. Without number there could be no harmonic proportionality to the soul. Medieval natural philosophy associated the element of numerical proportionality with the soul's ability to form the body and its actions. "*Ad infinite*," therefore refers to the element of number, though for clarification to the modern reader, it translates as "an infinite [number]," much like the Kemp/Walker translation of "an infinite number." Farago's translation as "to infinite [others]" can mislead with regard to the definition of "others."

The last portion of Leonardo's statement, "awaiting what nature ever [would have] created," makes two primary points. First, nature has yet to create some of what the painter can create. Secondly, whatever the painter creates, it is possible for nature to create. These points could be mistakenly avoided if one used Richter's "which nature never created." If nature never created some of what the painter created, one might argue that nature cannot create some of what the painter created. Nonetheless Leonardo clearly states that nature has yet to create whatever new number and new harmonious proportion the painter can create. Farago's translation, "which wait for nature ever to create them," may also be questioned on the basis that Leonardo's word "creò" refers more to "created" than to "to create them." For some of the painter's creations to "wait for nature ever to create them" suggests that nature may have plans to create what the painter created. Instead of this statement, Leonardo refers to what nature "ever [would have] created," since nature has not created some of the painter's number. But nature would be, and would have been, able to create that number.

So to further paraphrase the last line of Leonardo's Matthias chapter, one could expand it as follows: "For [painting creates] not only the works of nature, but an infinite number awaiting what nature ever [would have] created." To be more specific, "For she [/painting/true daughter of nature] embraces [(imitates and creates)] not only the [microcosmic] works of [the macrocosm of] nature, but [additionally (imitates and creates)] an infinite number [of proportional realities] awaiting what nature ever [would have] created [or may still create]." This interpretation, and that of preceding commentary, is not meant to challenge the work of previous translators, who have necessarily limited the amount of explanation on each statement. Instead, reinterpreting and extending the Matthias chapter's last line is meant to reify Leonardo's argument. As one of his earliest known writings, from around 1490, that chapter indicates a core belief in Leonardo's written *oeuvre*: that the painter can scientifically create elements of nature which move the soul by mirroring part of its own harmonious proportionality. This compliments Leonardo's concern about auto-mimesis; that the *scienza della pittura* creates "*ad infinite attende che la natura mai le creò*," (an infinite number awaiting what nature ever would have created). He therefore creates the self in a painting, if only in far more detail than would be recognized as simple auto-mimesis.

To conclude, I would like to return to the idea of the artist communicating directly through the surface rhetoric to the spectator. Leonardo most likely knew the traditional rhetorical duties noted by Cicero: to teach, entertain and move the spectator. For the painted surface to meet such rhetorical challenges, the creative *imaginativa*, combined with technical skill, could bring the surface to life. The Renaissance artist, as creator of a body of work, had to understand the causes and effects of various bodies to be re-made in the new body of work. About phenomena of the soul and body, Leonardo states: "Whoever would see in what state the soul dwells within the body, let him mark how this body uses its daily habitation, for if this be confused and without order the body will be kept in disorder and confusion by the soul."[60] He also stated that "art is produced by two masters, namely nature and man."[61] It was important to distinguish between nature and man, soul and body, because it was the interaction of these two phenomena that gave meaning to a body of artwork as well as to a person's body. Paintings, drawings, architecture and human or animal bodies all possessed, for Leonardo, distinct creative and technical elements. For this reason, he continually expressed the close interactivity between these two phenomena in any body. As Catherine Dunton writes:

> It is interesting to find that in Leonardo's diagrams demonstrating the perception of distance through perspective, the vanishing point is often referred to as the eye. This would seem to indicate not just the point that is visible to the artist's eye, but the point from which his own eye is visible. It appears, then, that the idea of the object both seeing and seen that was found in his optics is extended, equally, to his theories of perspective. This marks a significant departure from the traditional view of his system of perspective as corresponding to the static image as it falls on the retina; a concept of vision that is essentially passive in nature.[62]

Thus the painted subject is conceivably reflexive, referring back to oneself. Leonardo's proportional methods for expressing the active intellect form the basis of his approach to positing ourselves within the perspective landscape. For this kind of visually lush memory system, such as the *Annunciation hortus conclusus* – one in which we are capable of being imbedded – our human being is often nostalgic.

University of Oxford

NOTES

[1] A recital of Haydn, Beethoven and Mozart divertimenti was given by a string trio led by Tim Roberts, Director of Faneuil Hall, Boston.

[2] Cf. J. N. Findlay's comments in: Edmund Husserl, *Logical Investigations*, 2 vols, trans. J. N. Findlay (London: Routledge and Kegan Paul, 1970), Vol. I, p. 22.

[3] Cf. David Allen Brown, *Leonardo da Vinci: Origins of Genius* (New Haven: Yale U. Press, 1998), p. 75.

[4] For a standard approach to Renaissance memory systems, see: Frances Yates, *The Art of Memory* (London: Routledge and Kegan Paul), 1966.

[5] Martin Kemp, Leonardo da Vinci, *The Marvellous Works of Nature and Man* (Cambridge: Harvard, 1981), p. 98.

[6] Ibid.

[7] Alfred Gell uses a similar approach, whereby one examines the irrefragable linguistics, philosophy and psychology of a culture, in his: *Art and Agency, an Anthropological Theory* (Oxford: Clarendon Press, 1998). He argues that aesthetic theories construct passive perspectives on the art object, emphasizing its instrumental role rather than the actually complex intentionalities such art objects embody. For an earlier approach to the coherence of language, logic and world-view, see: Werner Jeager, *Aristoteles* (2nd ed., Berlin: Weidmann, 1955).

[8] *I Corinthians* 2: 11.

[9] Cf. Sylvia Berryman, "*Horror Vacui* in the Third Century BC: When Is a Theory Not a Theory?" in: Richard Sorabji, ed., *Aristotle and After* (London: University of London, 1997), pp. 147–57.

[10] Codex Atlanticus 76r.a, 219a. "Chi vuole vedere come l'anima abita nel suo corpo, guardi come esso corpo vsa la sua cotidiana abitatione, cioè se quella è senza ordine e confusa, disordinato e cofuso fia il corpo tenuto della sua anima." Jean Paul Richter, trans., *The Notebooks of Leonardo da Vinci*, 2 vols. (New York: Dover, 1970), Vol. II, p. 287, § 1143 (C.A. 75a).

[11] Though he wrote this around 1510, it reflects his earlier interests. Windsor RL 19115r. Also in: Charles D. O'Malley and J.B. de C.M. Saunders, *Leonardo on the Human Body* (New York: H. Schuman, 1952), p. 492.

[12] Triv. 78. "L'anima mai si può cirropere nella corruttio del corpo, ma fa nel corpo a similitudine del vento che causa il suono del organo, che guastadosi una canna, no resultava per quella del vento buono effetto." Trans. Richter, op. cit., p. 287, §1141. This relates to Galen's treatise: *That the Faculties of the Soul Follow the Mixtures of the Body.*

[13] Plato, *Philebus* 55c–65d, trans. Harold Fowler, The Loeb Classical Library (London: William Heinmann, 1925), pp. 357–393.

[14] Aristotle, *On the Soul*, II 412a 20–21, in: *The Complete Works of Aristotle*, ed. Jonathan Barnes, Bollingen Series LXXI, 2 (Princeton: Princeton U. Press, 1984), p. 656.

[15] Leonardo first visited the Sforza court around 1482–3, invited not as a painter or engineer, but as a musician. The Codex of Anonimo Gaddiano (Cod. Magliabecchiano) of c. 1506–1536, and Vasari's *Vite* of 1550 and 1568 (4th vol.), refer to Leonardo's trip to the Sforza court to play the *lira*. Harmony, *l'armonia*, was itself another word for sound, *suono*. Rather than refering to "harmonious sound," Vasari said that Leonardo made his lyre "acciocche l'armonia fosse con maggior tuba e piu sonora di voce" (in order for harmony [of sounds] to have great loudness and a more sonorous voice), in: G. Vasari, *La vite de più eccellenti pittori, scultori e architettori . . .* (2nd ed., 1568), in: G. Milanesi, ed., *Le Opere di Giorgio Vasari* (Florence: Sansoni, 1906), Vol. IV, p. 28. For the statement that M. Guastalino of Cesana "disturbed the air with dissonances and discords," see Paul Valery, *La Corte di Ludovico il Moro* (Milan 1929), p. 495. See also: E. Winternitz, *Leonardo da Vinci as a Musician* (New Haven: Yale University Press, 1982).

[16] Codex Atlanticus 138 r. b., trans. Edward MacCurdy, *The Notebooks of Leonardo da Vinci* (New York: George Braziller, 1958 [Reynal & Hitchcock, 1939]), p. 987.

[17] MS A 37 v., trans. Martin Kemp and Margaret Walker, *Leonardo on Painting* (New Haven: Yale, 1989), p. 56.

[18] MS A I v., trans. MacCurdy, op. cit., p. 992.

[19] Codex Urbinas 104r–v., trans. Kemp and Walker, op. cit., p. 119.

[20] Around 1505, Leonardo noted, "All natural powers have or are to be called pyramidal inasmuch as they have degrees in continuous proportion towards their diminution as towards their increase. Observe the weight which in each degree of its free descent is in continuous geo-

metrical proportion, and similarly for the force of levity." Codex Atlanticus 151ra., trans. Kenneth Keele, *Leonardo da Vinci's Elements of the Science of Man* (New York: Academic Press, 1983), pp. 93–94.

21 Codex Urbinas 107r and 157r, trans. Kemp and Walker, op. cit., p. 120.

22 Codex Urbinas 44r-v. Ibid. p. 204.

23 Windsor Royal Library 19115r. (c. 1510), trans. Charles D. O'Malley and J. B. de C. M. Saunders, *Leonardo da Vinci on the Human Body* (New York: Crown Publishers, 1982; reprinted New York: Wings Books, 1997), p. 492.

24 MS D 9v. (c. 1508), trans. Edward MacCurdy, op. cit., p. 228.

25 Ibid.

26 Ibid. pp. 228–9.

27 Ibid. p. 229.

28 Catherine Dunton refers to a number of times in which this object/subject condition occurs in Leonardo's writings, in her article, "Meaning and Appearance – a Merleau-Pontian Account of Leonardo's Studies from Life," *Art History* 22: 3 (September 1999), p. 345, fn. 7.

29 MacCurdy, op. cit., p. 229.

30 Merleau-Ponty, "Eye and Mind" in: Galen A. Johnson ed., and Michael B. Smith, trans. ed., *The Merleau-Ponty Aesthetics Reader: Philosophy and Painting* (Evanston, IL: Northwestern U. Press, 1993), p. 122.

31 Ibid. p. 142.

32 MS G 37r (c. 1510–15), trans. Kemp and Walker, op. cit., p. 53.

33 Martin Kemp, *Leonardo da Vinci, Marvellous Works of Nature and Man* (Cambridge, MA: Harvard, 1981), p. 128.

34 This point is also made in Kemp 1981, p. 128, and refers to *De Anima*, Bk III: 432a7, in: *Complete Works of Aristotle*, op. cit., p. 687.

35 *Works of Aristotle*, p. 687.

36 Themistius, *On Aristotle: On the Soul*, trans. Robert B. Todd, *Ancient Commentators on Aristotle* series, gen. ed. Richard Sorabji (London: Duckworth, 1996), pp. 7–8, n. 45.

37 Ibid, p. 142.

38 Hans-Georg Gadamer, *Dialogue and Dialectic: Eight Hermeneutical Studies on Plato* (New Haven: Yale, 1980), p. 161.

39 Cf: Annaliese Maier, *An der Grenze von Scholastik und Naturwissenschaft* (Essen: 1943; 2nd ed. Rome: 1952) p. 304. Edward Rosen, "Renaissance Science," in *The Renaissance: A Reconsideration of the Theories and Interpretations of the Age*, ed. Tinsley Helton, *Symposium on the Renaissance*, U. of Wisconsin, Milwaukee, 1959 (Madison, 1961, 2nd ed. Westport, CT: Greenwood 1980), p. 90.

40 Discussions of this Aristotelian coherence between language, logic and world-view appear in: Werner Jeager, *Aristoteles*, op. cit., and in: Charles Breslin, "The Ontological Foundations of Aristotle's Logic." Thesis, University of Louisville, 1967.

41 Gadamer, op. cit., pp. 156–193.

42 Ibid. p. 161.

43 *Trivulzianus* 7v (1485–87), trans. Raymond Stites and Pierina Castiglione, in: R. Stites, *The Sublimations of Leonardo da Vinci* (Washington: Smithsonian Institution, 1970), p. 170.

44 Plato, *Timaeus*, Vol. VII, trans. Rev. R. G. Bury, Loeb Classical Library (Cambridge, MA: Harvard U. Press, 1929), p. 247.

45 Codex Urbinas 1270, 15v, *Trattato della pittura* 27. Translated into English primarily by the following authors: Claire Farago, *Leonardo da Vinci's Paragone: A Critical Interpretation with a New Edition of the Text in the Codex Urbinas* (Leiden: E. J. Brill, 1992), pp. 234–7; Kemp and

Walker, op. cit., p. 46; Irma Richter, *Paragone, A Comparison of the Arts by Leonardo da Vinci* (London: Oxford University Press, 1949), pp. 67–68; A. Philip McMahon, ed. and trans., *Treatise on Painting by Leonardo da Vinci*, 2 vols., intro. L. Heydenreich (Princeton: Princeton U. Press, 1956) p. 28.

[46] Farago, op. cit., p. 237.

[47] Ibid. p. 356.

[48] *De Oratore*, III, x. 37, trans H. Rackham, Loeb Classical Library, Vol. IV (Cambridge, MA: Harvard U. Press, 1977), pp. 30–31.

[49] XI. i. 6, trans. H. E. Butler, Loeb Classical Library, Vol. IV (Cambridge, MA: Harvard U. Press 1979), pp. 156–157.

[50] XI, iii, 182, ibid.

[51] Farago, op. cit., p. 235.

[52] Ibid. pp. 236–7.

[53] Ibid.

[54] Ibid, pp. 234–5.

[55] Ibid. pp. 237.

[56] Ibid. p. 236.

[57] Richter, op. cit., p. 69.

[58] McMahon, op. cit., p. 17.

[59] Kemp and Walker, op. cit. p. 46.

[60] Codex Atlanticus 76r.a; 219a, see endnote 10.

[61] Urbinas 27v–26r, trans. Kemp, op. cit., p. 42.

[62] Dunton, op. cit., p. 345, fn. 7.

ALBERTO J. L. CARRILLO CANÁN

THE GARDENS OF VERSAILLES AND THE SUBLIME

In envisioning himself as the distinguished guest of Louis XIV in the king's *Manière de montrer les jardins de Versailles*, the architectural theorist Vincent Scully refers to his *surprise* at a point of the tour offered by the Sun King, and says:

> Then it all bursts out before us, the *tapis vert* that slants gently down through the trees that are brought tightly in to focus the view, the Bassin d'Apollon, the Grand Canal, and, most of all, the *overriding* sky. Louis tells us to *admire* it all. He was quite aware that it was a moment of vast *release*. The oval of Latona opens up and surprises us, *releasing us* to the burst of velocity that explodes at the middle of the garden. Our gaze moves rapidly down the *tapis vert*, but when it hits the water it literally takes off. It no longer adheres but slides across the water to the sky reflected on it. (. . .) We are *released to infinity*, or at least to *indefinitely expanding* space. (A 227n.)[1]

If they do provoke these effects at all, the Gardens of Versailles apparently do so by means of strict regularity. But characteristically, in a well-known passage of the *Kritik der Urteilskraft*, Kant refers to the "English taste in gardens" and associates it with his concept of beauty proper, namely "free beauty," whereas a few lines below he expressly rejects every "stiff" and "mathematical regularity," for, he says quite clearly, "it contravenes taste" (KU 72). Certainly, in asserting this, Kant refers to the "regularity" of a "pepper garden" instead of directly speaking of Versailles, but the criticism of "regularity" applies to the Gardens of Versailles as well. Now, Kant did not reduce the "aesthetical judgment" to the "judgment of taste," that is, to the "judgment on the beautiful," for according to Kant the other kind of "aesthetical judgment" is the "judgment on the sublime." The goal of this paper is to probe whether the "release to infinity" referred to by Scully falls under the Kantian judgment on the sublime. But in fact, this does not seem to be the case, for regularity is a main feature of the French classical garden, which not only contravenes Kant's criterion of beauty, but also his concept of the sublime. On the one hand, Kant does not refer the sublime to a man-made object like a garden, but to an "object of nature" (KU 115), such as the ocean, peaks, overhanging rocks, and even equally important, Kant does not refer the sublime to any form or regularity in nature but rather expressly to nature "in its chaos, or in its wildest and most irregular disorder and desolation" (KU 78). On the other hand, Kant relates the sublime either to the "immeasurabil-

A.-T. Tymieniecka (ed.), Analecta Husserliana LXXVIII, 47–58.

ity of nature" (KU 104), or to nature insofar as it is "represented as causing dread" (KU 102), and obviously any garden is both measurable, a finite region, and not dreadful but rather pleasant.[2] In any case, I must examine the Gardens of Versailles in view of Kant's concepts of beauty and of the sublime.

1. "FREE BEAUTY" AND KANT'S CRITICISM OF REGULARITY

Kant defines the beautiful in several ways, but the most salient feature for us now is the definition of beauty as "free beauty" or "self-subsisting beauty," which "presupposes no *concept* of what the object should be." (KU 48) The main point here is "mere apprehension (. . .) of *form*," and just without concept. To recall the meaning of such a Kantian idea in a simple way, one can refer to Kant's examples. He says: "Flowers are free beauties of nature. Hardly anyone but a botanist knows the true nature of a flower, and even he, while recognizing in the flower the reproductive organ of the plant, pays no attention to this natural end when using his taste to judge of its beauty." (KU 49) To speak of the "natural end" of "the true nature of the flower" amounts to speaking of a concept, according to which one could judge from the stance of knowledge but not from the stance of beauty. Kant thus goes on as follows:

> Hence no perfection of any kind – no internal finality, as something to which the arrangement of the manifold is related – underlies this judgement [about beauty]. Many birds (the parrot, the hummingbird, the bird of paradise), and a number of crustacea, are self-subsisting beauties which are not appurtenant to any object defined with respect to its end, but please freely and on their own account. (KU 49)

Kant's rejection of "concepts" as related to aesthetical judgement amounts to rejecting "meaning." He says: "So designs *à la grecque*, foliage for framework or on wallpapers, etc., have no intrinsic meaning; they represent nothing – no object under a definite concept – and are free beauties. We may also rank in the same class what in music are called fantasias (without a theme), and, indeed, all music that is not set to words." (KU 69)

From here one can easily understand Kant's further rejection, namely the rejection of mathematical regularity. He says:

> (. . .) geometrically regular figures, a circle, a square, a cube, and the like, are commonly brought forward by critics of taste as the most simple and unquestionable examples of beauty. And yet the very reason why they are called regular, is because the only way of representing them is by looking on them as mere presentations of a determinate *concept* by which the figure has its rule (according to which alone it is possible) prescribed for it. One or the other of these two views

must, therefore, be wrong: either the verdict of the critics that attributes beauty to such figures, or else our own, which makes finality apart from any concept necessary for beauty. (KU 70)[3]

Mathematical regularity is thus linked to some concept; for this reason it violates Kant's definition of "free beauty." The concept does not need to be a mathematical one. We do not need to refer necessarily to mathematical regularity; it may be regularity as mere symmetry. Kant says:

Where some purpose is perceived, as, for instance, that of forming an estimate of the area of a plot of land, or rendering intelligible the relation of divided parts to one another and to the whole, then regular figures, and those of the simplest kind, are needed; and the delight does not rest immediately upon the way the figure strikes the eye, but upon its serviceability for all manner of possible purposes. A room with the walls making oblique angles, a plot laid out in a garden in a similar way, even any violation of symmetry, as well in the figure of animals (e.g., being one-eyed) as in that of buildings, or of flower-beds, is displeasing because the form is contrary to the finality, not alone in a practical way in respect of some definite use to which the thing may be put, but for an estimate that looks to all manner of possible purposes. (KU 70)

Kant is thus very clear. He goes on as follows:

With the judgment of taste the case is different. For, when it is pure, it combines delight or aversion immediately with the bare contemplation of the object irrespective of its use or of any end. The regularity that conduces to the *concept* of an object is, in fact, the indispensable condition (*conditio sine qua non*) of grasping the object as a single representation and determining the manifold in the form of the object. This determination is an end in respect of knowledge; and in this connection it is invariably coupled with delight (such as attends the accomplishment of any, even problematical, purpose). Here, however, we have merely the value set upon the solution that satisfies the problem, and not a free and indeterminately final entertainment of the mental powers with what is called beautiful. In the latter case, understanding is at the service of imagination, in the former, this relation is reversed. (KU 70n., i. a.)[4]

Therefore, mere symmetry leads mostly to a concept and on its part this leads to Kant's rejection of symmetry:

With a thing that owes its possibility to a purpose, a building, or even an animal, its regularity, which consists in symmetry, must express the unity of the intuition accompanying the *concept* of its end, and belongs with it to cognition. But where the maintenance of a free play of the powers of representation (subject, however, to the condition that there is to be nothing for understanding to take exception to) is intended: in ornamental gardens, in the decoration of rooms, in all kinds of furniture that shows good taste, etc., *regularity in the shape of constraint is to be avoided as far as possible*. (KU 71)

After the rejection of regularity insofar as it is linked to some concept, it follows that what can be understood is an unuttered decision for the English garden against the classic French garden. Then Kant goes on as follows: "Thus English taste in gardens, and Baroque taste in furniture, push the freedom of imagination to the verge of what is grotesque, the idea being that in this divorce from all constraint of rules the precise instance is being afforded where taste can exhibit its perfection in projects of the imagination to the fullest extent." (KU 71)

Kant only mentions the gardens, but he immediately rejects explicitly any regularity connected with gardens. Kant says:

All stiff regularity (such as borders on mathematical regularity) inherently contravenes taste in that the contemplation of it affords us no lasting entertainment. Indeed, where it has neither cognition nor some definite practical end expressly in view, we get heartily tired of it. On the other hand, anything that gives the imagination scope for unstudied and final play is always fresh to us. We do not grow to hate the very sight of it. Marsden, in his description of Sumatra, observes that the free beauties of nature so surround the beholder on all sides that they cease to have much attraction for him. On the other hand he found a pepper garden full of charm, on coming across it in mid-forest with its rows of parallel stakes on which the plant twines itself. From all this he infers that wild, and in its appearance quite irregular beauty, is only pleasing as a change to one whose eyes have become surfeited with regular beauty. But he need only have made the experiment of passing one day in his pepper garden to realize that once the regularity has enabled the understanding to put itself in accord with the order that is the constant requirement, instead of the object diverting him any longer, it imposes an irksome constraint upon the imagination: whereas nature subject to no constraint of artificial rules, and lavish, as it there is, in its luxuriant variety can supply constant food for his taste. (KU 72).

We can conclude that Kant's criticism of "stiff regularity" applies to the geometric, in fact almost mathematical regularity of the Gardens of Versailles. Kant considers the English landscape garden nearer to beauty than the classic French garden.

2. REGULARITY, POWER, AND UNLIMITEDNESS

In fact, it is important to stress regularity as a main feature in Versailles. An actual guide to the gardens[5] refers to the rigor in their design, and goes on as follows: "The general layout is geometrical: The garden is ordered starting from a principal axis, with secondary axes, alleys out of stars, basins in circle and half-circle; the whole in symmetry and staged on several levels. The trees are rigorously cut, composing a true vegetable architecture."[6] The ideal of regularity and symmetry is not only proper to garden design but is also a main character of Baroque aesthetic as such. In this respect, Wölfflin says in the pioneering work *Renaissance und Barock*:

The Baroque does not submit itself to terrain, on the contrary the Baroque masters the terrain and tries by all means to impose on it a homogeneous character: a main motive from one end to the other; it tries to impose main perspectives, to refer each detail to the whole. The axis of the mansion is the same for the garden; pavilions and other structures are not hazardous or mottled in an edge, but on the middle line; over all reigns correspondence and *strict symmetry* (RB 149).

Such regularity, according to Wölfflin, characterises the transition from the Renaissance to the Baroque. He says: "Certainly, the Renaissance was concerned in giving architectural motives, but without giving themselves an architectural relationship, all remained without unity of composition. At this point, thus, the Baroque means progress toward architectural composition of space." (RB 148)

Theoreticians often tend to interpret this "progress" as a political fact, and not merely a neutral "composition of space," but rather as an expression of a will to power. For example, Michael Kelly says: "Gardens are often places that situate viewers in positions of visual control over vast landscapes, allowing an illusory symbolic control over the world external to the garden (. . .)" (K2 275.b). Directly referring to Versailles, Tom Turner says: "To Louis XV, Versailles was a symbol of his sunlike magnificence. Power radiated outwards into France from Versailles like the great avenues, but extended to the furthest corners of France. All Louis's subjects were drawn into the orbit of his power and crowds milled through the grounds at Versailles" (T).[7] The most characteristic feature revealing the will to power seems to be, in any case, even regularity. Especially the strict geometrical cutting of trees and shrubs, and also the topiary proper not only to Versailles, but to the Baroque garden in general, are interpreted as the will to master nature as the expression of power.[8]

But in the Baroque garden regularity is a feature, which in the ideal case is used only in combination, even submitted to great extension. In fact, as Turner says: "The French 17th-century garden, a manifestation of Baroque taste, required variety as well as *unlimited vista* and achieved it with fountains, parterres, and lesser gardens disposed within the boscages (wooded enclosures) that flanked the central axis." (T) The issue seems to be not regularity itself, but regularity over a great extension in order to achieve "unlimited vista," as pointed out by Scully.[9] I can pick out two characteristic elements to show what is at stake: the strict geometrically cut trees and shrubs on the one hand, and the fountains on the other. The first mainly show regularity, the second are in fact also regular – though not as the serpentine lakes in English landscape gardens – but such regularity is intended to better reflect the unlimited sky; such fountains are in fact nothing but *reflecting pools*. The fountains, thus, and not the geometrically cut trees, are means to achieve unlimited vistas or to stress unlimitedness, for they are indeed "an open specular field" (MI 87), provided they are not only regular but also monumental.[10]

Referring now to "Baroque architecture as developed by Bernini, Carlo Maderno, Francesco Borromini, and Guarino Guarini," one can say that it "emphasized massiveness and *monumentality*, movement, dramatic spatial and lighting sequences, and a rich interior decoration using contrasting surface textures, vivid colours, and luxurious materials to heighten the structure's physical immediacy" (B). Applying this to gardens, it seems then that not so much mere regularity as *monumental* regularity in garden design is the way to achieve "unlimited vista," and this, of course, using monumental regularity to create a "vanishing point disappear[ing] at the horizon, at infinity,"[11] and thus linking up to unlimitedness. And this gives us the link to the question of the sublime.

3. THE "INDEFINITELY EXPANDING SPACE" AND THE MATHEMATICALLY SUBLIME

Kant's well-known exposition about the sublime refers it to the formlessness of an object, so far as it immediately involves, or else by its presence provokes, a representation of limitlessness, yet with a superadded thought of its totality. In fact, the delight associated with the sublime is linked not with form, as delight produced by the beautiful is, but with quantity. And this is a special link: the beautiful implies furtherance of life and, thus, it is compatible with charm and playful imagination; yet on the contrary, according to Kant, the sublime implies not furtherance but hindrance of vital forces, and only after them, the sublime implies intense release. This interplay from restrain to release of vital forces produces not play but earnestness in the concerns of the imagination. Moreover, since the mind is not only plainly attracted but also repelled by the object, this interplay does not produce mere positive pleasure as *admiration* and *respect* (see KU 74–6). So far, these are some of the well-known characteristics of the sublime. But the point is how they are linked to quantity instead of form. Before I go further, I would like to remark that if the effect of the sublime is truly earnest passion, and specifically admiration and respect, then sublimeness is obviously the ideal goal to be attained by any ostentation of power by any State.

According to Kant, through the *quality* (form) the beautiful seems to be preadapted to our power of judgment, whereas through the *quantity* (limitlessness and totality) the sublime seems to be violent toward our imagination, and to override it. (Remember Scully talking about the "overriding sky".) The point here, according to Kant, is that the object appearing to be sublime does not possess any form. It is rather nature in chaos, or in its wildest and most irregular disorder and desolation, provided that through it "magnitude" and "might," which are "rational ideas," become visible somehow (KU 77n.).

It is thus noteworthy to stress that the point here is not necessarily chaos as opposed to regularity, but representation of magnitude and might. It may be that Kant was right in considering regularity as "contravening taste," but the point is not beauty, not "restful contemplation," but "movement of the soul" (KU 80). And this seems to be valid not only for Baroque gardening but for Baroque taste in general, or at least where it is linked to monumentality (or sumptuousness). Monumentality is somehow linked to magnitude and to might, and here these constitute the point concerning the sublime.

According to Kant, "[*s*]*ublime* is the name given to what is absolutely great" (KU 81), or equivalently, "*that is sublime in comparison with which all else is small*" (KU 84). In leading to "unlimited vista," to "sky," by means of a "vanishing point disappearing at the horizon," the main perspective of a regular garden's arrangement links it to the sky, and there is nothing more able to provoke the representation of "absolute magnitude," "in comparison with which all else is small," than the sky. The sky excites the idea of "absolute magnitude" through, as Kant states it, "a striving in our imagination towards progress ad infinitum" (KU 85).[12] The sky itself is, thus, in mere "subjective judgment," a good candidate to be Kant's "greatest," "beyond which no greater is possible" and which calls forth "emotion" (KU 86). But there is sky almost over all, so the function of monumental regularity in the main perspective of the garden, or of a monumental regular fountain, does not lie in mere neutral pointing to the sky, but rather in *releasing* the observer to it as "overriding sky." This release can be considered as Scully's paraphrase of the Kantian "movement of the soul."[13]

Kant attempts to explain the "emotion" by the sublime, saying that in talking imaginatively a quantum in the imagination involves both imaginative "apprehension (. . .) and comprehension"; and according to him, apprehension can progress indefinitely. But it is the incapacity of comprehension, that is, the imaginative taking together, that produces the "whole emotion" (KU 87n). It seems that just the geometric symmetry of the garden along a main axis leading the eye to the horizon is what produces Kant's imaginative apprehension, whereas the vastness of the sky seems to be what calls forth the impossibility of comprehension. It must be pointed out that not only the perspective with the "vanishing point disappearing at the horizon, at infinity," but also the magnitude involved here, create such an impossibility to comprehend. Comparing the Gardens of Versailles with the garden at Vaux-le-Vicomte, the other major French classical gardens, Allen S. Weiss says: "In both gardens, the vanishing point disappears at the horizon, at infinity. But the major difference from Vaux is that extremity of the Gardens of Versailles (. . .). The dimensions of the Gardens of Versailles are (. . .) *disproportionate* to human vision: the structural difference between Vaux and Versailles is situated between perceptual proportion and *disproportion*" (MI 47). Thus Versailles, according to this description, does violence to our perception and so to our imagination, to our capacity for comprehension.

Putting aside the regularity of perspective, Kant describes a similar situation. He says:

> The same explanation may also sufficiently account for the bewilderment, or sort of perplexity, which, as is said, seizes the visitor on first entering St. Peter's in Rome. For here a feeling comes home to him of the *inadequacy* of his imagination for *presenting* the idea of a whole, and hereby imagination attains its maximum, and, in its fruitless efforts to extend this limit, recoils upon itself, but in so doing succumbs to an emotional delight. (KU 88)

Here we are obviously not concerned with the sky, but with imaginatively non-comprehensible monumentality. Kant says: "Nature (. . .) is sublime in those of its phenomena which in their intuition convey the idea of their infinity." (KU 93) Clearly, the Gardens of Versailles make *nature*, in this case the sky, really "convey the idea of [its] infinity."[14] Weiss accounts for this fact in his own less precise way than the Kantian, by referring to "the garden's central axis, whose optical and geometric effects were the inclusion of real infinity into the garden's finite closure." (MI 73) Thus, a *manmade* and as such *finite* object, and even a *regular* one, links to the *formlessness*[15] and *endlessness* of the *sky*, exciting the idea of "absolute magnitude," and thus provoking the "feeling of the sublime."

4. THE SUBLIME GARDEN AND MIGHT

The sublimity of nature, in our case, of the "overriding sky" over Versailles is, according to Kant, not only linked to magnitude but also to "might." The relationship to magnitude constitutes, in Kantian terms, the "mathematically sublime," whereas a relationship to might constitutes the "dynamically sublime." On the other hand, mighty nature is sublime only when *fancifully* represented as a "source of fear"; we must *merely picture* the case of a might that would override us if we tried to oppose it (KU 106). To some extent surprisingly, Kant goes on to a relationship with God and His might: in fact nobody pretends to contravene His commandments. This pretension would not come about, but simply to imagine it is a "source of fear." The main point is, thus, that "(. . .) provided our own position is secure (. . .)", (KU 105) the fearsome object, for example God, becomes rather a source of "respect" and even "admiration" (KU 76). Now, is it possible to apply Kant's description of our reaction to a *merely imagined* "source of fear" to the Gardens of Versailles? In which sense are they sublime as such a display of might, i.e., are they not only "mathematically" but also "dynamically sublime"?

Of course, if the Gardens of Versailles are a display of might at all, they display not God's but the Sun King's worldly might. Applying Kant's description, the Gardens must in some way represent certainly not God, but the king as a source of "respect" or "admiration." Here again, the point is not the sky alone, but the Gardens and the sky together. Such a relationship, namely the continuity between the Gardens and the sky, is obviously able to *represent* [darstellen] as Kant puts it, such might, for which "we do not find our powers commensurable to the task" of opposing it, and, which is thus "a source of fear." And so long as we feel secure, the representation calls forth "admiration." Obviously, not the Gardens, either alone or in their relationship to the sky, but the Sun King, the master of the Gardens, is the object of admiration. One must really think of the Sun King in order for the effect to come about. That is the point: Sublimity is, according to Kant, the true way to the "presence of the Godhead" (KU 108n.). In our case, this means that the Gardens of Versailles can have "dynamically sublime" effects only in our thinking of or truly seeing the Sun King, or at least a king reigning in Versailles. The "feeling of the sublime" implies such a real or mere fictional *presence* of the king.[16] If the Gardens call forth such emotional thoughts of the king, then the Gardens of Versailles are surely "dynamically sublime"; without this they are only "mathematically sublime." The point is thus the imaginative self-embodiment in such a situation, in which there is a king who is the master of the Gardens.[17]

5. CONCLUSION: FRENCH AND ENGLISH GARDENS, AND SUBLIMITY

The sublime character of the French formal gardens, be it "mathematically" *and* "dynamically," or merely "mathematically" sublime, obviously rests on a key relationship between foreground and background. Such a relationship is also a principal issue for the English landscape garden. In its summit, the main theory and practice for it (Knight, Price, and Repton) prescribed a three-step transition from form to formlessness. Near the house, in the foreground, the garden should follow a formal layout and thus be "beautiful." In the middle ground, the garden must be "picturesque" and, as the third step, in the background it must become "sublime." The last can be achieved not through some relationship to the sky, i.e., not through the transition to formlessness and the unlimited sky, but rather by its being mere nature without any ruling. And at this point Kant's definition seems to fit better: "But in what we are wont to call sublime in nature there is such an absence of anything leading to particular objective principles and corresponding forms of nature that it is rather in its chaos, or in its wildest and most irregular disorder and desolation, provided it gives signs of magnitude and power, that nature chiefly excites the ideas of the sublime." (KU 78)

Because of its formality, the classic French garden does not fall under Kant's definition of taste, that is, of the beautiful, but in spite of Kant's rejection of regularity, the Gardens of Versailles fall under the other kind of aesthetic judgment: they are sublime. Is it possible to extend this conclusion to Baroque taste in general? In other words, is Baroque art a kind of art that aims to call forth the presence of the divine? Furthermore, is Kant's concept of the sublime as a kind of linkage between the human and the divine what the Baroque aspires to?

Puebla, Mexico

BIBLIOGRAPHY AND ABBREVIATIONS

B = *Encyclopædia Britannica*, "The History of Western Architecture: Baroque and Rococo."

KU = Kant, Immanuel, *Kritik der Urteilskraft*, Meiner, Hamburg, 1990.

K2 = Kelly, Michael, "Gardens as Art," in: *Encyclopedia of Aesthetics*, Vol. 2, Oxford University Press, New York, 1998.

A = Scully, Vincent, *Architecture: The National and the Manmade*, Harper Collins, London, 1991.

T = Turner, Tom, *English Garden Design*, internet edition 1998/(http://www.gardenvisit.com/t/c0.html).

MI = Weiss, Allen, *Mirrors of Infinity. The French Formal Garden and 17th-Century Metaphysics*, Princeton, New York, 1995.

RB = Wölfflin, Heinrich, *Renacimiento y Barroco*, Paidós, Barcelona, 1991.

i.a. = Italics from author of the quoted text.

NOTES

[1] For citations, see the list of abbreviations above.

[2] Panofsky's concept of "French garden" again opposes the possible relationship between such a garden and Kant's concept of the sublime. He says: "The 'formal' style of gardening, reaching its climax in Le Nôtre's Versailles, had proudly imposed upon the infinity and irregularity of nature the finiteness and order of a little universe conceived by man – a universe cut out of (and from) the great outdoors and rationally organized into a geometrical pattern of avenues suitable for the stately progress of carriages and the capering of horsemen rather than for solitary walks (. . .)" (ES 131). Here there is obviously no place either for "nature" as such, or for nature as "limitless," or as "dreadful."

[3] Italics within a citation are mine unless the contrary is expressly indicated using the abbreviation i.a.

[4] The author of the quoted text uses these italics.

[5] See the web page of Versailles at http://www.mairie-versailles.fr/.

[6] Such expressions as "extreme geometrization" (MI 14) are common in referring to the French garden.

[7] As a further example, I need perhaps only mention the title of Phil Agre's book, *Territorial Ambitions and the Gardens of Versailles* (Cambridge University Press, 1997).

[8] Remember the famous remarks in the Duke of Saint-Simon's *Mémoires* (writing on Versailles) about "tyrannizing" or "forcing nature" (See MI 15). Allen Weiss refers to "(. . .) the attempt to shape nature according to the human will, a task symbolized by the imposition of geometric formalism manifested in the French formal garden." (MI 75)

[9] This point is missing in Panofsky's concept of "formal garden" seen in a footnote above. But in fact, the vista *beyond the limits* of the garden is already a characteristic of the Renaissance garden: "The typical Italian Renaissance garden, exemplified by the Villa Medici (circa 1544) and the Villa d'Este (circa 1550), was constructed on a sloping terrain with the villa at the summit so as to provide a magnificent overall view of the gardens as well as of the landscape or the cityscape beyond. The garden was not a fully enclosed microcosm (. . .) providing a god's-eye view upon the princely domain." (MI 13)

[10] Referring to the pools in the Garden of Chantilly, A. Weiss says: "It is water, in the form of reflecting pools, that forms the archetypal mirror, one that expands the imagination." (MI 85)

[11] Referring to both Vaux-le-Vicomte and Versailles, A. Weiss says: "In both gardens, the vanishing point disappears at the horizon, at infinity." (MI 47)

[12] Here it is interesting to point out Weiss's above-quoted formulation: "In both gardens, the vanishing point disappears at the horizon, at infinity." Obviously, we are confronted with a description proper only when Weiss speaks of "the vanishing point disappearing at the horizon," but the added "at infinity" is not a description proper anymore; this expression only refers to the idea, in Kantian terms, of "absolute magnitude," which is "called forth" by the imagination in view of the perspective vanishing at the horizon.

[13] Compare with Weiss's description: Versailles is "structured according to the optical and symbolic signification of the central alley leading from the chateau to infinity" (MI 62n.).

[14] Weiss interprets the vanishing point in paintings using linear perspective as a link of "the spectatorial ego (. . .) to infinity" (MI 62), but obviously the effect of such a linkage can by no means be compared with the effect of symmetry of the Gardens of Versailles which we are considering.

[15] Weiss's description according to which the Gardens of Versailles are "disproportionate" is again interesting, for he speaks of "the profoundly *formless* aspect of [this] formal garden" (MI 47, i.a.) and of "the suppression of (. . .) entire visual logic" (MI 48).

[16] Weiss refers to the theological and psychological "phantasms" called forth by the gardens, and says: "They were articulated by the king's presence" (MI 72).

[17] See Weiss's question: "And today? What phantasms arise during our own visit to these gardens? (. . .) The question must be left open (. . .)" (MI 98).

RAYMOND J. WILSON III

GARDENS IN STOPPARD, AUSTEN, AND GOETHE

Three authors use landscape gardening in different ways. Tom Stoppard uses this activity to reveal the essence of different historical ages and to highlight the situation facing an intelligent woman in each era. Jane Austen uses attitudes toward landscape gardening to individualize her characters, and Goethe's use is even more complex.

A reference to Arcadia, an ancient image of a garden-like land, occurs in an early interaction between Tom Stoppard's characters. In Stoppard's *Arcadia*, thirteen-year-old Thomasina, the play's main female character, behaves in what her mother calls a "pert" fashion. Thomasina implies that her mother has incorrectly translated the phrase "*Et in Arcadia Ego*" as "Here I am in Arcadia" (12). The words "Et in Arcadia Ego" literally mean "and in Arcadia I," a translation which does not immediately reveal the expression's meaning. Even knowing that Arcadia is a lost Edenic garden of the Greeks and Romans does not add much to our understanding. We must know who is speaking to find the correct idiom. Later in the play, the sound of hunters' guns firing in the garden-like park leads Thomasina to say, "I have grown up to the sound of guns like the child of a siege" (13). Her tutor, Septimus, then gives the correct translation: "Even in Arcadia, there I am." Erwin Panofsky explains that Latin permits omission of the verb "to be" and that the "et" can be translated loosely – making for Septimus's interesting translation. Thomasina's next words prove she knew Septimus's translation and even knows who is speaking the words. She says, "Oh, phooey to Death!" (13). Thomasina identifies the speaker: Death. The meaning is that death is everywhere, even in Arcadia – the beautiful garden.

The phrase takes the image of the garden as a symbol of safety and happiness into stark contrast. It is derived from a sequence of paintings. The first "Et in Arcadia Ego" painting was by Giovani Guercino between 1621 and 1623.[1] A large skull, personifying Death, rests on a pedestal on which is carved the inscription "Et in Arcadia Ego." The skull delivers a message: death comes even to people who are happy and who live in the midst of plenty – not just to the poor and miserable.

Nicholas Poussin painted two "Et in Arcadia Ego" works. He completed the first one presumably around 1630. It is called "Et in Arcadia Ego." The pedestal has been replaced by a sarcophagus on which the words have been inscribed. However, a small skull resting on top preserves the allegorical

A.-T. Tymieniecka (ed.), Analecta Husserliana LXXVIII, 59–66.

dimension, as does the presence of the river god who is presumably pouring out the plenty which makes Arcadia a utopia. This is the form in which the motif often appears in literature.[2] In the painting, the shepherds can be interpreted as having just discovered the sarcophagus and to be interestedly exploring it.

Poussin's second Arcadian picture, "The Arcadian Shepherds," finished in 1635 or 1636, makes a transition. In this painting, the allegorical dimension has been transformed. The words "Et in Arcadia Ego" on the sarcophagus appear to be a message from the person interned within it. The words cry out to be translated "I, too, once lived in Arcadia"; allegorically, they would be interpreted as follows: "I, who am now dead, was once, like you, alive and happy." However, Panofsky demonstrates that this is not a translation allowed by Latin grammar; he points out instances where people educated in Latin misremembered the expression when they were referring to the meaning ostensibly proposed by Poussin's second painting on this motif; authors from Balzac to Dorothy Sayers misremembered the motif as "*Et ego in Arcadia*," a form which permits "Et" to connect with "ego" rather than with "Arcadia" ("Tradition" 307).[3] Some even remembered it as "Et tu in Arcadia vixisti," which unambiguously means "I, too, lived (or was born) in Arcadia" (Panofsky, "Tradition" 296). In Poussin's second painting, the figures are contemplative and sad; this painting is always said to be a much greater one than Poussin's first try.

The garden is artistically created space (or place), an insight to which Phenomenology can contribute. Space *as experienced by a human being*, in the phenomenological tradition, is called "place." "Heidegger credits Kant with anticipating the notion of the 'clearing' as the 'open' place within which beings can be articulated and identified, hence can appear. This circle-open-place-*clearing*-world is a relational place; it does not, like an empty container, preexist man and things"; rather, the opening "is precisely the relation *between* man and things" (Fell 190).

In fact, to understand a "place" or "clearing" as existing prior to humans (and to demand an understanding of that place) defeats the entire purpose of phenomenology. We understand any entity (figure) only in relation to something which is *not* that entity (ground). If, as a prelude to understanding phenomena, we demand understanding of the ground for phenomena, we repeat Plato's move (and Kant's repetition of Plato's move) of turning phenomena into a series of semblances whose source is more original or primordial. If we make this mistake, as Heidegger pointed out, we deny ultimate reality to the things among which "we know ourselves immediately at home, i.e., things as the artist depicts them for us" (Heidegger, qtd. in Fell 190).

The idea of artistically shaped space is explicated by Stoppard's characters in the play *Arcadia*. A stage direction sets up the exchange: "The sketch book is the work of Mr. Noakes. . . . The pages, drawn in watercolors, show 'before' and 'after' views of the landscape, and the pages are cunningly cut to allow the latter to be superimposed on the former . . ." (10). Referring to Noakes, the landscape gardener who drew the sketchbooks, Hannah says,

"He'd do these books for his client, as a sort of prospectus." (She demonstrates.) "Before and after, you see. This is how it all looked until about 1810 – smooth, undulating, serpentine – open water, clumps of trees, classical boat house." Bernard replies: "Lovely. The real England."

HANNAH: You can stop being silly now, Bernard. English landscape was invented by gardeners imitating foreign painters who were evoking classical authors. The whole thing was brought home in the luggage from the grand tour. Here, look – Capability Brown doing Claude, who was doing Virgil. Arcadia! And here, superimposed by Richard Noakes, untamed nature in the style of Salvator Rosa. It's the Gothic novel expressed in landscape. Everything but vampires. (25)

Stoppard's Hannah is writing a book on how the changing taste in gardens revealed the essence of each historical era, especially the transformation of the Age of Reason into the Romantic era. "The history of the garden says it all, beautifully," she claims (27). The garden was transformed in stages from around 1730, when the garden showed "Paradise in the age of reason," to its Gothic condition after 1810. For Hannah, the change represents "the decline from thinking to feeling" (27).

Bernard's sneering condescension toward her shows what a modern intelligent woman must face. Bernard is intent on proving that Lord Byron killed a man named Chater in the garden of the manor. Bernard entitles his paper, "Death in Arcadia." Further evidence proves him wrong, and demonstrates Hannah's clear-headedness. The reader understands that Thomasina has discovered a scientific theory that was only known to scientists hundreds of years later; she may have even solved Fermat's last theorem, but she dies in a fire on the eve of her seventeenth birthday. It is her death that proves to be the real death in Arcadia.

In Jane Austen's *Mansfield Park*, characters take varying attitudes towards the idea of "improving" the grounds. A curiosity about Austen's title is that it does not refer to a person, such as *Emma*, or to human qualities such as *Pride and Prejudice*. It is one of two of her titles denoting a place; perhaps this fits with the novel's emphasis on landscape gardening. The characters' attitudes exactly parallel the salient characteristics that Austen assigns to each, and, indeed, help her to assign these characteristics. Mr. Rushworth, Julia Bertram's suitor and eventual husband, is stupid and ignorant. Mr. Rushworth "had been visiting a friend in a neighboring county, and that friend having recently had his grounds laid out by an improver, Mr. Rushworth was returned with his head full of the subject, and very eager to be improving his own *place* in the same way" (46). It is interesting that Austen uses the word "place" in an almost phenomenological sense.

"'I wish you could see Compton [the friend's estate],' said [Mr. Rushworth], 'it is the most complete thing! I never saw a place so altered in my life. I told Smith I did not know where I was'" (47). The purpose of the older landscape design was to bring pleasure to the residents; the purpose shifts in the new plans to a design meant to impress strangers from a distance. "The approach *now* is one of the finest things in the country," continues Mr. Rushworth. "You see the house in the most surprising manner. I declare when I got back to Southerton [Mr. Rushworth's estate] yesterday, it looked like a prison – quite a dismal old prison" (47).

The facts about Julia are that she considers her suitor to be an ignoramus and marries him anyway, for money and social position. Julia's reply to Mr. Rushworth gives Jane Austen a chance to characterize Julia: "'Your best friend upon such an occasion [planning changes in the garden]' said Miss Bertram, calmly, 'would be Mr. Repton, I imagine'" (47). A note from the editor says: "The successor of Capability Brown who died in 1783, Humphry Repton was much attacked in the early years of the nineteenth century for his improvements, which, judged by the newly fashionable standards of the picturesque, were felt to be dull, vapid, and mechanical" (435n). Thus, Julia is not only tactlessly pointing out her lack of faith in Mr. Rushworth's competence in front of the others, but Austen is also, probably negatively, characterizing Julia's taste.

Mr. Rushworth thinks that "Repton, or anybody of that sort, would certainly have the avenue at Southerton down; the avenue that leads from the west front to the top of the hill you know" (48). The reference to cutting down so many trees gives Austen a chance to characterize Fanny, the heroine, who says, "Cut down an avenue! What a pity! Does it not make you think of Cowper? 'Ye fallen avenues, once more I mourn your fate unmerited'" (50).

Edmund, the hero, ruefully sides with Fanny, saying, "I am afraid the avenue stands a bad chance, Fanny" (50). Austen also characterizes Edmund: "'Had I a place to new fashion, I should not put myself into the hands of an improver. I would rather have an inferior degree of beauty, of my own choice, and acquired progressively. I would rather abide by my own blunders than by his'" (50). Edmund thus presents the position that the grounds would represent the taste of the resident, rather than serve as conspicuous consumption prepared by an expert in the current taste.

In Mary Crawford, Fanny's rival for the hero's hand, the author shows a different negative characterization. Mary remembers when her uncle was improving an estate of theirs: "For three months we were all dirt and confusion, without a gravel walk to step on, or a bench fit for use." Mary does not object to the new fashion, "but it must be all done without my care" (51).

Henry Crawford, the wrong man for the heroine, at first seems to be characterized positively, since he improved his estate according to his own plan: "I had not been of age three months before Everingham [his estate] was all that it is now. My plan was laid at Westminster [a secondary school] – a little altered perhaps at Cambridge, and at one and twenty executed" (55). However, we later see a different side of Henry. Henry Crawford uses the "improving" issue to tease Edmund Bertram. Henry tells how he came upon the parsonage where Edmund plans to live in a short time. "'Well,' continued Edmund, 'and how did you like what you saw?'" Pretending to refer to the "fun" Henry had while improving his property, Henry replies, "Very much, indeed. You are a lucky fellow. There will be work for five summers at least before the place is livable" (218). Saying this in front of Fanny, Henry is pointing out to her that he is much wealthier than Edmund and can give her a better material life.

Edmund replies: "No, no, not so bad as that. The farm yard must be moved, I grant you; but I am not aware of anything else" (218). Henry then gives a long list of supposedly necessary improvements, including buying up a great deal of the surrounding land – suggestions which he knows will never be within Edmund's means. Edmund disagrees, mildly, saying, "I think the house and premises may be made comfortable, and given the air of a gentleman's residence without any very heavy expense, and that must suffice me; and I hope may suffice all who care about me" (219). The reader knows that Fanny glows with approval and that Mary – who demands a more luxurious life – simmers with disappointed resentment toward Edmund. Thus, Edmund's expression of his attitude toward "improving" his property not only clarifies his character but also neatly sets off the difference between the two women who compete for his love: Mary and Fanny.

Johann von Goethe's novel *Elective Affinities* actually begins with one of the four main characters, Charlotte, supervising the renovation of a garden. Her husband, Eduard, asks his gardener where she is: "'She is over there on the newly laid out grounds,' the gardener replied. 'The summer house which she has been building against the rock wall opposite the castle will be finished today. Everything has turned out beautifully and will certainly please Your Grace.'" The gardener reveals a phenomenological perspective in that he assesses the landscape gardening based on what the people in it will experience. "'The view from there is remarkable: the village is below; a little to the right is the church, whose steeple you can almost look over; and, opposite, the castle and the park'" (3). As we will see, an analysis of the garden emphasis in this novel helps explain an aspect of its ending that otherwise seems out of place.

Charlotte has built the summer house on the assumption that she and Eduard alone will be using it. The direction of the plot receives a forecast by Eduard's reaction: "'I have only one criticism,' he added, 'The pavilion seems to me rather small.' 'Certainly large enough for two,' Charlotte answered. 'Yes, and there may be room even for a third person.' 'Why not? And for a fourth as well. If we have company, we can always make other arrangements'" (4). Goethe's entire project in *Elective Affinities* is to explore the interactions of four people, and the summoning of the third and fourth is intimately connected to landscape gardening. Eduard remembers that his friend the Captain is good at landscape design; Eduard asks if he can bring this friend to the estate. Eduard thinks that the Captain can improve on Charlotte's design; but more centrally, his friend is unemployed, and putting him in charge of the landscape design is a good excuse to give him a job.

Since the Captain and Charlotte later fall in love with each other, Eduard's decision marks an important complication point in the novel's plot. Charlotte's wisdom and the maturity and strong sense of morality of both people limit the potential harm of this development. However, other developments from the landscape gardening do not work out so well.

When he arrived, the Captain saw how to improve on Charlotte's plans. By removing a considerable portion of rock, they could build a gradually ascending path to the hilltop, to replace Charlotte's steep paths and steps. He told Eduard his ideas, but urged him to keep the idea "in strict confidence, between friends; any mention of it would confuse and annoy" Charlotte (27).

> . . . The Captain's criticism of the new grounds, which Eduard found to the point, weighed on his mind. He kept silent about it for a long time; but at last, when he saw her toiling up her little steps and paths again from the summer house to the top of the hill, he could no longer refrain and told her after some evasions, of his new ideas (27).

As the Captain had anticipated,

> Charlotte was completely taken aback. She was intelligent enough to see at once that the friends were right; but her own work, so differently planned, already existed and could not be undone; she blamed the two men for planning everything on too large a scale, for wanting to make important work out of a playful idea and pleasurable occupation, without considering the expense which a more elaborate plan would inevitably involve. She was upset, hurt, and annoyed. She was unwilling to drop her former plans, although she could not quite deny the advantages of the new ones; but, determined as she was, she stopped the work at once, in order to gain time to think over the whole matter and let it mature in her mind (28).
>
> Now Charlotte was missing her active pastime, the two men were more and more together and looked into everything, especially the nurseries and hothouses (28).

As a result, Charlotte decides to bring her niece Ottilie to the estate to keep her company. This proves to be a fatal turning of the plot when, later, Eduard and Ottilie fall in love. Both of these two have a strong childlike aspect to their characters. Thus, they are not able to make the right decisions to prevent pain to all concerned.

"Eduard and the Captain devised an easier ascent to the crest of the hill and also planned to build there a kind of lodge at the top of the slope and in front of a pretty little wood" (57). The Captain, "having planned and measured every detail with great care, now brought up the problem of the road to the village, the dam at the brook, and the raised road behind it. 'Building that convenient path to the hilltop,' he said, 'will produce exactly the quantity of stone required for the dam. As soon as one project links up with another, both become less expensive and can be carried out more quickly'" (58). The project is an external reality that parallels the emotional disasters of the plot.

At the celebration of the completed dam, several people fall in the water. One small boy almost drowns, but is saved by the Captain and revived by a physician. Later, Ottilie is caring for the baby boy of Eduard and Charlotte:

> The oar slipped overboard and when she tried to steady herself, the child and the book slipped into the water. She caught the child's clothes, but her awkward position prevented her from rising to a sitting position. She could not at first turn around and raise herself; but at last she succeeded. She pulled the child out of the water, but its eyes were closed, it had stopped breathing (262).

Thus, the baby boy died in the lake derived from the gardening preoccupation of all the characters.

Ottilie, who had never eaten enough, becomes acutely anorexic in her grief and remorse. She, too, dies, and a memorial to her becomes a salient aspect of the grounds. The Catholic-like outcome of peasants making pilgrimages to her chapel does not fit the prevailing Protestant culture of the novel, but the chapel where she is buried becomes the final piece in the remodeling of the estate, an ending that fits in well with the importance given to the concept of "garden" in this novel.

Loras College
Dubuque, Iowa

NOTES

[1] Dates for paintings are from Panofsky.

[2] Panofsky says that "in Evelyn Waugh's *Brideshead Revisited* the narrator, while a sophisticated undergraduate [and student of art theory] at Oxford, adorns his rooms at college with a 'human skull lately purchased from the School of Medicine which, resting on a bowl of roses, formed at the moment the chief decoration of my table. It bore the motto *Et in Arcadia Ego* inscribed on its forehead'" (311). Another example can be found in W. H. Auden's poem entitled "Et in Arcadia Ego" (Auden 544).

[3] Panofsky had applied the motif to Poussin and Watteau in a 1936 article ("Transience" 223–254).

REFERENCES

Auden, W. H. *Collected Poems*. Ed. Edward Mendelson. New York: Random House, 1976.

Austen, Jane. *Mansfield Park* (1814). Oxford: Oxford University Press, 1970.

Casey, Edward S. *Getting Back into Place: Toward a Renewed Understanding of the Place-World*. Bloomington: Indiana University Press, 1993.

Fell, Joseph P. *Heidegger and Sartre: An Essay on Being and Place*. New York: Columbia, 1979.

Goethe, Johann von. *Elective Affinities* (1808). Tr. Elizabeth Mayer and Louise Bogan. Chicago: Regenery Gateway, 1966.

Panofsky, Erwin. "*Et in Arcadia Ego:* On the Conception of Transience in Poussin and Watteau" (1936). Rpt. *Philosophy and History: Essays Presented to Ernst Cassirer*, ed. Raymond Klibansky and H. J. Paton. New York: Harper and Row, 1963, pp. 223–254.

Panofsky, Erwin. "*Et in Arcadia Ego:* Poussin and the Eligiac Tradition," *Meaning in the Visual Arts: Papers in and on Art History*. Garden City, NY: Doubleday, 1955, pp. 29–30.

Panofsky, Erwin. "The Creativity of Language: An Interview," *Dialogues with Contemporary Continental Thinkers*, ed. Richard Kearney. Manchester: Manchester University Press, 1984, pp. 29–30.

Stoppard, Tom. *Arcadia*. London: Faber & Faber, 1993.

SECTION II

MIYUKI KATAHIRA

APPROACHING ZEN GARDENS: A PHENOMENOLOGICAL-ANTHROPOLOGICAL APPROACH

Interpretative social scientists have recently come to view good ethnographies as "true fictions," but usually at the cost of weakening the oxymoron, reducing it to the banal claim that all truths are constructed. The essays collected here keep the oxymoron sharp. . . . Ethnographic truths are thus inherently partial – committed and incomplete. This point is now widely asserted – and resisted at strategic points by those who fear the collapse of clear standards of verification. But once accepted and built into ethnographic art, a rigorous sense of partiality can be a source of representational tact.

James Clifford, "Introduction: Partial Truths", *Writing Culture*

Both Figures 1 and 2 show the same material, the dry landscape in Daisen-in Temple, located in Kyoto, Japan. Figure 1 is focused, whereas Figure 2 is blurred. The focused photo provides a clearer view, and an explanation such as "the sand represents waves of the ocean, and the rocks symbolise sacred mountains", might have wide appeal. Thus seen in focus, the garden can be

Fig. 1. Dry Landscape Garden in Daisen-in Temple. (Focused)
Photographed by ASANO Kihachi.

A.-T. Tymieniecka (ed.), Analecta Husserliana LXXVIII, 69–84.

Fig. 2 Dry Landscape Garden in Daisen-in Temple. (Unfocused: Computerized)
Originally photographed by ASANO Kihachi.

read as a text, and we can question the meaning of the details that are visible within it. My attempt is, however, to deal with the blurry photo, the one lacking a focal point, which is therefore hard to *see* or *read*.

With the blurry shot of the garden, imagine that vision is no longer prioritised in the perceptual order. Imagine that the garden lies quietly at the foot of the mountain. Imagine yourself standing barefoot on a floor at some distance from the garden. Visiting the garden is not limited to visual engagement: it does not only equate seeing or watching; it involves the interrelations of a variety of senses. Visual information is but one of many sensory components that a person with normal vision is likely to experience during an actual visit to the garden. However, the overwhelming majority of narratives heretofore produced regarding such gardens only deal with how to understand what is seen. Reading the garden as a text is one possibility, but it is not the possibility that concerns me in this essay.

By using the example of the Daisen-in garden in Kyoto, the purpose of this paper is to criticise a dominant narrative in the anthropological discourse on material culture, that of the interpretative and structurist approaches. The discourse concerning material culture in anthropology often extracts meanings from objects, then applies the principles derived in the process to a wider and autonomous context of the objects themselves. According to the discourse, the garden is merely a *representation, symbol* or *manifestation* of something, such as an idea or system of thoughts. Consequently, scrutinising what the garden represents and symbolises is not only the best available narrative, but such a semiotic approach seems to be the only one. I find this disputable as it still does not answer the question of how such representations and symbols are communicated in and made part of the experience of the visitors to the garden. In other words, the notion of subjectivity is overlooked in this approach. I argue this based upon my own experience in the garden, although one problem that emerges here is that an experience as such belongs to the realm of private property and thus is often thought to lack academic relevance.

Nevertheless, I aim to scrutinise how a phenomenological approach has come to the forefront of the anthropological discourse in answer to the question above. First, I will summarise how an anthropological narrative, that of Joy Hendry, an English anthropologist, explains the dry landscape garden. Next, I will point out some problems with this narrative, parallel to describing how and why phenomenology has appeared within anthropological studies. Finally, I will reach an understanding of the garden in relation to its wider context, while still paying attention to the subjectivity and specifics of the garden. How does phenomenology outline the problems of the anthropological discourse? How can we discuss the garden in relation to a wider context without losing its particular actuality? How should anthropologists approach the garden at all?

Figures 1 through 6 display photos of the dry landscape garden of Daisen-in, which is one institution within the larger Daitokuji temple complex located in Kyoto. Joy Hendry accounts for this garden in her ethnography, which is marked by the most common way of accounting for material culture in the anthropological discourse: That is to reduce the objects under scrutiny to terms or meanings that eventually develop independently of the objects themselves. Such a narrative searches for another principle or reality, detached from the original object from which it was derived, and applies these principles or realities to other situations of other kinds.

Hendry's intention is to add to the many existing arguments regarding the social structure of Japanese society. She illustrates how the dichotomy of inside and outside, frequently presented as important concepts in studies of Japanese society, requires a further explanation: "The aim [of her study] is to break down a neat culture/nature distinction which we [Western readers of Japanese studies] find so hard to relinquish".[1] In examining the role of a fence that she observed in the Daisen-in garden, Hendry extracts a principle which she calls "wrapping".

Fig. 3. The northeast corner of the garden.
Daisen-in temple post card

Fig. 4. Drawing of the northeast corner.
Mirei Shigemori, Nihon Teien-shi taikei [History of Japanese Gardens].
CD-ROM no 1. (Tokyo: Shakai Shiso-sha, 1998)

This concept of "wrapping", she argues, is present throughout Japanese society in various manifestations. It is seen whenever something is "enclosed" in another item: a body wrapped in a kimono, a gift wrapped in cloth or paper, and even a miniaturised version of the world wrapped in the larger surroundings of a theme park. Her original term "wrapping", she claims, is more suitable than

Fig. 5. The Great Sea in the South corner of the Garden.
Photographed by ASANO Kihachi.

Fig. 6. The northwest rear of the garden.
Daisen-in temple post card.

"taming" or "cooking", both of which are employed with some frequency in analyses of Japanese society, because "wrapping" contains aesthetic implications. By studying the Daisen-in garden, Hendry appended her own term to the vocabulary used to examine the social structure of Japanese society. The following consists of a detailed summary of Hendry's approach.

HENDRY AND THE CONCEPT OF "WRAPPING"

In "Nature Tamed: Gardens as a Microcosm of Japan's View of the World",[2] Hendry argues as to the ways in which Japanese people perceive the world based on her observations in the dry landscape garden at Daisen-in, Kyoto. In the following passage, Hendry describes the role of gardens in general: "Gardens all over the world represent the efforts of human beings to create cultural versions of their natural surroundings. . . . They display local ideas about how the natural world is perceived, how it is constructed by that particular culture, how it relates to ideas about the supernatural, and how things in it are or should be ordered in a domesticated form".[3] In her view, gardens are representations of human ideas, through which one can reach understandings of ideas of nature, supernatural and social classification. In short, the Daisen-in garden is the means that allows her to grasp the Japanese view of nature and the world.

Hendry then summarises an oft-repeated argument about the Japanese view of nature, which can be found across disciplines. While some scholars insist that gardens are examples of how the Japanese live in harmony with nature, Hendry states that many others, on the contrary, argue that the Japanese prefer nature in idealised forms, such as gardens, and fear "raw" nature. Hendry agrees with the latter: "They fear 'nature in the raw', and the nature they love is 'cooked' nature, made palatable through culture such as literature and fine arts, through religious ritual, and through cultivation. A garden may be seen as a prime example of this 'cooked nature'".[4]

Hendry agrees that the concept of "raw" versus "cooked" is useful here, claming that gardens are a medium that "cooks" or "tames" nature. What she argues here is that gardens are rather a place where the "outside" world is transformed into "inside", they cannot be seen merely as an "inside" version of nature as opposed to the "raw" nature that is "outside". In other words, the original contribution of her argument is that gardens do not merely represent an "inside", but a complex way of mediating "inside" and "outside". Hendry describes the Japanese attitude towards nature in relation to gardens: "Building a garden is a way of mediating between the threatening, fearful aspects of the 'outside' world, of nature in the raw, and the safety and security of the 'inside' world of social and cultural life . . . an expression of a system of ideas, a system of classification and order".[5]

In other words, the gardens are a process of filtering the wild outside into a safe inside, according to Hendry. Through explaining the history of Japanese gardens, Hendry remarks that the most persistent characteristic of Japanese gardens is the notion of enclosure. She comments that a garden space is often enclosed by a bamboo fence, hedge and whitewashed wall with a tiled roof on top of it. In this enclosing practice, an aesthetic aspect for creating the ideal is required, and Hendry calls this process "wrapping". Instead of "cooking" or "taming", which are free of any aesthetic implications, "wrapping" suggests the implication of the aesthetics for creating the ideal.

"Wrapping" is comparable to layering, folding and creasing, according to Hendry: "The concept of wrapping in the way I have used it would seem to describe a process of cultural refinement very similar to the 'cooking' or 'taming' process . . . the term does carry more anthropological content to represent the complicated cultural process of mediating between 'raw' and 'cooked' nature".[6] Moreover, this aesthetic enclosure of "wrapping" is present in other situations such as enclosing the body, known as Kimono, and enclosing gifts in wrapping. "Wrapping", in her view, is a useful analytical tool to understand not just gardens but a variety of phenomena in Japanese society.

In order to confirm the effectiveness of the concept of "wrapping", Hendry applies it to the Tobu World Square, a theme park located near the Kinugawa Spa resort. The attempt is to illustrate how the Japanese tame a fearful world outside. This theme park features 102 exquisite one twenty-fifth miniature replicas of world famous architectural works and ancient monuments from 12 countries. In addition, the park also displays miniatures of several UNESCO world heritage buildings and locations. The Sphinx and Pyramid of Egypt, the Taj Mahal of India, the Great Wall of China and others are all exhibited as exact replicas. What strikes a visitor is likely to be the accuracy and detail given in creating these figures; for instance, the replica of Antonio Gaudy's Sacred Family Church in Barcelona is displayed as "under construction", incomplete, with 200 miniature replica people working on it. Likewise, the miniature Central Tokyo Station contains a crowd of people hurrying to work. Moreover, Rokuonji temple, known as a golden pavilion with its famous pond and surrounding forest in Kyoto is in miniaturised form.

Tobu World Square is, according to Hendry, a "wrapping" of the world within an amusement park, and it thus provides another example of how the Japanese tame the world outside. In finding the notion of "wrapping" within gardens, kimonos, and the miniaturised world in the theme park, Hendry seems content to argue that she has grasped the way that the Japanese perceive nature and the world. With the concept of "wrapping", the argument departs from the garden; in other words, Hendry takes it beyond the bounds of the garden to find other situations wherein the concept is relevant. Just as the garden "wraps" raw nature, so Hendry finds that the theme park "wraps" the world outside. Both gardens and the theme park are representations of the Japanese view of "outside", thus they become comparable, if not structurally interchangeable, with one another. Other particulars, in which one might find divergences irreducible to the dichotomy under consideration, do not find their place in Hendry's analysis.

CRITICISM OF STRUCTURAL REDUCTION

It is efficient to read a meaning and an idea behind an object in order to reach a broader understanding of its context. Hendry's work represents an attempt to read the ideas behind the garden of Daisen-in so as to develop an insight useful in reconsidering the social structure of Japanese society. However, I would argue that one can efficiently read a meaning behind objects only when the assumption of reading symbolic meanings behind the garden consists of only one of many ways of explaining it. Yet, approaches such as Hendry's are so dominant that they sweep other variant narratives away entirely. In these cases, the meaning departs from its situation and becomes its own autonomous significance. In other words, such a meaning loses its particularity to the original situation and no longer exists in relation to the garden. Reading symbolic meanings is one possibility, but such an approach does not represent the best choice of narratives. Regrettably, it seems rather to have become the only one conceptually available in the anthropological discourse.

Likewise, Dorine Kondo, an American cultural anthropologist, accounts for the tea ceremony in Japan with similar grammar. She notes that during the ceremony, the participants leave the tearoom in order to get some water in the garden. Then they return to the room with the water to prepare tea. Kondo describes this act as the "symbolic journey from the mundane to the ritual".[7] In her conclusion, she relates this journey to the dichotomy of inside/outside. Here, the variations and details of the act of getting water are not given sufficient attention; instead, Kondo reads meanings into the ceremony, as if the phenomenon itself were so rigid and codified that the details of who is doing it and how the ceremony is practised become irrelevant. Moreover, in a fashion similar to Hendry, Kondo neglects to explain how the notion of "symbolic journey" is communicated to the participants in the tea ceremony.

Instead, Kondo examines how the structural dichotomy of "inside/outside" is being perpetuated through the practice.

There are two major problems with approaching gardens in the way that Hendry does. First and foremost, the actuality and specificity of the garden are overlooked. Secondly, the role of spectator/audience is absent; in other words, the role of subjectivity is taken as transparent or just simply missing. Such approaches do not pay enough attention to how the ideas of the garden are communicated in or made part of the experience of the actual visitors to the garden. When approaching objects or phenomena, anthropologists can only explain them with already existing grammar, which overlooks their specificity and subjectivity. Are these the limits of conceptual availability that anthropologists face? Can't anthropologists scrutinise this otherwise?

Prioritising particular features of the garden, and paying attention to how people actually visiting the garden experience them, would provide one route to reaching an understanding of Japanese society. In this regard, Bruce Kapferer's approach, which combines phenomenology and anthropology, seems to offer significant possibilities. Kapferer, a social anthropologist of the Manchester school, points out how major anthropological approaches often read human activities as texts, and that the main concern of most anthropologists is usually to reach a meaning, especially since the emergence of post-modern criticism. Kapferer attempts to overcome such a dominant narrative by incorporating phenomenology into ethnography as a theoretical framework.

Kapferer locates phenomenology in his study on demon rituals among Sinhalese Buddhists in Sri Lanka as follows: "The phenomenological orientation I took was intended to give full status to the importance of exorcist understanding and practice which did not subordinate them to metropolitan 'authority'. . . . A phenomenological approach involves a radical suspension of the analytical categories of normal social science".[8] Phenomenology, in Kapferer's definition, is precisely what is needed to expand the limits of anthropology.

In his framework, Kapferer investigates the social ordering and formation system among the Sinhalese Buddhists of Galle, the capital of Sri Lanka's Southern Province. Kapferer describes their demon rituals, paying special attention to the details of the demon rituals as experienced by Sinhalese Buddhists. In other words, rituals are not merely a representation of ideas, and so Kapferer places emphasis on the details of the rituals, their participants and forms of practice, as a dynamic operating source. Specifically, Kapferer explores how a social ordering is constructed in relation to the belief in demons and the nature of the demonic through exorcism rituals within the Sinhalese Buddhist community. According to Kapferer, the Sinhalese believe that misfortunes, including suffering, illness and disease, are related to demonic power. In order to grasp the complexity of the Sinhalese understanding of demons and their related system of thought, Kapferer concentrates on the demon ceremonies wherein the Sinhalese exorcise these miscreant demons.

Combining phenomenology with anthropology, Kapferer approaches the relation between artistic form, meaning and experience in the Sinhalese exorcism rituals. His analysis places the aesthetic elements of the rituals, such as music, dance and drama, in a central role. Influenced by *The Phenomenology of Aesthetic Experience*, written by French phenomenologist Mikel Dufrenne,[9] Kapferer defines the significance of the aesthetic of Sinhalese exorcism in relation to the participants:

> Exorcisms place major aesthetic forms into relation and locate them at points when particular transformations and transitions in meaning and experience are understood by exorcists to be occurring or are to be effected. Demon ceremonies are then occasions where the communicational, meaningful and experiential possibilities of specific aesthetic forms within a particular culture can be subjected to examination.[10]

Through phenomenological insights, Kapferer gives detailed accounts of the sound of the music, dance movements and dramatic voices in the demon ceremonies. He does not simply *read* the ritual as *a text*; on the contrary, he emphasises the significance of hearing, moving and speaking in detail. The demon rituals are more than a mere visual engagement, in which subjectivity for cognising the aesthetics is manifested in practice.

As Kapferer points out, a phenomenological approach is successful in illuminating the problems of anthropological discourse. Taking this into account, I will reconsider the Daisen-in garden, the same garden that is the starting point for Hendry's study, using a similar phenomenological approach. Central to my quest are two complex questions: How is the garden communicated or made part of one's experience? What is the relation between the garden and the spectator?

APPROACHING THE PHENOMENOLOGICAL GARDEN

At one point, Joy Hendry agrees that the garden evokes human emotions. As stated above, she argues that with the concept of "wrapping", one can reach "a deep understanding of the human condition".[11] Although the scholars, an art historian and an architect, differ on some details, they both seem to accept the notion that the stones in the garden evoke human emotions. They argue that the garden manages to form a relationship between the stones – that is, the physical reality outside – and inner emotions of the people in the garden.

Hendry agrees about how gardens evoke human inner emotions; nevertheless, she provides no further exploration in detail for the disposition of the dynamics between people's emotions and stones. The important issue of the ways in which the inner self is related to the outer world of stones in the gardens is not discussed thereafter. The garden remains as a text throughout her argument that a "human relationship with that cultural version of nature can be abstract and profound, that elements of nature, suitably arranged (or wrapped) in an artistic form, can be *read* as expressions of the deepest human concerns".[12]

Inspired by Kapferer's proposal, I would argue against the position that gardens are representations of the Japanese view of nature, that they symbolise the ways in which the Japanese engage with the outside world. Symbolic meanings of gardens are nothing more than just that: meanings of the garden. In other words, they cannot be "explored as a thing in itself, either apart from the life worlds of its production or as divorced from the body and contained in abstract categories of thought removed from concrete reality".[13]

In this light, dry landscape gardens are not just mere representations of a Japanese view of nature; they are places where visitors are bodily engaged with gardens. The experience of being in a garden does not simply involve seeing or reading the garden as a text. Other senses are evoked so that vision is no longer prioritised, and thus, vision in the garden seems more blurred. In the following, I will give a short ethnographic overview of the Zen garden in Kyoto. Moreover, I shall aim to provide a narrative for the experience of the garden, using the elements of both phenomenology and anthropology.

ETHNOGRAPHY OF THE DRY LANDSCAPE GARDEN IN KYOTO

There are more than 1600 Buddhist temples and 270 Shinto shrines in Kyoto, in addition to an imperial palace, several villas, and numerous other gardens.[14] The city is located in the midwest of the main island of Japan, in a valley surrounded by hills and mountains in all directions except the south. The four seasons, all different from one another, are frequently celebrated in paintings, poetry and many other forms of art, where we find descriptions of the flowery spring, the hot and humid summer, the colourful autumn, and icy winters with snow.

Kyoto's city plan was designed based upon Chinese geomantic science, a field of study focusing on cosmic energy (*Ch'i* in Chinese, *Ki* in Japanese) and centered on the notion that the earth is a living body. For *Ch'i* to flow properly, it is important that city designers pay attention to the four directions, each of which is associated with an element: north with water, south with fire, east with wood and west with metal. The south is considered the most auspicious direction, and the northeast corner is a "demon gate". The northwest corner of Kyoto is a high mountain, Mt. Hiei, which blocks and guards the demon gate, whereas the south is not blocked by mountains, and thus open for fortune to enter the city. It is said that many Buddhist temples are laid out according to these principles as well.[15]

Daitokuji, the temple complex containing Daisen-in, is located in the north of Kyoto at the foot of a mountain, just as many other temples are also situated at the bases of mountains. Daitokuji was built in the early sixteenth century during 100 years of civil war. Daisen-in, one temple within the Daitokuji complex, was originally constructed for the retired abbot of Daitokuji. The designer of the garden is unknown, yet he is often identified as Soami, a famous Noh drama performer.[16] One might also apply the principles of geomancy explained above to the Daisen-in temple. In the following text I will describe in some detail the structure of the garden at Daisen-in, using some of the traditional vocabulary and thought associated with landscape gardening in Japan.

Upon entering Daisen-in, visitors must take off their shoes. Walking inside, one might become slightly disoriented because the inside of the building is only dimly lit, in contrast to the open space outside which is naturally bright. The sunlight does not filter directly inside; instead, only the light reflected off the ground finds its way into the room. One uses this indirect light to find one's way gradually through a hallway leading to the garden. Surrounding the abbot's quarters located in the centre of the structure, the garden consists of a series of sand patches that mimic the movement of water. Each of these patches of sand is on a different side of the building.

In the northeast corner of the garden are three rocks, standing for Mount Shumisen, the sacred mountain of Buddhism. From this corner, the visual high point of the garden, the remainder of the garden extends toward the north and east, thus forming an L shape. The width of the L-shaped dry landscape is only about nine feet (three meters), in which stones and rocks are placed in white sands. Beside Mount Shumisen, a stream of white sand forms a waterfall that flows both north and east.[17] (See Figures 1, 3, and 4.)

Along this flow of white sand in the L-shaped veranda, one finds water, several rocks and stones that represent a bridge, an islet, a waterfall, mountains and so forth. There are other stones in the shape of a crane and a turtle, both of which are symbols of longevity. It is frequently said that the whole composition of this part of the garden was designed to bring the sequence of a landscape painting to mind.

The stream of sand in the L-shaped part of the garden further moves on to flow beneath the stone bridge and under the rebuilt corridor, before running into the southern garden. The composition of the southern garden is placid and furtive, in contrast to the northeast veranda. The southern garden has two mountains and a great sea, all depicted using only white sands within the bounds of a rectangular space (Figure 5). To the north and northwest is a rear garden (Figure 6), which contains a sea of sand within which there are several rocks. Like the southern garden, it is simple in design. The whole scene, all around the abbot's quarters, reminds one of a painting on a scroll, unfurling as the scroll unrolls before the visitors.

The readers might have noticed that the above description of the garden uses the language of semiotics. I described the stones, rocks and sands in the garden as *representations of* things, such as religious symbols. As a consequence, it is inevitable to read meanings behind the garden. Needless to say, in this process the garden starts to lose its actuality and becomes a collection of signs *of something*. Moreover, the above narrative was carried by the medium of *writing*, thus what was written was, namely, what was seen. The process of describing excludes my own experience or subjectivity, evoked in the auditory, olfactory and tactile. Although the blurry shot of the garden (Figure 2) is more likely to match the experience in the garden, scrutinising the garden nevertheless indicates the focused photo (Figure 1), as if there were no choice otherwise.

What if the description placed emphasis on the auditory, olfactory and tactile, and not so much on seeing? What if explaining the garden payed enough attention to the subjectivity of the visitors? In that process, does the actuality of the garden give way to a "broader understanding" of Japanese society? Next, I will scrutinise the garden, free of the authority of seeing. Then, I aim to explore how the actuality of the garden is communicated in the visitors' experience, and how meanings are carried into that sequence.

Daitokuji is large and surrounded by a quiet neighborhood near the mountain. Although it is not far from a bus stop and wide, busy streets, the fast-paced scene along the road gradually gives way to serene greenery as one progresses along the path to the temple. When passing through the main entrance gate, the white walls of many sub-temples appear, and cars and people dwindle in number. The asphalt street gives way to a path of pebbles, which make a sound with each step. As the surroundings begin to offer less visual stimulation, the visitor's auditory engagement with the place becomes more pronounced. Hardly a noise is audible, and in the quietness, one finds that one's ears respond with more sensitivity than usual to each sound that greets them.

After several minutes of walking, one arrives at the entrance to Daisen-in, where one takes off one's shoes and leaves them neatly behind. Upon setting one's bare foot on the floor, everyone becomes acutely aware of the temperature and humidity or dryness of that particular day. One's first encounter with Daisen-in is, in short, tactile. The hallway extends through a rather dark passage; therefore, when the garden appears at the end of this darkness, it stands out brilliantly under the bright light of the sky. Reflecting the sunlight is the white sand of the garden, which makes the whole scene even more dazzling to visitors. If one should visit in the summer, the whole garden resonates with the sounds of cicadas during the day, and the voices of birds in the late afternoon and at sunset. One hears these sounds even though the animals themselves remain hidden from view by the trees, white wall, and garden roof.

As for the stones and rocks themselves, it is forbidden to touch them, but they seem cold and still, as if either absorbing all the sound or merely letting the surrounding sounds echo off their surfaces. Since one cannot touch the stones, one cannot of course weigh them, but they give the visitors a sense of ponderous weight. This impression begins with sight, but seeing leads one to imagine the auditory effects and tactile sensations that they possess.

Sitting for a while on the edge of the abbot's quarters and examining the garden, one becomes acutely aware of how the garden changes slightly from moment to moment. The wind blows through and alters the sound as well as the temperature, and the shadows move, allowing the sun to form different patterns on the surroundings. The whole scene also invites one to imagine the garden in other settings: how it looks in the early morning or at night under the moon, under the crimson and gold leaves of autumn, or coated in winter snow. The garden is not rigidly structured; on the contrary, it is flexible and "unwrapped" so that the imagination can play with it. The visitors communicate with the garden, finding in its flexibility and changing states various different experiences. In such a relation, the shifting meanings and emotional reactions brought about by the garden, as well as the visitor's intuition about it, become part of the overall experience.

The endeavour not to define the garden as a mere representation leads one to take notice of oneself as a visitor and of the various ways in which one engages with the garden in terms of the senses. Although a fuller description of a visit to the garden might require one to dissect the whole garden into describable individual components, thus inviting the use of a semiotic language of symbolism, I would stress that such symbols are governed by the manner in which the garden is felt or sensed. The garden in its physical actuality is sensed and felt concretely by each visitor; it is not there only to be read as a text. Moreover, the relation between the visitors and the garden is more than visual, and a visual description of its components is not enough to exhaust the manifold ways in which the garden is experienced.

CONCLUSION

Bruce Kapferer's phenomenological insights, when combined with his anthropological approach, did much to clarify the problems of the dominant approach in anthropological discourse. In the last section, I tried to describe the ways in which one might reach a broad, sensory and bodily understanding of the garden without losing its specificity.

When paying attention to the details of the garden, instead of merely reducing them to ideas to be extracted, it becomes apparent that seeing is not necessarily authoritative in the perceptual order. Audition and tactileness are as relevant as vision. The actuality of the garden is felt and sensed as much as it is there to be read as a text with the eyes. Openness to sensory engagement and felt actuality reveals the ways in which a meaning is given, communicated to, and becomes part of the visitors' experience. Exploring the visitors' sensory perceptions also raises the complex question of how humans form and constitute their reality. By approaching the garden phenomenologically, one opens oneself to different experiences than by using the dominant methods of dealing with material culture in the anthropological discourse. By keeping more to sensory experience – blurring one's sight and bringing one's subjectivity into scrutinising it – one can approach the garden in different ways, ways that allow innovative anthropological approaches.

NOTES

1 Ibid. 103.

2 Joy Hendry, "Nature Tamed: Gardens as a Microcosm of Japan's View of the World," in *Japanese Images of Nature*, ed. Pamela J. Asquith and Arne Kalland (London: Curzon 1997), pp. 83–105.

3 Ibid. 83.

4 Ibid. 84.

5 Ibid. 84.

6 Ibid. 85.

7 Dorinne Kondo, "The Way of Tea: A Symbolic Analysis," in *Man* 20 (1985), 287–306.

8 Bruce Kapferer, *The Celebration of Demons* (Bloomington: Indiana University Press, 1983), pp. 18–19.

[9] Mikel Dufrenne, *The Phenomenology of Aesthetic Experience* (Evanston, Illinois: North-Western University Press, 1973).
[10] Kapferer, ibid. 11.
[11] Hendry, ibid. 103.
[12] Ibid. 97.
[13] Bruce Kapferer, *The Feast of the Sorcerer* (Chicago and London: The University of Chicago Press, 1997), p. 25.
[14] Japan National Tourist Organization, *Japan* (brochure, 1995).
[15] Mitchell Bring and Josse Wayembergh, *Japanese Gardens: Design and Meaning* (New York: McGraw-Hill Book Company, 1981), pp. 1–7.
[16] Daisen-in Brochure, pp. 30–32.
[17] Ibid. 18–29.

REFERENCES

Bring, Mitchell, and Josse Wayembergh, *Japanese Gardens: Design and Meaning* (New York: McGraw-Hill Book Company, 1981).
Clifford, James, and George E. Marcus, *Writing Culture: The Poetics and Politics of Ethnography* (Berkeley: University of California Press, 1986).
Dufrenne, Mikel, *The Phenomenology of Aesthetic Experience* (Evanston, Illinois: North-Western University Press, 1973).
Sigemori, Mirei, *Nihon Teien Taikei* [The History of Japanese Gardens]. (Tokyo: Shakai Shiso-sha, 1998).
Hendry, Joy, "Nature Tamed: Gardens as a Microcosm of Japan's View of the World," in *Japanese Images of Nature*, ed. Pamela J. Asquith and Arne Kalland (London: Curzon, 1997).
Japan National Tourist Organization, *Japan* (brochure, 1995).
Kapferer, Bruce, *The Celebration of Demons* (Bloomington: Indiana University Press, 1983).
——, *The Feast of the Sorcerer* (Chicago: The University of Chicago Press, 1997).
Daisen-in Brochure.
Kondo, Dorinne, "The Way of Tea: A Symbolic Analysis", in *Man* 20 (1985), pp. 287–306.
Ohasi, Haruzo, *Nihon no Niwa Katachi to Nagare*, vols. 5.1 and 2 (Tokyo: Creo, 1999).

PHOTO CREDITS

Figures 1, 2, 5: Photographed by ASANO Kihachi.
Figures 3, 6: Daisen-in post card.
Figure 4: Mirei SHIGEMORI, Nihon Teien-shi Taikei [The History of Japanese Gardens] (Tokyo: Shakai Shiso-sha, 1998).

ANNE-MARIE BOWERY

HATHA YOGA: A PHENOMENOLOGICAL EXPERIENCE OF NATURE

We have forgotten who we are.
We have alienated ourselves from the unfolding of the cosmos.
We have become estranged from the movements of the earth.
We have turned our backs on the cycles of life.
We have forgotten who we are.[1]

1. INTRODUCTORY REMARKS

In the insular world of academia, we often find ourselves cloistered in windowless rooms. Electricity, air conditioning ducts and heating shafts surround us. Our chosen environment separates us from the world of nature. We often leave in the still dark hours, venturing forth from temperature controlled houses, travelling vast distances in temperature controlled cars, to classrooms and offices heated according to budgetary necessity rather than comfort or season. We live within walls (albeit ivy-covered) of brick and plaster. Our intellectual socialization rarely includes more than a brief walk across campus to the local brewery or coffeehouse or a foray across town to a restaurant recommended in a conference program. Nietzsche's meandering walks through Europe and Socrates' journeys beyond the city walls of Athens remain amusing anecdotes. We, like much of the post-industrial society, have lost touch with the natural world that grounds our intellectual experience.[2] We seldom stop to admire a sunset, pick a flower, or smell the air around us. As Chellis Glendinning notes, "You and I are not people who live in communion with the Earth, and yet we are people who evolved over the course of millions of years through savannah, jungle, and woodland – to live in communion with it" (1994, ix).

Instead, we live in a civilization that separates us from the world. Unfortunately, we cannot exist healthfully in this created environment. Our vital energy slowly dissipates. We become "dislocated from our roots by the psychological, philosophical, and technological constructions of our civilization, and this alienation leads to our suffering: massive suffering for each and every one of us, and mass suffering throughout our society" (Glendinning 1994, x). We suffer from environmental illness and "sick-building" diseases. We experience seasonal affective disorders. Plagued by chronic sinus

A.-T. Tymieniecka (ed.), Analecta Husserliana LXXVIII, 85–92.

problems from air pollution and by stomach upset from stress and unhealthy eating patterns, we swallow the quick fixes provided by prescription drugs and over-the-counter medications.[3] Unfortunately, we often mindlessly affirm civilization and the creature comforts it brings. We ignore the violence to nature that these comforts entail. A dismal scene lies just beyond our view: "Mowing down forests. Blanketing valleys and mountains with deadly poisons. Spewing garbage into rivers. Building machines to exterminate life" (Glendinning 1994, x).

Despite everything that civilization produces to separate us from our ground as natural beings, we nonetheless search for this ground. We try to create what comes naturally. We use industrial-strength fixes for natural problems. We tan in ultra-violet coffins; we use sunlamps for mood enhancement; we take a slew of vitamins rather than eating whole foods. We even search out oxygen bars where people in Japan, Canada and the United States pay for air and the energy that comes with pure breath. One can also see the call to nature in the increased popularity of backpacking and mountain climbing. Sadly, however, we are often woefully unable to cope with the dangers and delights of such experiences. Technology fails us, but we depend upon it rather than upon ourselves as natural beings.[4] We remain tied to the civilized technologies that separate us from ourselves. We bring all the conveniences of technology into the natural world. It is easier to pack a cell phone than what we will need to survive in the snow-covered woods for a week. Civilization is just a phone call away. A Mount Everest climber's tragic phone call shows the triumph of technology over nature. Just as clearly, however, it reveals the fleetingness of the triumph, the ultimate futility of our choice. "Here we have the 'postmodern mind': a rootless, undigested perception of life whose hallmark is the absolute relativity of all human-made experience – the very opposite of the primal matrix's caring respect for the nature-inspired differences among cultures and the penetrating sense of archetypal patterning that binds them" (Glendinning 1994, 118).

At the same time, one also notices the rise in popularity of spiritual practices. One thinks of the New Age meditation groups, "Rebirthing" techniques, Jacquelyn Small's work on integrative breathing. Think also of the pervasive interest in Buddhism and the popularity of "12-step groups," not to mention the rise of religious fundamentalism and movements like the Promise Keepers. Like our attempts to reconnect with nature, these spiritual practices represent another societal attempt to find our lost ground. They represent our search for connection with what we have lost, namely the awareness of ourselves as a part of nature. Glendinning sensed a link between "two of the most important social issues of our times: the psychological/spiritual challenge each of us is facing, and the ecological crisis besieging our planet" (1994, x).

For many people, the ancient practice of Hatha Yoga becomes a spiritual quest that unifies the disparity between self and world. I will focus on three particular aspects of Hatha Yoga which provide a possible solution to the dissociation from the natural world that we so often experience. First, the poses themselves. Second, the associated breathing practices of *pranayama*. Third, the changed mental attitude that arises out of the practice of hatha yoga. Finally, I will suggest how the practice of yoga might lead to greater social action to prevent further harm to the environment.

II. POSES (ASANAS)

According to B. K. S. Iyengar, "*Asanas* keep the body healthy and strong and in harmony with nature" (1994, 21). The names of the various poses (*asanas*) suggest this direct link to the natural world. Think for instance, of the all-important Sun Salutation (*Surya Namaskar*). It forms the basis for Asthanga yoga, the form of Hatha Yoga with the most direct link to the ancient *Yoga Sutras* by Pantanjali. Traditionally practiced at dawn, the body rises to greet the sun. As the sun energizes the world, providing light and warmth, the body moves through an intricate series of poses that warm the blood. The movements energize the body. They offer energy to the body just as the sun gives the possibility of growth and life to the world. The basic mountain pose (*Tadasana*) provides another obvious example. You simply stand with your weight evenly balanced between your feet. Your sacrum is aligned over the arches of your feet. Your shoulders align over your hips and your head balances between your shoulders. Yoga practitioners often imagine themselves as a mountain. A mountain draws strength from being grounded to the earth as it simultaneously ascends to the sky. Just as the mountain remains through external changes of the seasons, wind, snow, hikers, so your strength endures through the countless permutations of life. In the Tree Pose (*Vrkshasana*), the arms and legs become branches. One popular video encourages practitioners precariously balancing on one leg, "If you start to sway, don't give up. Trees sway."[5] The visualization works. The foot becomes solidly grounded to the earth. It grounds the arms as they reach above the head, and the opposite leg as it rests on the inner thigh.

In Sanskrit, the word Hatha itself means sun-moon. In an established practice, you balance a solar phase of the high energy poses like cobra (*Bhujangasana*) and locust (*Salabhasana*) with a corresponding lunar phase of relaxing poses like the child's pose (*Garba-pindasana*) or corpse pose (*Savasana*). As the stretch of the solar phase deepens, the ability to experience relaxation, rest and release in the lunar phase increases dramatically. Just as the sun follows the moon and the moon follows the sun, energy emerges out of relaxation and relaxation flows from the energy. You mimic the very processes of nature.[6] In doing so, you join them. Fittingly, the word yoga itself means yoking or joining together. According to Iyengar, the yogi "knows that throughout the whole gamut of creation, from the lowliest insect to the most perfect sage, there breathes the same Universal Spirit, which assumes innumerable forms" (1994, 42).

As if contemporary Americans were so divorced from nature that they would not see this integral connection, many popular yoga videos are filmed in natural settings. Rodney Yee offers us the varied landscapes of Yosemite National Park. Eric Schiffmann leads us to the stark beauty of White Sands, New Mexico. These settings underscore a simple fact. Yoga will place us in the natural world. Similarly, enticing advertisements for yoga workshops and retreats grace the pages of *Yoga Journal*. The experience of nature awaits. One ad offers "Yoga in Paradise." You can "practice under blue Bahamian skies and bask in the gentle sunshine." Another offers the possibility of "relax[ing] on pristine beaches and refresh[ing] yourself in the clear warmth of the Caribbean."[7] Certainly, a morning sun salutation on the beach of Isla Mujeres will enhance your affinity with nature, but the possibility of that affinity exists in the practice of yoga itself.

The animal names for the poses also illustrate yoga's relationship with the natural world: Fish (*Matsyasana*), Dog (*Adho Mukha Svanasana*), Cat (*Bidalasana*), Rabbit (*Sasangasana*), Camel (*Ustrasana*), Scorpion (*Virschikasana*), Crow (*Kakasana*), Eagle (*Garuda Asana*). Often, one can easily image the attributes of the animals reflected in the poses. You stretch each vertebra of the spine as the cat does. In another pose, your spine curves like a scorpion. You look up to the sky like a dog who howls at the moon. You float like a crocodile in the river. Your chest puffs up as a pigeon's does and your legs move softly like butterfly wings. Rick Fields sees the practice of these poses as leading from Hatha Yoga to the practice of Bhakti Yoga, the yoga of devotion. They "help reconnect us through imaginative embodiment with our animal kin" (Fields 1998, 68). This aspect of yoga parallels the nature-based religions of Native America.[8]

III. BREATHING (PRANAYAMA)

In the practice of Hatha Yoga, the breath integrates the mind and the body. Without focus on the breath, the poses are just exercises, beneficial on a physical level, to be sure, but the practice of breathing is what cultivates higher levels of awareness. Richard Freeman notes, "An intimate relationship exists between our breathing and our attention" (1997, 60). Similarly, the breath furthers the connection that one feels with nature: "The breath moves like the wind through the inner landscape."[9] Breath becomes a vehicle for transporting us into the natural world from whence we came, just as it reintegrates the mind and the body. Rodney Yee suggests that "the mind follows the exhalation and surrenders to the earth."[10] The mind exists completely with the breath.

Richard Freeman explains, "Because we are so attracted to or repulsed by things in our inner and outer worlds, our breathing often becomes erratic, shallow, and uneven." (Freeman 1997, 60). We feel separate from our basic life force (*prana*). However, regulated breathing or "*Pranayama* is a remarkably effective means of deconditioning the body and the mind so that something as familiar as the breath regains an extraordinary vitality and mystery" (1997, 70). Interestingly, Freeman is clearly aware of this separation from nature. He describes the process of Ujjayi breathing, "During the practice, make a soft aspirant sound with the flow of the breath – similar to the wind in the trees, the surf from a distance, or (for the more urban among us) water running through pipes in an adjacent apartment" (1997, 62).

Plato may have sensed this possibility. If Jeffrey Gold is correct, Plato advocates breathing practices as a means of centering the mind. Gold believes that "the esoteric physiology at the end of the *Phaedo* provides a specific technique of meditation and *pranayama* that one can use as a means of acquiring liberation. According to Eliade, Yoga teaches techniques of meditation that are indispensable tools used in the acquisition of liberation. Paramahansa Yoganada also describes meditation as a technique for liberation. I am suggesting that Plato makes the same connection" (1996, 25).

IV. MENTAL ATTITUDE

This emphasis on the breath eventually focuses the mind. As one's practice intensifies, feelings of calmness, peace, and acuity greatly increase. One has "attention [that] becomes timeless, like the granite faces" of a steep rock ledge. Iyengar describes the mental focus in this way: "The still waters of a lake reflect the beauty around it. When the mind is still, the beauty of the Self is seen reflected in it" (1994, 24). This aspect of Hatha Yoga helps us understand the natural world by cultivating alertness, an awareness of one's body, one's surroundings, and the interrelationship of all things. Frances Harwood describes this harmony as a "transparency of psyche, a wide-openness that emanates a space and time far beyond what we associate with the individual" (Glendinning 1994, 19). Similarly, the title of Sharon Salzberg's 1999 book, *A Heart as Wide as the World*, reflects this desired unity with the soul and the world.

Images of nature often help us cultivate this sense of spiritual affinity. In his video *Yoga of Meditation*, Rodney Yee talks of the space he sees before him: "When I see the beauty of Yosemite, I see you." He sees spaces, spaces carved by the valley. He sees skies, skies expanding above open fields. Space in mind. Space in body. The skies radiate openness. Your heart opens in response. "Your inner landscape" becomes part of the space of nature.[11] When I returned to the open skies of Texas after years in rural Pennsylvania, enclosed by rainclouds, piles of snow, and relentless grayness, I felt a palpable openness in both my mind and my body. Shortly afterwards, I began practicing yoga. I frequently envision the clear blue skies. I allow my mind to become part of the vast expanse. Glendinning argues that "To recover from trauma . . . we need to be psychologically open, to attune ourselves to the flow of the world around us and the flow of feelings and images within us" (1994, 131). Hatha Yoga facilitates this process. Hatha: Sun-Moon. The movements of Hatha Yoga bring balance. Yoga: Yoking together. Hatha Yoga connects the self with something larger. It heals the spilt between human being and world just as it heals the spilt between mind and body on an individual level. Rick Fields writes, "And so earth yoga: the yoga that leads us to realize the unity, the oneness of human and earth, civilized self and wild Nature: the great Mystery we have never been, and cannot ever be, apart from" (1998, 68).

V. ENVIRONMENTALISM

Many people view yoga as an ascetic practice that denies the needs of the body and ignores its physical surroundings. Indeed, there are stories of yogis meditating high in the Himalayas where the intensity of their body heat melts the surrounding snow. As such, it would seem to have little to offer the environmental movement. Those who mischaracterize it in this way overlook its power to reconnect us with the earth. Fortunately, for most practitioners, Hatha Yoga extends beyond asceticism. As Christopher Chapple writes, "The ultimate philosophical goal of yoga, involves the cultivation of higher awareness, which, from an environmental perspective, might be seen as an ability to rise above the sorts of consumptive material concerns that can be harmful to the ecosystem" (1995, 184). While yoga teaches its practitioners the art of satisfaction, "the yogi believes that every creature has as much right to live as any person" (Iyengar 1994, 32). Furthermore, yoga teaches the ability to become mindful of one's surroundings. The "Yogi looks upon creation with eyes of love" (Iyengar 1992, 32). Yoga fosters loving acceptance of the present moment without requiring resignation or withdrawal from undesirable circumstances. Rick Fields further explains this possible parallelism between yoga and environmentalism. He writes, "Just as in yoga we relax and stretch muscles constricted by the stresses of civilized life, so would the practitioner of earth yoga extend his or her practice to include the body of an earth attacked by the multiple stresses of civilization. From the viewpoint of earth yoga, the body of the earth is, like our own, a complex living organism, and like our own, sacred" (1998, 67). Indeed, if one easily imagines the "lower ribs like wings of a bird riding on the currents of breath," and sees the "exhalation as a tranquil pool and the inhalation as the morning mist rising to the sky,"[12] the earth becomes an extension of the self. Caring for the earth becomes a means of caring for oneself. In this way, Rick Fields extends the conception of yoga to the spirit of environmentalism itself. He writes that "a growing recognition that we need a spiritual response to the ecological devastation of our planet is taking shape under many banners: spiritual ecology, deep ecology, earth-based spirituality, eco-psychology, feminist ecology, creation spirituality, Gaia consciousness, and Dharma Gaia . . . they all can be considered as a kind of yoga of the earth" (1998, 67).[13] Fields continues, "But while earth yoga embraces action, it is based on the peace of contemplation and the calm eye of wisdom at the center of the hurricane" (1998, 68). Hatha Yoga offers us access to the calm eye of wisdom. "Our primal matrix grew from the Earth, is inherently part of the Earth, and is built to thrive in intimacy with the Earth" (Glendinning 1994, 16). Yoga brings us home.

LIST OF SOURCES CITED

Chapple, Christopher Key. "Nonviolence to Animals, Earth, and Self in Asian Traditions." *Indo-Iranian Journal* 38 (1995): 184–204.

Fields, Rick. "Earth Yoga. The Time Is Now." *Yoga Journal* 138 (1998): 67–68.

Freeman, Richard. "Triumphant Breath." *Yoga Journal* 137 (1997): 60–70.
Glendinning, Chellis. *My Name Is Chellis & I'm in Recovery from Western Civilization*. Boston: Shambhala, 1994.
Gold, Jeffrey. "Plato in the Light of Yoga." *Philosophy East & West* 46 (1996): 17–32.
Iyengar, B. K. S. *Light on Yoga*. New York: Schocken Books, 1994.
McGraw, Ali. *Yoga, Mind & Body*. Director Claudio Droguett, with Erich Schiffmann. (Videotape) Satellite Films, 1994.
Roberst, Elizabeth and Elias Amidon. *Earth Prayers from Around the World*. San Francisco: Harper Collins, 1991.
Salzberg, Sharon. *A Heart as Wide as the World*. Boston: Shambhala, 1999.
Yoga Journal's Yoga Practice for Meditation. Dir. Steve Adams, with Rodney Yee. (Videotape) Living Arts, 1996.

NOTES

[1] United Nations Environmental Sabbath Program. Roberst 1991, 70.

[2] Glendinning argues that our problems began long before the industrial revolution. Rather, we became separated from our natural primal matrix when we moved from a hunter-gatherer to an agrarian lifestyle. "It is a severance that in the western world was initiated slowly and subtly at first with the domestication of plants and animals, grew in intensity with the emergence of large-scale civilizations, and has developed to pathological proportion with mass technological society – until today you and I can actually live for a week or a month without smelling a tree, witnessing the passage of the moon, or meeting an animal in the wild, much less knowing the spirits of these beings or fathoming the interconnections between their destinies and our own" (1994, 64).

[3] Glendinning mentions several signs of trauma in society and in the individual that I think yoga can help address. Hyperactivity (89), recurrent intrusive recollections of trauma, flashbacks, and nightmares (90), psychic numbing, constriction of feeling, warding off of such intrusions (91), a sense of powerlessness over one's destiny, a sense of futurelessness, and surrender patterns (92), arrested genetic and/or psychosocial development (93), narcissism (94), and thinking disorders (95).

[4] "Essential to the existence of deconstructive postmodernism is the predominance of technological society over all other ways and cultures" (Glendinning 1994, 121).

[5] Ali McGraw's *Yoga, Mind & Body*.

[6] Iyengar elaborates on this point. Some poses "are named after vegetation . . . some after insects . . . some after aquatics and amphibians . . . some after birds and quadrupeds" (1994, 42).

[7] *Yoga Journal* 137 (1997): 178.

[8] Divinity within, Divinity in the natural world. A kind of pantheism. Look in Joseph Campbell's sacrifice and bliss for this paper. He talks about nature and its role, and the access to the supernatural.

[9] *Yoga for Meditation*.

[10] Ibid.

[11] Ibid.

[12] *Yoga for Meditation*.

[13] Actual environmental work, cleaning up litter etc., would fall under the category of "karma yoga, the yoga of work and action. And just as yoga restores our body, earth yoga restores the body of the earth: renewing streams that are no longer safe for spawning salmon, replanting clear-cut hillsides, reintroducing wolves and buffalo" (Fields 1998, 68).

MITRA K. MARTIN

IN SEARCH OF PARADISE: GARDENS IN MEDIEVAL FRENCH AND PERSIAN POETRY

INTRODUCTION

A man decides to build a garden. His purpose is to create a space that gives him pleasure. First, he chooses a site and builds a wall around it to separate it from the disorganized world beyond. Next, the gardener selects those elements from nature that please him, such as fragrant flowers, fruit trees, and running water. He arranges these elements within the garden according to a design that allows him to appreciate each of them to the utmost. Thus a garden is a special space full of pretty things from nature arranged in a pleasing way.

After making his garden, the gardener must maintain it with utmost vigilance. He fears anything that threatens his garden's order: reckless, stupid animals; violent storms; the coming of winter. Without his constant regulation, the natural elements of the garden would become overgrown and disorderly. Flowers would bleed out of their precise rectangular beds; weeds would grow at random and choke the rosebushes; rodents would nibble at leaves and gnaw at tree trunks. To preserve his garden as he likes it for any significant amount of time, the gardener must actively work *against* nature. In building a garden, then, the gardener has four tasks related to nature: separation, selection, imposition of order, and maintenance of order.

Gardens are inherently artificial. Their spaces are artificially bounded; their elements are artificially selected and arranged; their appearance is artificially maintained. The universe naturally moves toward disorder, and gardeners must expend energy in working against this. Depending on the gardener's habitat, his job may be easier or more difficult. In hot, arid Iran, for instance, creating a cool, pleasant space is a challenging task. There is a stark contrast between the inside and the outside of a Persian garden. "On the outside [of the garden's wall] is the desert . . . the harsh reality of life on the Iranian plateau. Within the wall are flowers, fruit, shade, water and life."[1] Within the garden, "protection was afforded from the encroaching sand, and from the noise and bustle of the city. High walls, trees and shrubs afforded privacy. . . . Outside was the barren landscape, the heat and the glare; inside was shade and vegetation. . . ."[2] Because of the harsh Persian climate, collecting garden elements there demands more energy than it might in a

A.-T. Tymieniecka (ed.), Analecta Husserliana LXXVIII, 93–138.

temperate, fertile land. The Persian garden is artificial in its design as well as its collection of elements. This "man-made geometry"[3] is dictated by tradition:

> The plan is rectangular . . . the main axis is formed by the water course, often higher than the surrounding ground, flanked by paths, bordered with trees and sometimes crossed by one or more secondary axes at right angles. These secondary axes either also carry water or are simply walks. Flowers and trees are often located in the quadrants.[4]

Though the constituents of a garden come from nature, they are artificially arranged and maintained. In building a garden, the Persian gardener imposes his will on disordered nature in a way that pleases him. The Persian garden is finely wrought artifice at its best, and Persians treasured their hard-won, man-made gardens.

The purpose of every garden's artifice is to engender sensual pleasure. Garden elements stimulate all five senses. Flowers are fragrant; trees bear ripe, succulent fruit; the soothing sound of running water delights the ear; grass is soft to the touch. Each of the garden's component parts, as well as the whole arrangement, are visually stunning. The best gardens even have virtuoso songbirds and are cooled by a gentle breeze.[5] Sensual delectation, however, is not the only kind of pleasure the garden affords. The artificially crafted garden also engenders aesthetic excitement. While all gardens are artifice, the best gardens are works of art.

Oscar Wilde writes:

> The more we study Art, the less we care for Nature. What Art really reveals to us is Nature's lack of design, her curious crudities, her extraordinary monotony, her absolutely unfinished condition. . . . It is fortunate for us, however, that Nature is so imperfect, as otherwise we should have had no art at all. Art is our spirited protest, our gallant attempt to teach Nature her proper place.[6]

By *Nature*, Wilde means "natural simple instinct as opposed to self-conscious culture,"[7] which is *Art.* According to Wilde's definition of art, gardening is the quintessential art form. This invites us to compare gardening with other art forms.

Like gardening, poetry is a careful selection and arrangement of elements designed to produce a profound effect on the beholder. Words of a poem are chosen as painstakingly as flowers in a garden, and both are arranged with care. Just as no one would consider a random torrent of words a poem, a chaotic area of vegetation and undergrowth could hardly be called a garden. The choice and arrangement of garden elements contribute to the aesthetic enjoyment of the garden just as the choice and arrangement of poetic images appeal to the reader's aesthetic sense. Poetry and gardens both appeal to a 'sixth sense' which responds to what is aesthetically pleasing. This is what Wilde refers to when he writes, "There is in us a beauty-sense, separate from the other senses and above them. . . ."[8]

However, whereas gardens provoke a sensual response in addition to this aesthetic response, the immediate sensory appeal of a poem is limited to its music. Real fruits and flowers actually appeal to sight, hearing, smell, taste, and touch; words on a page cannot. No matter how evocative the sensory imagery of a poem is, it cannot enthrall the bodily senses as garden elements can. Whereas a garden delights all six senses, a poet directs his energies towards pleasing the sixth.

I've discussed how gardens are similar to and different from poems. Now, I will consider how real gardens compare with gardens in poems. Terrestrial gardens are, of course, imperfect. But the gardens depicted in poetry are the best gardens imaginable. At once sensually and aesthetically enthralling, they represent paradise. Fantastical creations made by poets, these gardens don't exist anywhere on earth but in terrestrial poetry. And unlike transient earthly gardens, in poems gardens are immortal. Garden poetry addresses the tension between real and ideal, the paradox of terrestrial paradise.

Nowhere was garden poetry more abundant and rich than in the medieval period. I have chosen two examples of medieval garden poetry, each with a radically different cultural heritage. Manuchihri Damghani, a Persian poet of the early eleventh century, wrote panegyrics for his court patrons in Ghazna, Persia. I shall examine his treatment of the garden in a handful of his panegyric poems. I shall continue my exploration with a study of the gardens in the *Roman de la Rose*, a mid-thirteenth-century French narrative poem written by Guillaume de Lorris and continued by Jean de Meun. With such fertile material, this analysis promises to yield fruit.

It has been suggested that the *Roman de la Rose* was influenced by Persian poetry, in which the rose is a central symbol:

> The enthusiastic pan-Iranist, Pizzi, has endeavoured, but without adequate evidence, to show influence of the Persian Rose motif in the mediaeval *Roman de la Rose*, whose symbolism is reminiscent of this. But in the absence of definite documentary evidence and of preparatory studies we cannot exclude the opposite hypothesis, namely that a Hellenistic motif may have penetrated into both cultures, derived from that civilization which in various ways and forms fertilised them both, *i.e.*, late Hellenistic symbolism.[9]

In my discussion I shall assume no explicit connection between the two texts.[10]

I shall address two themes in my exploration of the different ways in which poets with dissimilar religious and cultural backgrounds use the garden as setting and symbol. First, I shall examine the contrasting social and religious goals of these poets' differing representations of gardens as artifice. Next, I shall explore how each poets' representation of the garden enables him to fulfill these goals. It will become clear that the intentions and aesthetics of even apparently secular poetry are dramatically affected by religious belief.

In Chapter One, I establish the cultural context of the garden poetry of Manuchihri Damghani and his goals in representing gardens as artifice. Chapter Two answers these questions with respect to the *Roman de la Rose*. These chapters deal with the garden as a setting for human activity, and the contrasting poetic representations of the gardens advocate very different models of behavior. In Chapters Three and Four, I examine each poet's conception of good poetry, and discuss the way it reflects his conception of the divine art of Creation, represented in microcosm by the garden. These chapters address the aesthetics of garden poetry.

CHAPTER ONE: MANUCHIHRI'S GARDEN KINGDOMS

Between 1033 and 1041, Manuchihri's patron was the Ghaznavid Sultan Mascud. As the Sultan's court poet, Manuchihri was expected to compose poems for all court occasions, from royal births and deaths to bacchic festivals. In medieval Persia, the centerpiece of every palace was a formal court garden, in which the monarch traditionally held court during the first day of spring.[11] The dynamic of the courtly milieu is the driving force of Manuchihri's poetry. "One cannot understand the poetry of this period without placing it in the context of the aristocratic and courtly tradition."[12]

Manuchihri composed poetry in two fixed forms, the *qasidah*[13] and the *musammat*. In this study, I shall focus only on his *qasidahs*. The Persian *qasidah* is generally between fifteen and thirty *bayts* (distichs) long, although it can be somewhat shorter or much longer. Complex rules govern its rhyme and meter. In composition and in meaning, each *bayt* is independent from those which surround it. The *qasidah* consists of three parts: it begins with the *nasib*, an exordium which frequently describes *Nowruz*, the first day of spring at the vernal equinox. This is followed by the *madh*, the panegyric proper, in which the poet enumerates the good qualities of his patron. The *qasidah* ends with the *duca*, a prayer for the poet's patron.

Like all court poets, Manuchihri depended upon his royal patron for his livelihood. Accordingly, his poetry was designed to engender and retain the favor of Sultan Mascud. Julie Meisami writes, "The court poet enjoyed considerable privilege and influence in return for ensuring his patron's repute by means of his eloquent verses. . . ."[14] As we shall see, however, his poetry was more than just a device to sustain his personal livelihood.

Instruction by Example

Manuchihri's poetry is overtly panegyric. Modern readers may doubt the sincerity of the effusive flattery in each *madh*. It seems transparently contrived to garner the king's favors; no monarch could be perfect, as is the one this poet describes. However,

> Panegyric preoccupations were never far removed from didactic ones. The ethical dimension of court poetry has received little serious attention from critics eager to associate court power in general, and panegyric in particular, with flattery and bombast, despite abundant evidence . . . for the view that an important function of poetry is moral instruction by example.[15]

This didactic function is exemplified in *qasidah* XIII. Here, Manuchihri's praise of his king's virtues should not be read as empty flattery, but rather as a characterization of the ideal king. Here is the *madh* in its entirety:[16]

> Ce premier élu du Sultan, il sera un jour l'élu du Maître du monde, du maître de la majesté, le premier et le dernier.
>
> Il sait dompter ses passions et celui-là est le meilleur qui sait dompter ses passions.
>
> Il n'est dur ni pour les autres ni pour lui-même; celui qui est dur pour les autres le sera pour lui-même.
>
> Chez lui la générosité s'allie à l'illustration, celui qui démolit l'avarice élève l'édifice de la libéralité.
>
> Sa naissance est pure et son charactère est encore plus pur; lorsque le caractère est pur le galbe du corps sera plus pur.
>
> Je vois quel est son empire sur sa colère, il faut qu'un homme sache maitriser sa colère.
>
> Son unique pensée est de triompher de ses ennemis et dès qu'il en a triomphé il sait pardonner.
>
> O homme fort par le jugement, plus puissant par l'esprit, je ne connais personne qui soit, comme toi, fort par le jugement et puissant par l'esprit.
>
> Tu possèdes beaucoup de richesses mais plus encore de reconnaissance envers Dieu; celui-là grandira en richesses qui sait être reconnaissant envers Dieu.
>
> Chez toi la raison et le corps commandent, c'est la passion qui obéit aux ordres reçus; la raison et le corps seront esclaves, si la passion commande.
>
> Par la garde que tu exerces sur toi-même tu ne fraies pas avec les hommes dissolus, et certes celui qui s'assied avec les dissolus deviendra lui-même dissolu.

Lorsque l'on est en paix avec toi la mauvaise fortune même apporte des avantages, le jour de ta colère le bonheur lui-même est nuisible. (XIII, 12–23)

The verses are instruction, not flattery. The traits of the ideal ruler are projected upon Manuchihri's kingly patron, who in reality did not reflect this high standard. The projection of the traits of the ideal monarch upon the real king at once praises the king, and invites him to emulate this ideal. Traditionally, it was the poet's voice that expressed the community's norms and conventions, and in each *madh*, Manuchihri articulated the Persian cultural conception of the ideal monarch. Similarly, the gardens in Manuchihri's *nasibs* are idealized for didactic and panegyric purposes.

The Garden Idealized

To carry out these purposes, Manuchihri first establishes the royal ownership of the garden by describing its elements as rare and priceless treasures. Flowers and trees are compared to precious gems and metals, and exotic perfumes: "Les tulipes tout épanouies devant les pétales des jasmins sont des bouches de corail qui chuchottent des secrets à leurs blanches oreilles" (XXIV, 5). "La manche de la rose de Chine est remplie de boules d'ambre gris; le pan de la robe de l'amandier est rempli de perles éclatantes" (XIII, 5). "Les branches *d'arbres* font surgir des perles . . ." (XXX, 5). No one but a king could own such expensive, rare treasures. Manuchihri dramatizes the king's possession of the garden elements when he compares a flower to a prostrate subject: "Voyez cette violette penchée vers la terre, c'est comme l'ennemi du Khadjeh, la tête retombée sur les genoux et le visage enduit de bleu" (XXIV, 7). Personified, the garden elements know that the garden has a kingly master, and that they are subject to his power.

Poem seventeen is richest in metaphors that explicitly compare garden elements to royal treasures. Narrating the "conquest" of spring over winter, Manuchihri characterizes *Nowruz* as a king: "Le Nowruz était parti de ce pays en voyage comme un Roi; en effet d'illustres monarques entreprennent des voyages" (XVII, 6). During his absence, Winter plunders *Nowruz*'s rich kingdom: "[L'Hiver] brisa boîtes d'or et de perles des arbres fruitiers" (XVII, 10). When the north wind sees Winter's mischief, it reports to *Nowruz*:

"Dans ta royale maison, tout ce qu'il y avait de luxe et d'ornements, l'avant-dernière année passée,

"tout cela, l'armée de l'Hiver a saccagé de fond en comble et pilla tant ton trésor que tes perles royales;

"tes petites bien aimées, rose, fleur de grenadier et jasmin, il leur a arraché les bracelets des bras et les boucles d'oreille des oreilles;

"tes musiciens, le ramier et le rossignol, il leur brisa la flûte entre les mains et le tambourin sur le sein" (XVII, 14–17).

In response to this outrage, *Nowruz* mobilizes a decorative army and tells his messenger to warn the wintry king of his imminent arrival, upon which *Nowruz* plans to present the reigning monarch with many rich gifts:

"J'irai là avec de bons augures, avec d'immenses richesses, avec l'éclat d'une heureuse étoile, avec un choix fortuné.

"Avec cent mille coupes d'un vin rouge au parfum de musc; avec cent mille pétales de la rose Kamkar.

"Avec tes chers petits rossignols à tête rouge pinçant la harpe, avec tes chers petits jasmins aux joues de corail et versant des parfums de musc" (XVII, 34–37).

Manuchihri describes *Nowruz* as a monarch who owns many beautiful and rare treasures: flowers, trees, and birds. Through a conceit, he compares these natural treasures to treasures a Persian king would own: precious gems, metals, lovely slaves and musicians. The poem skillfully conflates *Nowruz* with the royal patron as it elides from *nasib* into *madh*.

At their most literal level, Manuchihri's *nasibs* describe the palace garden, owned by the king. His comparison of garden elements to royal treasures emphasizes the king's ownership of the garden. But, as the *qasidah* quoted above demonstrates, the king's ownership of the garden has an allegorical level as well: Manuchihri compares the court garden to the king's whole realm.

While treasure images establish the king's ownership of the actual court garden and the allegorical garden kingdom, they also conflate the court garden on *Nowruz* with the Islamic conception of celestial Paradise. The Koran states: "For them [the believers], the gardens of Eden,[17] under whose shades shall rivers flow: decked shall they be therein with bracelets of gold, and green robes of silk and rich brocade shall they wear, reclining then therein on thrones" (Koran, 18:31). Elsewhere in the Koran, Paradise is similarly luxurious: "But God will bring in those who shall have believed, and done the things that are right, into gardens 'neath which the rivers flow. Adorned shall they be therein with golden bracelets and with pearls, and their raiment therein shall be of silk" (Koran, 22:23). On the Koranic depiction of Paradise, Hanaway writes:

Two points should be stressed here: first, that Paradise will be the eternal and unchanging abode of the believers . . . ; and second, the elite nature of the Koranic images. Green silk and brocade garments, golden bracelets, brocade-lined couches, houris as lovely as rubies and coral all suggest royalty, affluence, and ease.[18]

Luxury and immortality are the essential features of celestial paradise, and Manuchihri's poetic representations of gardens stress both of these qualities.

I've identified many images that describe the garden's elements as precious gems and metals or rich cloths, emphasizing the royal ownership of the garden. At the same time, these images conflate the earthly garden with celestial Paradise. In particular, the *nasib* of *qasidah* XLV, which is entirely devoted to description of the garden as a treasure-trove, beautifully weds Koranic images of Paradise with images of royalty:

> Le nouveau printemps est venu comme une nouvelle lune; et chaque plaine est devenue comme le paradis de l'Éden.
> Sur la tête du narcisse brille une pleine lune et six étoiles brillent à côté de chaque lune,
> ou bien c'est comme six lunes admirables onduites d'argent entouraient, en boucles, de l'or pur.
> Au point du jour on voit dans les airs un arc-en-ciel semblable au pan d'une robe royale.
> Il y a là cinq pièces d'étoffe de soie de toutes couleurs; et le pan de chaque étoffe séparé,
> Partout où tu tourneras il y aura une aire d'émail; partout où tu iras il y aura une tente de brocart.
> Dans la prairie, le frais narcisse est comme une fossette d'or au menton . . . (XLV, 1–7).

Furthermore, treasure imagery makes the terrestrial garden more durable, associating it with immortal Paradise. Rosebuds die; rubies don't. While Christian theology and art stress the earthly garden's transience, Manuchihri's idealization of the garden implies permanence. In his poetry, the garden never fades, because it is always depicted at its freshest moment, the moment of its renewal. Following established poetic tradition, the garden exists in Manuchihri's poetry only on *Norwuz*. "Le Nowruz renouvelle toutes les choses, et par ses jardins il renvoie au loin les jardins d'Irem" (XLVII, 1). Manuchihri begins another *qasidah* with the celebratory words, "C'est la saison du printemps; c'est la saison de la rose vermeille, le monde est paré comme le séjour éternel. Le monde était devenu décrépit, dos vouté et visage rébarbatif; regarde comme il est devenu frais et rajeuni!" (X, 1–2)

Hanaway summarizes the aesthetic of Manuchihri's garden *nasibs*:

> First, the poets modelled their garden descriptions on a divine archetype, adding from their own visual experience elements lacking in the Koranic model. Second, while they observed nature carefully and mentioned many sorts of flowers and trees, the power of the ideal model was a prominent factor in their idealization of the actual garden into a paradise of eternal spring with brilliant, jewel-like colors, the water of life flowing in its channels and houris on hand to delight the nobles taking pleasure within its bounds.[19]

Clearly, the power of the ideal is the driving force in Manuchihri's depiction of gardens. Because the garden is an allegory for the king's realm, idealizing the garden reflects on the monarch. Manuchihri praises the king when he upholds the royal ownership of the ideal garden. Furthermore, by projecting the qualities of the ideal garden on the kingdom, Manuchihri instructs the king to make his realm as luxurious as a paradisal garden.

But luxury alone does not a good kingdom make. In a garden, man makes the disorganized forces of nature conform to the artificial order that he dictates. Hanaway writes, "In the arrangement of the Persian garden, [the] chaos of individual flowers is framed and controlled by the larger order imposed upon it."[20] Manuchihri's gardens represent an ideal of order, which a king should strive to exercise over his kingdom. Predictably, the poet projects this ordering ability onto his monarch: "Voici le Nowruz qui t'arrive et avec lui la fête du Nowruz: Maître, reprends à nouveau ce nouvel ordre du monde" (XXIV, 1). Like the garden, the ideal kingdom under the divinely sanctioned monarch exemplifies order: "C'est un démon que celui qui désobéit à ses [monarch's] ordres; démon est en effet celui qui se montre rebelle aux ordres de Dieu" (XXX, 28). In one *qasidah*, Manuchihri praises his patron for his justice, order, and mercy, and then asserts that "Notre Dieu n'a pas créé ce monde pour l'injustice, ni pour le désordre, ni pour les rancunes et les vengeances" (XXX, 40). The order of the ideal garden is a model for the order a good kingdom should exhibit.

By relating the garden to the state, Manuchihri praises the monarch for bringing order and prosperity to the kingdom. At the same time, he instructs by example, invoking the exemplary paradisal kingdom characterized by divine order and extravagant luxury. As A. Bartlett Giamatti notes, this poetic fusion of garden, kingdom, and Paradise is the etymological root of the English word "paradise": "the word 'paradise' derives from the Old Persian word *pairidaeza* – formed on *pairi* (around) and *diz* (to mould, to form) which meant the royal park, enclosure, or orchard of the Persian king."[21] Radically ordered and full of finery, the gardens in Manuchihri's *nasibs* are undeniably artificial. This artificiality enables Manuchihri to carry out his panegyric and instructional goals.

Manuchihri praises and instructs this king. In addition, he instructs his audience. Bursting with sensual garden imagery, Manuchihri's garden *nasibs* are designed to encourage sensual experience of the world's luxuries. As settings, his poetic gardens have visceral sensual appeal. Aside from their visible sparkle, the gardens fairly exhale sweet scents: "Le vent répand le musc, le nuage verse des gouttes de rosée" (XXX, 5). The reader is entreated, "Sens la chevelure de la violette et bois le bienheureux rubis, saisis l'anse de la lyre et penche-toi vers la coupe" (XXX, 2). Songbirds contribute to the auditory appeal of the garden. Clinton writes:

> At the frequent feasts and drinking bouts which were so familiar a part of the royal court and so well-beloved of Mascud, he was called upon to improvise on subjects set as a competition between the poets, to enliven the company with romantic and bacchic lyrics, and probably to prove himself always a witty, delightful companion.[22]

Manuchihri's purpose in idealizing gardens is threefold: he both praises and instructs his king, and he encourages his audience to enjoy the world's luxuries.

In the next chapter, I shall discuss the goals of the *Roman de la Rose*. Unlike Manuchihri's poetry, the *Roman* is not designed to extol and instruct a specific royal patron. However, it does convey a message to a public audience, just as Manuchihri's poetry does. But whereas Manuchihri encourages his audience to experience the sensual world, the *Roman* instructs its audience to fear and avoid the sensual world. Since Christian poets of the *Roman* believe this is their God's message, they honor Him in communicating it effectively. Their effective communication of this message carries out the panegyric purpose of the *Roman*.

CHAPTER TWO: NATURE'S ARTIFICE

The court milieu is the driving force for Manuchihri's poetry. However fantastic the garden depicted in his *nasibs* are, they are based on Manuchihri's actual experience of the palace garden. Before analyzing the gardens in the *Roman de la Rose*, we should consider its cultural context. Is its central garden modeled on a royal court, like the gardens in Manuchihri's *Divan*? In his *Allegory of Love*, C. S. Lewis claims that in medieval European poetry, the garden traditionally "means Love; in Guillaume it is changed slightly and made to mean the life of the court, considered as the necessary sphere or field for love's operations."[1] What was court life like in thirteenth-century France? Diametrically different from the luxury-loving court of the Ghaznavids, the French court life under Louis IX "consists largely of sermons, Scripture study, and deep theological consultations between the young saint who sits beneath the canopy of fleurs-de-lys and his courtiers. The royal family . . . debates questions of abstract morality."[2] Citing the piety of Louis's court, and the historical improbability that Guillaume de Lorris ever was a part of that court, John Fleming refutes C. S. Lewis's claim that Guillaume's garden refers to court life.

The view that Guillaume's paradisal garden means the life of the court – any court, let alone the extraordinary French court under Louis IX – seems to me untenable, mistaking not only the nature of court life in the thirteenth century but also the nature of the terrestrial paradise from which the image of the garden of Love is taken.[3]

It is unlikely that Guillaume's sensual *Jardin* reflects any actual experience of Louis's meditative court.

The Garden's Artifice and Amant's Response

Like Manuchihri's gardens, the Jardin de Déduit of the *Roman de la Rose* is artificial in its order and luxury. Everything about its contents and design implies regulation by a controlling intelligence: exotic flora and fauna, collected together within the garden walls, exhibit impeccably sustained order. Even the garden's inhabitants, from its tall, straight trees to its comely women, suggest artifice. In the introduction, I delineated the four tasks of the gardener: selection and separation of elements, and imposition and maintenance of order. The maker of the Jardin de Déduit has carried out these tasks masterfully.

The garden is emphatically described as a collection, which implies artificiality; there are no pretensions to spontaneous assembly of the garden's elements. The exotic trees didn't just spring forth from the garden's soil; they were "imported from Saracen land and planted throughout the garden" (590).[4] The garden includes an assortment of fruit and nut trees, a menagerie of exotic flowers, birds and animals, a variety of unusual spice plants and, in addition, "there were many singing birds, collected together, throughout the whole garden" (655). Collections are inherently artificial, and the garden's artifice is emphasized by Amant's extended catalogs of the contents of each collection. Its artifice is further highlighted by the "transplantations arbitraires affranchies de toute logique climatique"[5] that were necessary to create the garden. All the trees Amant lists – Saracen trees, domestic fruit trees, laurels, pines, olive trees, cypresses, elms, hornbeams, beech, hazels, aspen, ash, maples, firs, cedars, and oaks, to name a few – do not spontaneously grow in the same climate and terrain.

Although the garden houses an extraordinary variety of species, the gardener has not included every natural specimen in existence. Instead, he has carefully selected the natural elements for his garden with judgement and discriminating taste. "Except for some trees which would have been too ugly, there was no tree which might bear fruit of which there were not one or two, perhaps more, in the garden" (1324). The chosen collections are separated from the world beyond the garden by "a highly crenelated wall, sculptured outside and laid out with many fine inscriptions" (132). Garden elements are arranged in an orderly geometrical design within the garden, which is "a completely straight, regular square, as long as it was wide" (1323). For instance, trees are arranged in regular intervals. "Know too that these trees were spaced out as they should be; one was placed at a distance of more than five or six fathoms from another" (1360). Like Manuchihri's gardens, the *Jardin* is at once highly ordered and sensually delightful. Although the garden's artifice enthralls all the senses, it is primarily designed to stimulate Amant's sense of sight, with its explosive color, visible order and elaborate surface detail. "The earth was very artfully decorated and painted with flowers of various colors . . ." (1410).

Like the garden itself, the fair inhabitants of the garden are consistently described as if they were works of art. Each careful assemblage of well-formed traits appears to have been crafted by an artist with an unfailing sense of human beauty. Déduit was "handsome, tall, and straight . . . his mouth [was] fine, his nose most carefully shaped. Somewhat broad in the shoulders and narrow in the waist, he was so elegant and full of grace, so well formed in all his limbs, that he looked like a painting" (801). When Amant sees Joie's nose, he claims that "no one ever made a better one out of wax" (840); similarly, upon seeing Beauté, he affirms that "her nose, her eyes, her mouth, were all well made. So help me God, my heart is touched with great sweetness when I remember the arrangement of her every limb . . ." (1010). Indeed, all the carollers are described as if they were paintings or sculptures. Like the garden itself with its beautiful contents and design, each lovely feature of each of its inhabitants is arranged to create an exquisite whole. These enchanted garden-people are incarnations of the enchanted garden: their visible beauty comes from the artistry of their creation.

The superficiality of the carollers' beauty is emphasized by descriptions of their garments. Ingenious craftsmanship went into their creation, and they are rendered with the same close attention to surface detail that characterizes the descriptions of the bodies and faces of the dancers: "[Déduit's] body was richly clothed in samite decorated with birds and beaten gold. His dress was highly ornamented; in several places it was cunningly slashed or cut away. He was shod with great skill in low-cut, laced shoes" (825). Similarly, "[Joie's] hair was laced with a golden thread, and she had a brand new chaplet of gold embroidery. I have seen twenty-nine of them, and never had I seen a chaplet so beautifully worked in silk. Her body was dressed and adorned in a samite worked in gold" (860). All the carollers' garments are wrought by craftsmen skilled in embroidery, jewellery, and other arts. The garden's luxury manifests itself in these carefully designed, rich clothes which echo Koranic luxury in their craftsmanship.

This rampant artifice dazzles the foolish dreamer, and lures him deeper into the garden to explore its visible charms:

> Thus these people, along with others of their households, danced their carols there. All together they were warm, open people, well-instructed and beautifully trained. When I had seen the appearances of those who led the dances, I then had a desire to go see and explore the garden . . . (1279).

The manifest artifice of the enchanted garden is sensually desirable, and in particular visually striking. In the medieval literary world, it was universally accepted that visual beauty led to love. The *Roman* upholds this close association. "The images of the opening part of the poem . . . emphasize the role of sense impressions. Sight images predominate and underline the idea that love begins with sight."[6] Indeed, Amant's love of the rose begins with his sight of it. But it is Oiseuse who first charms his vision. The sight of luxurious Oiseuse leads him into the garden, to the fountain where he sees the rose.

Oiseuse exemplifies female artifice. Her features are sculpted: she "had a straight, well-made nose . . . her neck was of good proportion, thick enough and reasonably long . . . her body was well-made and svelte" (539). Her hair is compared to a man-made "copper basin," (530) her face is "white and colored" (538), as if painted. Each of Oiseuse's traits is described separately, implying that her beauty results from a carefully designed arrangement of beautiful component features: "The space between her eyes was not small but very wide in measure" (535). Emphasizing the superficial, Oiseuse's clothing, like her body, is exquisitely crafted. Oiseuse spends all her time maintaining her charms: "By the time that she had combed her hair carefully and prepared and adorned herself well, she had finished her day's work" (570). Based on his superficial appraisal of Oiseuse's superficial beauty, Amant concludes she has a "gracious heart" and leads a "good life"; moreover, he follows her into the garden. I shall address the tension between the garden's appearance and its reality in Chapter Three. For now, it is important only to realize that "the psychological process by which Amant is to become entirely drunk and befuddled with the beauty of *visibilia* begins with his easy capitulation to Oiseuse. . . ."[7] The visually stunning artifice of Oiseuse, reflected also in the rest of the garden and its inhabitants, leads directly to the rape of the rose.

The Identity of the Artificer

The *Jardin* is luxurious, like Manuchihri's garden. But while Manuchihri idealized the garden by comparing natural elements to manmade luxuries, Guillaume implies that the luxurious garden is actually the creation of an absent artificer. Not only is the luxury of the lovely women and their garments scrupulously crafted, but the artificial order and exotic collections of the garden also reveal its artifice.

Inevitably, the reader wonders who is responsible for creating the artifice that leads Amant to love. Surely this kind of refinement couldn't occur spontaneously in nature. Yet Amant asserts that Nature *did* create the garden's artifice. He makes this claim at the Fountain of Narcissus, and again when he first looks at the rose.

Like everything else in the garden, explicit artifice characterizes the Fountain of Narcissus. Amant describes a complex mechanism by which the water in the fountain retains its purity: "The water is always fresh and new; night and day it issues in great waves from two deep, cavernous conduits" (1533). Such a unique system of hydraulics is clearly the work of some intelligent designer, not the result of untamed natural forces. But in the next moment, despite the manifest artifice, Amant explains that "Nature, with consummate skill, had placed the fountain under the pine within a marble stone, and in the stone, on the border of the upper side, had cut the small letters saying that there the fair Narcissus had died" (1430).

Amant echoes this assertion in the next moment, when he sees the rose. He again refers to "nature's skill" when he describes the work of art of a flower too symmetric to possibly be natural:

> It glowed with a color as red and as pure as the best that Nature can produce, and she had placed around it four pairs of leaves, with great skill, one after the other. The stem was straight as a sapling, and the bud sat on the top, neither bent nor inclined. (1660)

Everything in the garden is like the pretty rose: artificially designed to delight the senses. While the women and their dresses are meant to please the eye, streams are made specifically to arouse auditory pleasure: "In little brooks, which Diversion [Déduit] had had made there as channels, the water ran along down, making a sweet and pleasing murmur" (1384). Mouthwatering fruits hang in clumps from branches of artificially transplanted exotic trees. Even the birdsong is a rehearsed choral performance, designed to please the ear: "The birds went along performing their wondrously sweet and pleasing service, in which they sang love lays and elegant songs, one high, one low"

(701). The garden is also delightful to touch: the "thick, short grass" is "sweet and moist," and Amant compares it to a manmade feather bed. Nature itself has been artfully crafted to serve the purpose of the garden.

In the introduction, I distinguished between nature and artifice. Citing Oscar Wilde, I showed that Nature and Art work at cross purposes, and I concluded that gardens are more like Art than like Nature: through the art of gardening, man dominates nature to create an artificial space he considers beautiful. Although wary of the ambiguous concepts "nature" and "artifice," Pierre Grimal summarizes the meanings of the words as I have used them so far in my discussion of gardens:

Si, donc, les hommes ont voulu, depuis des temps immémoriaux, se donner des jardins, c'est que, dès l'origine, ils étaient enclins à ne pas accepter cette Nature qui leur était offerte, ou imposée. Même lorsque l'art des jardins se réclame du "naturel", et refuse (en théorie) les "artifices" . . . il n'en reste pas moins "artificiel" dans bien de ses aspects. Il faut s'y résigner: la création n'est pas spontanément "belle" (selon nos critères) ni totalement satisfaisante pour notre sensibilité.[8]

Grimal maintains that gardens are artificial even though "le jardin utilise les matériaux que lui fournit la Nature."[9] In modern vocabulary the word *nature* denotes that which exists spontaneously and organically without external human interference. When he states that the Fountain and the Rose are *Nature's* creation, Amant undermines the modern opposition between nature and artifice. His claim that Nature creates Art – that natural and artificial are identical – is worth examining.

The capital "N" suggests that Amant refers to a personification of nature.[10] Although Guillaume never introduces this character, Jean de Meun builds Amant's unlikely assertions in Guillaume's part of the poem into the complex figure of Nature, who works in a forge, where she "put all her attention on forging individual creatures to continue the species" (15892). This immediately identifies Nature as a skilled craftsman.

When Nature, sweet and compassionate, sees that envious Death and Corruption come together to put to destruction whatever they find within her forge, she continues always to hammer and forge and always to renew the individuals by means of new generation. When she can bring no other counsel to her work, she cuts copies in such letters that she gives them true forms in coins of different monies. (16005)

Through her craftsmanship, Nature makes forms that continue the species.

Thus Nature is the mysterious artificer who wrought the stunning *Jardin*, which incites Amant's love for the rose and results in the rape of the rosebush. It is appropriate that what Nature creates leads to procreation, for her task is "forging individual creatures to continue the species" (15891). She creates circumstances to ensure that procreation takes place. This is why she makes the garden so luxuriously sensual, and the rose so irresistibly beautiful. Beautiful things please the eye, engage the rest of the senses, and make reasonable people unable to resist falling in love with them. Visible beauty leads directly to sex, which continues the species. To ensure that procreation will occur, Nature crafts beauty that ravishes the heart.

Nature attends to the other side of the equation as well. Not only does she form visually appealing artifice, but she also engenders man's delight in it. This is confirmed by Reason:

> Nature wills, since father and mother disappear, that children rise up to continue the work of generation, and that one's life may be regained by means of another. For this purpose Nature has implanted delight in man because she wants the workman to take pleasure in the task. . . . Thus Nature uses this subtle means of gaining her end. (4403)

Clearly, "the bonds by which man is joined to Nature are those of an unenlightened sensuality,"[11] and the first sense Nature engages is vision.

The Sin Caused by the Artifice

But in medieval Christian theology, love that begins with sight is sinful. Lady Reason characterizes Amant's love for the rose as follows:

> Love, if I think right, is a sickness of thought that takes place between two persons of different sex when they are in close proximity and open to each other. It arises among people from the burning desire, *born of disordinate glances*, to embrace and kiss each other and to have the solace of one another's body (4376, my emphasis).

The love Reason describes is carnal love, or "cupidity: a love of an earthly object for its own sake, not for the sake of God,"[12] as contrasted with charitable love. Amant's devotion to the rose begins with the beautiful vision he sees reflected in the fountain's crystals.

> There crystals enable the dreamer to see the garden on one side at a time. For they are the eyes of the flesh, the dreamer's own eyes, and to see all of the garden through them, he must turn his head. Having succumbed to *concupiscentia carnis*, the dreamer is now ready to indulge in *concupiscentia oculorum*, which leads, as St. Paul tells us, to pride of life. The wrong love, it will be remembered, proceeds, "ex visione."[13]

Seeing is the first step in the progression of carnal love. The successive stages are *visus*, followed by *alloquium, contactus, osculum*, and finally *factum*.[14]

Alain de Lille warns, "Sicque castitatem expugnat armis suis luxuria, quae sunt: 'Visus et alloquium, contactus et oscula, factum.'"[15] These *gradus amoris* correspond to the five branches of the iconographic sinful tree whose crowning fruit is *luxuria*, or lechery, which leads man to *cupiditas*. There is a corresponding tree *sapientia*, or reason, which leads man to *caritas*. Robertson writes:

Any tree may have implications for the individual, for society, or for the afterlife. Thus the Tree of Life variously represents *sapientia*, the Cross, Christ, or the good Christian. The Tree of Knowledge of Good and Evil was not evil in itself, since God put nothing evil in Paradise, but eating the fruit of the tree represents turning away from God in pride.[16]

Manuscript illustrations of these two trees can be found in *De fructibus carnis et spiritus* with the works of Hugh of St. Victor, printed by Migne. In the course of the drama, Amant systematically climbs the five branches of the lecherous tree, beginning with *visus*, the sinful sight of the rose. "It should be clear that the moral context into which Amant's career must be placed is that of *luxuria*, or what the Middle English text of the *Romaunt* calls Lechery."[17]

Saint Bernard's paradigm for carnal love also stresses vision as its starting point. He suggests that each of man's five senses corresponds to a different type of love. Charles Dahlberg reconstructs Bernard's argument in the following passage.

In ascending order, the first [love] is an *amor pius* or *naturalis*, directed between parents and children; it corresponds to the sense of touch because it is the love closest to the body. . . . [He goes on to characterize the charitable loves that correspond to taste, smell, and hearing.] Finally, there is the love that is *sanctus* or *devotus*, the love of God. It corresponds to the sense of sight, which is still better and more perceptive than is hearing. These five loves are kinds of charity; the first four concern the love of one's neighbor and the fifth the love of God. But when one reverses the order of the senses, going from sight to touch, one traces the course of cupidinous love via the bodily senses. . . . *Touch is the goal of cupidity and the starting point of charity; sight is the starting point of cupidity and the goal of charity, the beatific vision. . . .*[18]

Cupidinous love begins with sight and proceeds to engage the rest of the senses, one by one. "Eyes, ears, mouth, hands, and *oeuvre vilaine* agree with *visus, alloquium, osculum, tactus*, and *factum*, a concordance between each *gradus* and a specific organ of sense perception."[19] Amant's love clearly follows the cupidinous route, beginning with the sight of fair Oiseuse. She is visually irresistible, and Amant doesn't resist.

But why is love beginning with sight sinful? The definition of sin is the rejection of reason in pursuit of sensual gratification. Vision begins a progressive pleasurable engagement of the senses causing the sinful lover to overrule reasonable behavior. This is why sins tend to happen in intoxicatingly lush menageries of sensual *choses*: gardens. Like Amant's folly, man's first sin happened in a *paradis terrestre*, with a crystal fount at its center and where "the trees were pleasing to the eye and good for food" (Gen., 2:8). The drama set in Eden is the paradigmatic sin: man's desire, inflamed by a beautiful sight, seduces his reason. Despite God's warning, "when the woman saw that the fruit of the tree was good for food and pleasing to the eye, and also desirable for gaining wisdom, she took some and ate it" (Gen., 3:6). This sounds suspiciously similar to Amant's flawed rationale for all the foolish things he does within the garden. Amant reenacts the fall when he falls in love with the precious rose, the apple of his eye.

Through the descriptions of the garden that Amant himself presents, certain details alert the Christian reader that the garden's beauty is not to be trusted. Familiar with the iconography he adopts, Guillaume's contemporary readers would realize the perils of the garden. Oiseuse's comb, mirror and gloves are iconographic clues to her true nature: these are the typical accoutrements of the iconographic figure of *Luxuria*, representing lechery. Through these clues, Guillaume conveys that following Oiseuse is utterly foolish.

> Amant describes Oiseuse as beautiful but clearly indicates, to the wary reader, that she is to be shunned. There is here a kind of irony, but no confusion of the values and no refusal to arbitrate claims on the reader – in short, no double vision.[20]

Medieval Christians know that Oiseuse is not to be trusted, and that Amant is foolish to follow her into the garden. More importantly, however, the text suggests that Amant, too, knows that Oiseuse and her garden are dangerous. Consider, for instance, this description of birdsong:

> These birds that I describe to you performed a lovely service: they sang a song as though they were heavenly angels. Know well that I was filled with great joy when I heard it, for mortal man never heard so sweet a melody. It was so sweet and beautiful that it did not seem the song of a bird; one could compare it rather with the song of the sirens of the sea, who have the name *siren* on account of their clear, pure voices. (661)

Amant correctly identifies the songbirds as the temptress sirens of antiquity, but nevertheless continues to enjoy their music. All the warning bells the word "siren" should sound off simply don't register with him. Amant knows the garden is dangerous. He just ignores it.

Amant's awareness of the dangers of Narcissus's fountain support this conclusion. The tension between morally desirable behavior and Amant's actions peaks in this comic example of Amant's utter inability to act reasonably when presented with visible beauty. After his lengthy retelling of Narcissus's tragic end and his expostulation of the fountain's dangers, it is singularly inane for Amant to even consider looking into the fountain. Amant's "reasoning" is delicious; he doesn't even offer a counterargument:

> When the inscription had made clear to me that this was indeed the true fountain of the fair Narcissus, I drew back a little, since I dared not look within. When I remembered Narcissus and his evil misfortune, I began to be afraid. But then I thought that I might be able to venture safely to the fountain, without fear of misfortune, and that I was foolish to be afraid of it. (1511)

It is psychologically unsatisfying that no agency is presented in the text to explain this central incident. Like Eve in Eden, Amant appears to erroneously judge superficial beauty and the prospect of sensual gratification as a valid reason for disregarding palpable danger. But this pattern of reasoning is not even made explicit. Amant's behavior at Narcissus's fountain, and his ensuing pursuit of the fair rose, dramatize his rejection of reason, by which he falls into sin.[21] It is the garden's artifice that causes Amant's sinful insanity; it is created by Nature, who delights the senses as she carries out her task.

Nature's "Sin"

Nature's fundamental intention in making the garden's beautiful artifice is to ensure procreation. Yet in so doing, she incites Amant's sinful cupidity. Should we condemn Nature for her work, since it is her artifice that leads Amant to sin? Lest we forget, sin is serious business. Mary Ellen Faughnan reminds us that "the final picture [of the *Roman*] is one of horror rather than happiness. The Lover has succeeded in turning completely away from God, a scene which carries with it the threat of eternal damnation."[22] It is tempting to convict Nature for pandering: through her arts, an innocent man is led into lechery.

Although she never states it explicitly, we can infer that this is the mysterious sin Nature is confessing in her lengthy, unusual "confession." She is troubled by the rampant sinning that accompanies her efforts at continuing the species, for it is contrary to Reason. But when she "confesses" this "sin" to Genius, the penance he gives her is suspiciously light:

> The valiant priest Genius absolved her and gave her a penance that was suitable and good, one that accorded with the magnitude of the fault he thought that she had committed. He enjoined her to remain within her forge and labor as she was accustomed to do when she had no sorrow . . . (19411).

If it is Nature's *sin* to encourage cupidity through her arts, then why would her (albeit dubious) confessor send her right back to her forge to continue perpetrating this wicked crime? There, she will resume making the alluring roses that make reasonable people reject Reason, whose efforts she purportedly upholds.

In Nature's confession, there is tension between the character Nature and "man's natural ways": his habits, inclinations, instincts and impulses. Man's natural ways are inscribed in the stars. Without regulation, they lead him to folly. Nature states,

> I know very well that it is quite true that however *the heavens work to give them those natural ways that incline them to do things that draw them to this end*, obedient to the material that goes about to bend their hearts in this way, even so, they can . . . obtain another result, *provided that, like intelligent people, they have bridled their natural ways*. (17059)

Since the folly these "natural ways" cause is "very harmful to Nature" and "make[s] [her] waste all [her] pains to lead [man] to a natural death" (17027), Nature advises man to regulate them. Restriction of man's natural ways is the office of reason, or "free will." Nature asserts that "free will" has power over the natural ways inscribed in the heavens: "Free will is above all destinies, no matter how destined they are" (17725). When man exercises his free will, the species is safer. Therefore Nature strongly encourages man to adopt Reason, although its function is to restrict man's natural ways. Nature, whose goal is procreation, divorces herself from the "nature" represented by man's unreasoned instincts, impulses and desires. Her goal is the continuation of the species, and Reason is necessary to help her realize that end.

Reason is part of procreative Nature. Nature's comments on nobility characterize the relationship between Nature and Reason:

> When I cause likenesses to be born, they acquire a fresh nobility if they have so much ability and if they want to be noble with a nobility other than that which I give to them. This nobility is a very beautiful one. Its name is natural freedom, and I have put it into everyone equally, along with the reason that God gives them. . . . (18870)

The nobility "other than that which I give to them" is Reason. Nature's nobility, "natural freedom," is only given to man *provided he has the ability and desire to be reasonable*. To be truly noble, one must acquire both nobilities, and Nature goes on to state that "no one is noble unless he is intent on virtue, and no man is base except because of his vices, which make him appear unbridled and stupid" (18610). If Nature withholds her nobility until man is noble by Reason's criteria, then natural freedom cannot be awarded to any man who lacks reason. Paradoxically, only reasonable men have natural freedom.

The logic of Nature's discourse reflects the paradox of her relationship with Reason. Speaking like a professor of philosophy, she appears to present well-reasoned arguments in an orderly way. "I would like very much to explain everything, to raise more objections, reply to them, and give many illustrations for them . . ." (17730). It is part of Nature to be reasonable. "I would never yet have spoken of this question [of free will], if I should have kept silent about it, but it belongs to my subject" (17737). Yet often her logic is circular, paradoxical, or self-undermining. Although it is a defining trait of her character, Nature's reason is not rigorous enough to understand that when she champions Amant's cause, *she encourages the folly that she has condemned.* She states unequivocally, "To enter into sin is evil" (19320). But though she understands the evil of sin, Nature does not take it upon herself to punish sinners: "Of course, I leave to God all sins with which the wretch [man] is stained; let God take care of them and punish when He pleases" (19323). Nature wants man to consciously and reasonably procreate his divine self; this would both eliminate sin and continue the species, encapsulating the ideal relationship between Reason and Nature. However, Nature's highest law is continuation of the species, and not elimination of sin. "Nature functions by a morality which goes no further than her own ends. Her morality is determined by the universal laws of creation. If these laws are not violated, she is content."[23] Since lovers procreate, Nature denounces those who get in their way:

> But of those of whom Love complains (for I have indeed heard the complaint), I complain myself, as much as I can, and I should do so because they have denied me the tribute that all men have owed me, and always do and always will owe me as long as they receive my tools. (19323)

Nature does not distinguish between carnal lovers and charitable lovers. Both kinds of lovers engage in sexual intercourse, which has the outcome that Nature desires. "It is Nature who wishes to carry out the command of her Creator: 'Be fruitful and multiply.' She is God's aid and only, unknowingly, does she help the Lover in his sinful quest. Nature, as a procreatrix, has not been successful in this poem."[24]

Ironically, Amant's carnal love affair will *not* continue the species. By and large, the activities of carnal lovers who pursue sex for selfish sensual pleasure have the by-product of creating a child. But we cannot ignore the fact that at the most literal level, the object of Amant's love remains a rosebud, not a woman. This emphasizes what we already know: that Amant's desire is carnal, and that he seeks only sensual gratification, not the continuation of the species. Fortunately for Amant, Nature is entirely removed from the action of the romance and has no immediate knowledge of Amant's love affair.[25] Her complaint "of those of whom Love complains" is a general statement; nearly always, what thwarted lovers desire leads to procreation, be it deliberate or accidental. But, like Pygmalion's lunatic adoration of a stone statue, Amant's love for the rose is sterile. Ultimately, Pygmalion did have a child, Paphus, by the beloved statue animated by Venus. This appears to undercut the idea that Pygmalion's cupidinous love is infertile. "Nature's purposes appear to be served by the procreation resulting from Pygmalion's passion. But the fate of his progeniture cannot be ignored." Paphus's son Cynaras begat Myrrha who was turned into a tree for the incestual liaison that produced fair Adonis. At its highest remove, carnal love is in fact destructive to the species, despite the immediate procreation it may prompt.

In the *Roman*, Guillaume creates a fictitious garden. This sensually appealing space, full of order and luxury, is made by the artificer Nature. Because of the sensual artifice, Amant falls madly in love with the rosebud. Through his poetry, Guillaume depicts the proper setting for carnal love, and dramatizes the process by which man falls in carnal love. This process begins with vision. The poem's medieval audience would know that this kind of love is sinful according to Christian doctrine. Knowing this cultural context, we can conclude that the poem has a moral: it shows that sensuality threatens Christian morality. It demonstrates that good Christians should fear and avoid real gardens that are like the *Jardin*, for sensual artifice causes lunacy in a reasonable man. The invocation and dedication of the opening lines of the poem support the validity of this Christian moral. "God grant that she for whom I have undertaken it may receive it with grace. It is she who is so precious and so worthy to be loved that she should be called Rose" (42). The rose is a familiar symbol for the Virgin Mary. Written in her honor, the poem is panegyric. The Virgin epitomizes pure charitable love, and the poem praises her by instructing man to follow her example.

Both Manuchihri and the authors of the *Roman de la Rose* conjure up sensually appealing gardens characterized by luxury and order. Although the gardens themselves are similar in their contents and design, they convey diametrically different messages, which can be understood through poetic analysis hand-in-hand with an understanding of the cultural context of the poetry. Manuchihri's poetic representation of gardens is primarily designed to carry out a socio-political goal: extolling and instructing his king. This is recognizable in light of the court tradition of medieval Persia. The *Roman* lacks a political agenda. Its central purpose is to warn Christians away from sensual experience.

As I mentioned at the close of the last chapter, Manuchihri's poetry stimulates the opposite behavior in his audience. Instead of warning his audience away from the garden, Manuchihri explicitly encourages his audience to experience its sensual delights. This reflects the Islamic conception of morally desirable behavior, which clashes with proper behavior according to Christianity. Both cultures' gardens are full of sensual delights. But gardens are frightening for Christians because, according to Christianity, sensual appeal tempts man to turn from God. On the other hand, in Islam gardens are innocent of the peril Christians associate with them. In the next two chapters, I shall examine the religious origins of these contrasting conceptions of sensuality, and explore how they influence each poet's aesthetic.

CHAPTER THREE: ALLAH'S DIVAN

Gardens are works of art, at once sensually appealing and aesthetically pleasing. This invites us to explore the garden as a metaphor for poetry. In this chapter, I shall examine what Manuchihri's representations of the gardens imply about his conception of his own art. In Chapter One, I suggested that the rigorous order of the Persian garden is a model for the order a kingdom should exhibit, and that Manuchihri idealizes his king as the man who brings perfect order to his realm. The presence of order implies a similarity between the ideal kingdom and poetry. In Persian, the word for poetry, *næzm*, means *order*, while *næsr* (prose) implies a lack of order. The *qasidah* is an example of highly ordered language; strict rules govern its composition.

Thus garden, kingdom, and poem exhibit order. Recall the line, "Notre Dieu n'a pas créé ce monde pour l'injustice, ni pour le désordre . . ." (XXX, 40). Whereas disorder is associated with injustice and suffering, order is divine. Allah has created everything in nature, and poets, kings, and other artists emulate His ordered Creation.

The Garden's Luxury

I've showed that gardens are luxurious as well as orderly. The ideal kingdom should be prosperous and replete with luxury, like a lush garden. Manuchihri compares the garden full of natural luxuries to a rich carpet: "Le nouveau printemps endosse ce tapis aux cent couleurs . . ." (XIII, 11), and the narcissus to precious metals: "c'est comme si six lunes admirables enduites d'argent entouraient, en boucles, de l'or pur" (XLV, 3). The rainbow is like a splendid royal garment: "Au point du jour *on voit* dans les airs un arc-en-ciel semblable au pan d'une robe royale. / Il y a cinq pièces d'étoffe de soie de toutes couleurs" (XLV, 4–5). Even the rose exhibits luxury: "La rose à deux faces est comme un cercle sur une soie rouge, lorsque tu aurais fait passer à son envers un cordonnet d'or" (LIII, 12). To sum up, "Partout où tu te tourneras il y aura une aire d'émail; partout où tu iras il y aura une tente de brocart" (XLV, 6). Clinton writes, "Nature is for [the court panegyrists] the gorgeous and luxurious world of the palace garden and they depict it as a concoction of rare stuffs and precious gems, redolent of costly perfumes."[1]

Each image above conveys nature's luxury by comparing natural elements to fine artifice in the form of rich treasures created by skilled craftsmen. Carpet-making, for instance, is a learned craft, and carpet images reinforce the artifice of the springtime garden: "Les branches des fleurs sont comme des échecs d'argent et de cornaline qui, à la tombée de la nuit, jouent aux échecs sur un tapis de verdure" (XXIV, 3). Indeed, actual Persian gardens were designed to resemble carpets.[2] "Since paths are usually at a level higher than the surrounding garden, the geometrical character of the layout is emphasized, the surrounding vegetation is prevented from obscuring the architecture and one has the illusion of walking at the level of a floral carpet."[3] Artifice is emphasized by comparing the garden to hand-woven cloths or fine brocade: "De tous les côtés, le vent, comme un voleur, enlève quelque brocart, et le jardin fleuri est comme l'échoppe d'un marchand d'étoffes" (XIII, 9); "Le vent est devenu un porte-drapeau, le nuage est devenu un drapeau noir et l'éclair est comme une ou deux broderies d'or sur ce drapeau" (XXX, 9). Elsewhere, Manuchihri compares natural elements to well-wrought armored mail, another example of man-made artifice: "Le vent est devenu un fabricant de cottes de mailles; l'eau en fait des mailles, le nuage s'est fait confectionner de tentes et le brouillard fait chainer des tentes" (XXX, 15).[4] Manuchihri consistently compares nature to manmade crafts:

> On voit bien la nature exercer l'art ou le métier du peintre, du scribe, du teinturier, du tailleur, du frappeur de monnaie, du forgeron, du fabricant de papier et même du messager, de la nourrice; les réalités naturelles imitent l'art et l'œuvre humaine et les transcendent.[5]

The Jardin de Déduit also exhibits artifice, not only in its natural elements, but also in its eminently manmade elements like the stone fountain and the dancers' garments. The artificer responsible for these creations was Nature, who made beautiful artifice to encourage procreation. Manuchihri, on the other hand, compares only truly natural elements to human artifice; we know that roses and rainbows come from nature, not from man. But by comparing them to artifice, Manuchihri implies that nature itself is the work of some Creator. Nature is created by God, the artisan who crafts the ordered, sensual art we find in the world. "Dieu est le créateur de cette nature, mais il est au-delà."[6] In Manuchihri's poetry, God does what the character Nature does in the *Roman*. There, God is distinct from Nature, whom Jean describes as "God's chambermaid." This separation of Nature from God conveys the Christian belief that Nature's law, to enthusiastically encourage procreation, does not always coincide with God's will. However, there is no separation between God and Nature in Manuchihri's poetry, reflecting the Islamic belief that God has created natural luxury for man to delight in, and that He communicates with man directly through sensual experience of the world. This is the difference between medieval French and Persian poetry that Charles de Fouchécour gestures toward when he writes,

> Un example très typique de description de la nature est celui de la description du verger par Guillaume de Lorris dans le Roman de la Rose; le caractère stylisé de la description et les réalités que voit le poète rappellent bien ce que nous connaissons dans notre poésie [la poésie lyrique persane du XIe siècle]; mais la progression de la description, la façon dont le poète se met en cause et le contexte général où s'inscrit cette description, rendent les rapprochements difficiles.[7]

In his *nasibs* Manuchihri describes the garden as orderly, reflecting God's desire. The garden also contains natural luxuries, which are compared to artifice, implying divine craftsmanship. The palace garden as Manuchihri depicts it is a metaphor for God's ordered, luxurious Creation, the highest example of aesthetically pleasing and sensually appealing art. In his poetry, Manuchihri emulates God's art.

God's Intention and Aesthetic

Islam recognizes a parallel between God's art and literate arts such as poetry. It shares with other religions a basic belief in the natural world as a book in which the believer can read divine mysteries. Known as *cebrat*, this concept is presented in Surah 16 of the Koran:

> And He hath subjected to you the night and the day; the sun and the moon and the stars too are subjected to you by his behest; verily, in this are signs for those who understand.
>
> And all of the varied hues that He hath created for you over the earth: verily, in this are signs for those who remember.
>
> And He it is who hath subjected the sea to you, that ye may eat of its fresh fish, and take forth from it ornaments to wear – thou seest the ships ploughing its billows – and that ye may go in quest of his bounties, and that ye might give thanks.

> And He hath thrown firm mountains on the earth, lest it move with you; and rivers and paths for your guidance.
> And way marks. By the stars too are men guided. (Koran, 16.12–16)

This passage suggests that God's intention in making such a beautiful Creation is that man might experience its delights. It is an infinitely benevolent God who makes the fish and the ornaments of the sea, for the purpose that man might eat them and adorn himself with them. Similarly, roses and fruit in lush gardens should be smelled and tasted. Enjoying the earth's luxury is not sinful, according to Islam. On the contrary, to remain blind, cold, and unresponsive to nature's manifest luxury would be a sin.[8] Sensual luxuries are created by God, and therefore loving them represents loving God. Thus sensual love is holy love.

This contrasts with the Christian idea that it is sinful to pursue that which delights the senses; there is no tension between "caritas" and "cupiditas" here, as there is in Christianity. The sensually apprehended characteristics of God's natural creations convey God's intentions in creating them: beautiful, harmonious, fragrant, succulent delights are themselves a divine invitation to look, hear, smell, and taste, and man cannot go wrong if he accepts this invitation. The passage even implies that man can understand divine mysteries better through sensual experience of God's lush Creation than through intellectual exegesis. Thus God's message in his Creation is that sensual pleasure is holy, and his intention in creating the world's luxuries is to draw man into sensual experience of the world. He communicates his message and realizes his intention through direct, immediate sensory communication with man. In his Book of Nature, God adopts a direct aesthetic.

Manuchihri's poetry has the same message, intention, and aesthetic as the poet's perception of God's Creation. In Chapter One, I showed that Manuchihri's poetry is designed to encourage sensual experience of the world's luxuries. As court poet, he was often called upon "to enliven the company with romantic and bacchic lyrics."[9] Manuchihri conveys this goal directly, with straightforward directives such as: "Sens la chevelure de la violette et bois le bienheureux rubis, saisis l'anse de la lyre et penche-toi vers la coupe" (XXX, 2). Like God, Manuchihri conveys his message of encouraging sensuality directly. The text of the poetry that contains his message is innocent of irony, allegory, or other subtle literary devices that might complicate or negate the message.

Thus, like God's Creation, Manuchihri's poetry is straightforward, emulating the Islamic conception of the divine aesthetic with its directness. The poem's purpose of encouraging sensual experience is communicated sensually as well as literally. Form contributes to function, for the poem itself is sensually appealing in its music. Manuchihri even compares his poetry to birdsong: "Le rossignol au doux language, perché sur un noyer se fait conteur de contes; ce rossignol tisseur de contes zend, liseur du zend se fait poète" (XIII, 7). In this line, Manuchihri places himself within the garden as court poet, personified as a singing nightingale. The sensual music of his poetry delights the ear, reinforcing and advancing its exhortation to enjoy sensual beauty, just as the sensuality of natural luxury conveys God's intention that man delight in the world's beauty. Thus Manuchihri's aesthetic emulates his conception of the divine aesthetic.

The full significance of God's Creation does not stop with its sensually apprehended meaning. It goes even deeper: accessible to the curious mind is the Book of Nature's rich subtext which reinforces sensually apprehended conclusions. In this subtext are "signs for those who understand." Like God's Creation, Manuchihri's poetry also has an allegorical subtext, the meaning of which compliments the literal import of the text. His poetry holds figurative language, which comprises "signs for those who understand."

Manuchihri's poetic representation of the garden as a microcosm of Creation reflects the Islamic conception of the divine, both in form and in function. In his poetry, Manuchihri emulates his conception of God's aesthetic to convey his conception of God's intention. In the next chapter, I shall examine the relationship between human and divine art with reference to the *Roman de la Rose*, which also suggests a parallel between the art of poetry and the divine art of nature in the garden. Like Manuchihri's poetry, the *Roman* invites the reader to interpret nature within the garden as a metaphor for God's Creation. After examining the Christian conception of God's intentions and of His method of communicating them, I shall show how the *Roman* also adopts a divine aesthetic to convey a divine message.

CHAPTER FOUR: POETIC ALCHEMY

In Chapter One, I proposed that gardens are by definition artificial, and that artifice is opposed to nature, according to modern usage of the word "nature." It has become clear that the medieval definition of nature does not rule out artifice; neither Manuchihri nor the authors of the *Roman* conceive of "nature" and "artifice" as opposites. In the *Roman*, Nature is literally an artificer. The conclusion that nature is artifice is reinforced by observation. Nature's creation of the human eye, for instance, is an elegant study in adapting form to function with economy and grace. The human body is a well designed and highly ordered organism with "liaison entre ses membres" and "harmonie," qualities which Manuchihri states are essential to a good poem (XXIV, 13). What we see in nature is apparently designed by some intellect. William Paley writes: "There cannot be design without a designer; contrivance without a contriver; order without choice; arrangement without anything capable of arranging."[1] If one found a watch in the middle of a meadow, Paley argues, one would assume that

> there must have existed . . . an artificer or artificers who formed it for the purpose which we find it actually to answer, who comprehended its construction and designed its use . . . [this observation] may be repeated with strict propriety concerning the eye, concerning animals, concerning plants, concerning, indeed, all the organized parts of nature.[2]

Manuchihri and the authors of the *Roman* had a very different idea of "nature" from the conception of nature that modern, secular minds implicitly accept. We can ascribe this inconsistency to the intervening centuries which brought many scientific and philosophical discoveries. Darwin's theory of evolution debunked Paley's compelling Argument from Design, and caused twentieth-century thinkers to reconsider the conception of nature as divine design. Rather, the theory of evolution suggests that apparent designs in nature result from random genetic variation, not divinely crafted artifice. But in Manuchihri's *Divan* and the *Roman de la Rose*, both written centuries before Darwin, nature is presented as high artifice, created just so by the highest Artificer.

In the *Roman*, Nature is the immediate artificer of the *Jardin*, but her artifice exists only to maintain God's original Creation. Nature states, "'When God himself, through His grace, had placed His other creatures according to His plans, He so honored and valued me that He established me as chambermaid'" (16768). What Nature makes is ultimately created by God. This implies that the lush artifice crafted by Nature that fills the garden is, at its highest remove, the work of a divine Creator.

The Irony of the Roman

In the last two sections of Chapter Two, I addressed the irony of the *Roman*. On the surface, the poem is a celebration of carnal love, which is presented as a noble, worthy pursuit. As a result of his love, the Lover becomes "so rich that (I declare it as *truth*) there was no wealth as rich" (21765, my emphasis). Yet the poem's subtext makes it clear that Amant's behavior should not be replicated. The iconography of Oiseuse's mirror and comb and the allegorical reference to Ovid's Narcissus myth demonstrate to the pious Christian that Amant's behavior is un-Christian. Through subtle literary devices, Guillaume conveys that carnal love is a fearful madness that can and should be avoided. He suggests that the affliction is most likely to be escaped by staying away from sensuous artificial spaces like gardens. Although it never states it explicitly, the poem conveys the basic Christian teaching that carnal love should be avoided. But if the reader misses the subtle irony and the indirect allegorical references, he will glean the opposite meaning from the text.

As a setting within the *Roman*, the *Jardin* is Nature's creation, designed to entice lovers to procreate. The plot of the poem narrates the drama by which Nature's luxurious artifice leads to Amant's rape of the rosebush. Nature's artifice has no subtext. Because she wishes to continue the species, she creates conditions which make procreation probable, and there is no mystery in her means of achieving this end. Amant need not intellectually interpret the garden's artifice and draw the conclusion that its "meaning" is to make him desire to continue his species; he simply inhales the garden and falls in love with its rosebud. But whereas the meaning of Nature's garden can be communicated directly through Amant's senses, the reader can't know what Guillaume and Jean intend simply by holding the book or smelling its pages.[3] Nor even is understanding the literal meaning of the text enough; if the reader took the poem literally, the poets' true intentions would still elude him. The intended meaning of the text that describes the garden can be discovered by identifying and analyzing subtle and even deceptive literary conventions like Ovidian allegory and irony. The opacity of the *Roman* is heightened by these devices, for in order to understand it rightly, one must be familiar with a vast body of other texts. Guillaume's poetic representation of the *Jardin* is characterized by an aesthetic of concealment. Guillaume's concealing, deceptive method of communication differs dramatically from Nature's openness. Although Guillaume's poetic garden is artificial like Nature's work, its meaning is masked.

Nature vs. Art

Late in the poem, Jean makes some interesting comments about art that ultimately support the hypothesis that the *Roman* operates with an aesthetic of concealment. At one point, he makes indisputable reference to the poetic artifice with which Guillaume depicts the Jardin de Déduit:

For Art, no matter how hard she tries, with great study and great effort, to make anything whatever, no matter what shapes they have – whether she paints, dyes, forges, or shapes . . . beautiful birds in green groves; or the fishes of all waters; all the wild beasts that feed in the woods; all plants, all the flowers that little boys and girls go to gather in the spring woods when they see them in bloom and leaf; tame birds and domestic animals; balls, dances, and farandoles with beautiful and elegantly dressed ladies, well portrayed and well represented, either in metal, wood, wax, or any other material, in pictures or on walls, with the ladies holding handsome bachelors, also well represented and portrayed, in their nets – even so, Art, for all her representation and skillful touches, will never make them go by themselves, love, move, feel, and talk. (16035)

Jean evokes Guillaume's poetic representation of the garden as an example of Art, and asserts that it is inferior to Nature's artifice, the actual garden. Jean's forceful declaration that Nature is an artist whose work is superior even to Art continues with this passage:

From [Nature's true forms], Art makes her models, but she does not make her forms as true. However, with very attentive care, she kneels before Nature and like a truant beggar, poor in knowledge and force, she begs and requests and asks of her. She struggles to follow her so that Nature may wish to teach her how with her ability she may properly subsume all creatures in her figures. She also watches how Nature works, for she would like very much to perform such a work, and she imitates her like a monkey. But her sense is so bare and feeble . . . (16020+)

Humorously, this emphatic declaration is immediately undercut. Jean makes a small exception to his vehement rule that Art is inferior to Nature. "Nevertheless, it is a notable thing that alchemy is a *true art*" (16083, my emphasis).

Whoever worked wisely in [alchemy] would find great miracles; for, however it goes with the species, the individuals, at least, when they undergo intelligent operations, are changeable into many forms. They can so alter their appearances by various transformations that this change puts them into entirely different species and robs them of the original species. (16083)

Nature's law of continuing the species is broken when an individual member of a species is transformed into another by the miracle of alchemy, such as the transformation of fern into ash and glass that Art can effect (16100). Nature claims, "All things by their dispositions, through natural directions, have a rule which neither fails nor lies: everything goes back to its beginning. This rule is so general that it does not want to fail in the case of the elements" (18970). However, this law does "fail in the case of the elements" when an alchemist transforms an inferior metal into gold. Unlike ineffectual artists, alchemists can transmute Nature. This is what makes alchemy a "true art": it is *super*natural.

Inferior works of art are depicted within the poem. Amant describes the images on the wall of the Jardin de Déduit as good likenesses of their subjects. Regarding the painting of Papelardie, Amant states, "The image resembled her very much; it was made in her likeness" (420). When he sees Villainie, he salutes the absent painter: "He who could produce an image of such a truly contemptible creature knew how to paint and portray" (160). Clearly, Nature far surpasses painterly imitation of this variety. No matter how well these images copy their subjects, they are inferior to Nature's work, for they cannot "go by themselves, love, move, feel, and talk" (16020). The forms Art makes emulate Nature's creations, but cannot surpass them: "They will never catch up with Nature" (16148).

Is Jean's poetry like these ineffectual paintings, or is it a "true art"? In the next passage, Jean's excessive humility about his poetic abilities seems to suggest not. "I would willingly describe [Nature] to you, but my sense is not equal to it. My sense! What have I said? [etc.]" (16165). His apologetic effusion should alert the reader to question its sincerity. This is an example of false modesty, a familiar topos.

Such "modesty formulas" achieve an immense diffusion, first in pagan and Christian late antiquity and then in the Latin and vernacular literature of the Middle Ages. Now the author protests his inadequacy in general, now he bemoans his uneducated and rude speech.[4]

After Jean's inflated protestations that he cannot describe nature, he does; by stating what Nature is not, he conveys what she is. Amant presents such a negative description when he says, "I make no tale either about her body or her face, so pleasing and beautiful that no lily at the beginning of May, no rose on its twig nor snow on a branch is so red nor so white" (16240). Furthermore, Nature's lengthy confession serves as her description; if the poet were genuinely incapacitated by his awe for Nature's beauty, characterizing her would be entirely too audacious.

Guillaume is similarly modest. The poet narrator is reluctant to describe the beautiful things he sees in the *Jardin* and denies his poetic ability: "There was never a girl more elegant or better arrayed; nor would I have described her right" (550). "It is useless to speak of how quaintly they danced" (771). "I don't know what to tell you about her nose: no one ever made a better one out of wax" (848–849). "Why should I go on telling you? She was beautiful and beautifully adorned –" (853–854). "There were so many different trees that one would be heavily burdened before he had numbered them" (1358–1361). But just like Jean's affected humility, Guillaume's modesty is false. Each of the clumsy excuses is followed by an evocative description of what the poet claims he cannot describe. "I fear I should be grievously burdened in describing this dress, since it was not of silk but of tiny flowers . . ." (874–878). Sentences such as this one undermine the avowals of artistic inadequacy.

Jean also refutes his self-undermining pose by blatant contradiction. He states that Plato, Aristotle, Algus, Euclid, Ptolemy, Pygmalion, Parrasius, Apelles, Miro, and Polycletus don't have the skill to describe Nature, and that "*Even Zeuxis* could not achieve such a form with his beautiful painting" (16196, my emphasis). This implies that no human could possibly succeed at the task of capturing Nature in a sculpture, painting, or book. However, he then notes that "Tully [Cicero] recalls the story to us in this way in the book of his *Rhetoric*, a very authentic body of knowledge" (19198), which indicates that Cicero's writing in fact can depict Nature.

Alchemical Language

All the claims about the inferiority of human artists and poets in particular are undercut. Although they profess that they cannot, Guillaume and Jean do effectively capture Nature on paper. Although the literal level of the text asserts the impotence of art, the poets demonstrate that the art of writing *can* effectively convey truth. Therefore writing is a true art, analogous to alchemy. A successful poet has the alchemical power to "make himself subtle enough to prepare the spirits so that they had the force to enter into bodies and not fly out again once they had entered . . ." (16130). This alchemical activity is a metaphor for the writer's craft of making effective metaphors. Sophists such as Plato, Aristotle, Algus, Euclid, and Ptolemy write in a straightforward, unmetaphorical manner, and this is why "those who work with sophistry would never do [alchemy]." (16146) Furthermore, Pygmalion, Parrasius, Apelles, Miro, Polycletus and Zeuxis are plastic artists; purely representational artwork, like the paintings on the garden's wall, is inferior to Nature for it cannot transmute forms in the manner of alchemy. Susan Stakel writes:

> At first glance, Jean seems to contrast all-powerful alchemy and ineffectual poetry, for the high praise accorded to alchemy is followed immediately by Guillaume's admission of his inability to describe nature. The pose itself is a feint. The feelings of inadequacy expressed therein are part of the medieval *topos* of "affected modesty." Poetry recovers its prestige very quickly.[5]

Alchemy as a metaphor for poetry has many implications. A wise alchemist can perform magic: he can transform inferior metals into gold. Similarly, poets transform raw language into poetry, by a process we can make explicit. A poet has a number of tools at his disposal. Poets make use of figurative language such as similes, metaphors, symbols, and personification. In addition, they use musical elements such as rhyme, meter, assonance, consonance, and alliteration. But there are other even subtler means a poet might use, such as irony, allegory, or the *topos* of affected modesty discussed above. Poetic alchemy includes any and every operation a poet might choose to employ to transform ordinary language into poetry. His goal is to make raw language into the equivalent of an alchemist's gold: language that is at once beautiful and capable of communicating truth. "The poet is thus an instrument of transmutation, not of metal but of words."[6]

This explains why Jean upholds Cicero's writing. His assertion that Cicero's *Rhetoric* supports authentic communication implies that Cicero is not a sophist, but an alchemist like Jean. He approves of Cicero's style; in fact, he mimics it: Curtius notes that Cicero originated the device of affected modesty: "We have traced the modesty topos as far back as Cicero."[7] In *De Inventione*, Cicero himself affirms the value of this device: "Prece et obsecratione humili ac supplici utemur."[8] Jean's poetry follows Cicero's advice and his example. Thus both writers are poetic alchemists.

When an inferior metal is transformed by alchemy, the result is a substance identical to naturally occurring gold. That it is gold is obvious from its appearance. With this observation, alchemy as a metaphor for poetry appears to break down. Unlike alchemical gold, the meaning of alchemical poetry is not immediately obvious. The literal level of poetry is most often puzzling, and seldom in accordance with its actual meaning. In the *Roman*, the literal level of the text is downright deceptive; it appears to suggest that the pursuit of carnal love is noble, but on examination the subtext conveys that such a quest is actually sinful. When poetic tropes are identified and examined, the well-masked meaning is revealed. And although the poetry seems cryptic and deceptive, those elements that make it up so unite to convey what the poet wishes to communicate in an even more compelling way, rewarding the study of a patient scholar. Alchemical gold is a metaphor for the truth made possible only through poetic alchemy. This is Jean's eloquent answer to the question that has tortured writers from Plato to Kafka: "How is Art to convey Truth when its medium is fallen language?"[9]

Nature's gloss on the moon is an allegory for poetry and supports this theory of Jean's aesthetic of concealment. Moonspots were believed to be evidence of the moon's imperfection. "It seems to men that the moon may indeed not be clean and pure, because in some places it shows up dark" (16833). However, Nature explains these apparent defects: "It is because of its double nature that it appears opaque and cloudy in some places, shining in one part and ceasing to shine in another, because it is both clear and opaque" (16833). Like the moon, poetry has a "double nature." In some places the transparency of the text makes the meaning of the language clear. This is true when the intended meaning of a text equals the literal meaning of the words on the page. Most often in the *Roman*, this is not the case. Whenever a poet uses alchemical language, his poetry becomes opaque; its literal meaning hides its true meaning, which is found only beneath the surface. "But the opaque part, which can resist the rays well and overcome its light, *shows the light*" (16833). Taking light as a metaphor for understanding, this passage states that the enlightened intellect, upon contemplating Jean's opaque poetry, will be rewarded with illumination. Moreover, pure transparency in poetry is undesirable, because it cannot communicate effectively:

> What makes its light fail is the fact that the clear part of its substance cannot reflect the rays that the sun throws out toward it; instead they pass on through and beyond . . . transparent glass cannot show shapes, since the eye-beams cannot encounter there anything that may retain them. (16840)

Just as nothing in the clear part of the moon stops the light to shows its shape, nothing arrests a reader's intellect when reading sophistry, for it is innocent of alchemical language. Poetic alchemy is a necessary obstacle to easy comprehension, rendering poetry opaque but richer. It enables the poetry to contain and convey truth. "In the collective myth of literature as alchemical undertaking, the crystal in its complexity and transparency is at once jewel and conductor of light, end and passage, goal of the cosmic reverie and operator of perfect understanding."[10]

Jean's poetry is more like lead than glass; even more, it resembles "some polished object which could reflect light," which is "opaque of its own nature or through some other material" and therefore can "return an image." This "image" is the truth Jean could not convey without the apparent obstacle of alchemical language. It is evidence of Jean's mastery of poetic alchemy that he can make these complicated ideas clear through a treatise on moonspots, language dense as lead. "Alchemical language, meant both to hide and to reveal, is based on plurality and enigma, often a double image of a double significance,"[11] such as the double image of the moon. The passages on alchemy and moonspots support the hypothesis advanced earlier: that Jean's aesthetic is grounded in the theory that some element of deception or concealment of truth is necessary for effective communication.

Understanding poetic alchemy resolves the much debated invocation and dedication of the opening lines of the poem. Guillaume writes, "God grant that she for whom I have undertaken it may receive it with grace. It is she who is so precious and so worthy to be loved that she should be called Rose" (42). Medieval Christian readers, familiar with the close association of the rose with the Virgin Mary, would undoubtedly interpret this line as a dedication to the Virgin. In making this leap, readers interpret the text's symbolism, alchemical language they normally expect in poetry. Given the substance of the *Roman*, however, this dedication seems strange. At its literal level, the *Roman* celebrates a sin: the gratification of carnal love. This is the antithesis of the Virgin, a symbol of purity and holy love. Upon realizing this, Christian readers will now claim that deluded Amant made the dedication in ignorance of Christianity. Thus, they will reverse their interpretation of the rose as a symbol for Mary and reinterpret the statement as the poet-dreamer's dedication of the poem to a pretty woman of his acquaintance named Rose. Orthodox Christians would thus infer that Guillaume is either ignorant or irreverent, and denounce him as a heretic.

But despite the literal celebration of sin, the scholar who sees through the deceptive alchemy of the poem's language is rewarded with an orthodox Christian truth. For all its apparent celebration of sin, the *Roman* in fact advocates Christian charity. Therefore, Guillaume's invocation of God's grace and his dedication to the Virgin Mary are completely in order. Although we are usually wise to seek the irony and subtext in the *Roman*, this rare moment is innocent of it. It is perhaps the only sentence we are meant to interpret literally (taking into account, of course, the transparent symbolism of rose as Virgin). Guillaume is uncharacteristically direct; the words mean what we think they should mean. For once, the intended meaning and the literal meaning of the poetry overlap. In a poem so dense with subtext, such surprising honesty is moving.

Significantly, however, the statement is deceptive in its very honesty. Jean suggests that to convey a truth, one cannot state it directly. Only poetic alchemy makes true communication possible through the imperfect medium of language. Defective language causes the truthteller to resort to deception – alchemical language. This nearly drove Franz Kafka to despair. He even found it difficult to construct metaphors, which are basic to both ordinary and figurative language, because of the "feeling of falsity" they gave him.[12] Luckily, misgivings such as these paralyzed neither Kafka nor the poets of the *Roman*.

Divine Alchemy

Just as Manuchihri's aesthetic reproduces his conception of God's aesthetic, the poetic alchemy adopted in the *Roman* is a reflection of the Christian conception of the divine aesthetic. Jean concretely characterizes God's work as poetic alchemy. Within the *Roman*, God's work is characterized in explicitly alchemical terminology. Nature cites Virgil's *Fourth Eclogue*, which predicts the incarnation: "'Already a new line is sent down to us on earth from high heaven to guide the people, who have lost their way. With this line the age of iron will perish and the age of gold will spring up in the world'" (19175). Christians interpret this "new line" as Jesus Christ, who performs the collective alchemical transformation of the souls of mortals. Similarly, the virgin birth, which Nature cannot comprehend – "It could never be through me that anything might ever be born of a virgin" (19170) – is an example of the kind of "great miracle" only a master of alchemy is capable of performing. God is a master alchemist. When human artists write poetry, they emulate his work with poetic alchemy.

Christianity, like Islam, conceives of God's Creation as a Book of Nature through which God's intentions can be read. In Chapter Three, I demonstrated that Manuchihri's aesthetic mirrors his conception of a divine aesthetic, just as his poem's function is to convey a divine message. I've shown that the well-masked intention of the *Roman de la Rose* is to communicate a Christian message. Now I shall demonstrate that its poets replicate their conception of the divine aesthetic in communicating this message.

God has created a beautiful, seductive garden full of enticing luxuries: the world we inhabit. However, rather than simply responding to the world's charms in a non-interpretive, sensual manner as Moslems do, Christian believers endeavor to read the world as if it were a book, seeking its subtext and its tropological implications. In the last chapter, I cited a passage from the Koran that explains that man can learn from the world he inhabits, and there is a congruent assertion in the Bible. In his famous letter to the Romans, the apostle Paul writes: "For since the creation of the world God's invisible qualities – his eternal power and his divine nature – have been clearly seen, being understood from what has been made, so that men are without excuse" (Rom., I:20). The meaning of the Book of Nature, however, is crystal clear only to the blessed; most often, God's creation requires rigorous exegesis to unravel the divine secrets it encrypts. In *De genesi ad litteram*,[13] Saint Augustine writes, "It is not by way of assertion, but by way of inquiry that we have to treat the hidden matters concerning the natural things which we know were made by God, their almighty maker."[14]

The lecherous Amant acts against God's intentions, because he responds to the sensuality of the garden. The garden is a microcosm of Creation, and Amant represents all men who are tempted by the beautiful things the world holds. Christianity exhorts man to deny the sensual invitation of Creation's beauty and instead to understand its subtext, God's true intention. This enables the Christian to love charitably rather than lasciviously. Amant ought to delve below the surface beauty of the natural artifice that encourages his lechery, and he should instead seek its hidden, divine purpose.

In the same way, man should resist the superficial meaning of the poem in search of the concealed subtext that conveys its true message. The sensual world is like alchemical language: deceptive, but interpretable. One can glean its true import only through interpretation and analysis. Dedicated to the Virgin Mary, the *Roman* advises Christian behavior, which opposes what the text literally encourages. The reader must be aware of deceptive poetic conventions like irony and false modesty to discover the truth of the poem. In the Christian world, the superficial can deceive the naïve reader of poetry or of Creation, for it masks the hidden truth. The reason for this necessary deception lies in the basic Christian belief that theology is "not a mere intellectual or conceptual knowledge, but a prayerful activity in search of God in his very presence – for our way of finding God is precisely to seek him."[15] By prolonging and complicating our search for spiritual truth, poetic alchemy helps us experience it more completely.

Relating the congruent Islamic and the Christian beliefs of Nature as a great Book through which the believer can understand the divine, Julie Meisami writes,

> The medieval conception of the garden was as a place wherein the soul might read the most profound spiritual lessons. The idea of the Book of Nature, "written" by God to provide signs of Himself, is found both in Christianity and in Islam: the Augustinian view that "beauty is not mere spectacle but God's rhetoric in the book of creation, a call to the movement of the heart" is echoed by the Islamic, that "Nature is a fabric of symbols, which must be read according to their meanings."[16]

Meisami implicitly equates "the garden" with "Nature," supporting the conception of the garden as a microcosm of Creation, designated by "Nature." Moreover, Meisami correctly indicates that both Islam and Christianity espouse the idea of nature as a guide to the divine. However, her statement is misleading because it implies that both cultures' religions teach that God communicates his intentions through nature in the same way. Nothing could be further from the truth.

Whereas deception is central to the Christian conception of the divine aesthetic, it is alien to Islam. The medieval Persian believer would be unable to fathom Nature's appealing luxury as sinful or remotely perilous; he would simply interpret it as wonderful though unnecessary evidence of God's infinite wisdom and benevolence. This does not imply that the Islamic conception of the Book of Nature lacks a subtext. On the contrary, as Meisami correctly notes, Nature (God's Creation) is richly symbolic. Interpreting its web of symbols enables the believer to understand divine intentions. In this study, for example, I have focused on the complex symbol of the garden and its elements. Although Islamic poetry is rich in figurative language, it is never deceptive. The poetry means *more* than the literal meaning of the words, but it never means the *opposite* of the literal meaning of the words. Deceptive devices such as irony and false modesty are absent, and so are subtle literary conventions like allegorical references to other texts. God's Book of Nature doesn't mislead, and therefore a devout Muslim poet's *Divan* must not.

In every culture, art represents man's pursuit of the ideal, which is culturally defined by the accepted religion. The apple was beautiful, but it deceived Adam and Eve. As a fundamental Christian teaching, this is the basis for the belief that apparent beauty perceived by the senses does not denote true worth. In Christianity, conclusions based on information discerned by the eye and the other senses are most often wrong. The aesthetic adopted by Jean and Guillaume's Christian poetry reflects this conception of God's aesthetic. If a Christian poet wants to say something, he must at all costs avoid telling it in a straightforward way. Rather, he emulates God's roundabout method of conveying truth. Since truth cannot be effectively communicated through direct sophistry, the poet must resort to alchemical arts to convey his message. These arts deceive in order to enlighten. This translates into Jean de Meun's aesthetic: he utilizes deceptive literary conventions to convey truth through language.

In the last two chapters, I have examined the garden as a microcosm of Creation, in Manuchihri's *Divan* and in the *Roman de la Rose*. Nature within Manuchihri's gardens is sensually appealing, aesthetically pleasing artifice, crafted by a divine Creator. In Islam, God communicates with man sensually; behavior encouraged by sensual experience of Creation is good and holy. Manuchihri's poetry parallels his conception of God's work. Not only does his poetic creation convey the divine message to enjoy sensual pleasures, but it also emulates the divine aesthetic by communicating this message directly, through the text's literal meaning and musical appeal. Thus Manuchihri's *Divan* reflects both his conception of God's intention, and his conception of God's method of communicating this intention. The garden of the *Roman*, which is also a sensual artificial space, is maintained by Nature but originally created by God, suggesting that it is also a microcosm of Creation. In Christianity, what is sensually perceived should be mistrusted. Within the *Roman*, God's work is compared to the poet's alchemy. The *Roman* conveys the Christian message that man should avoid and mistrust the sensual garden in a subtle, deceptive way, emulating the Christian conception of the divine aesthetic.

CONCLUSION

Oscar Wilde writes, "It is through Art, and through Art only, that we can realize our perfection."[1] An artist implicitly builds his pursuit of perfection on his conception of the divine. This is true of secular as well as religious art. According to both Islam and Christianity, God represents the unity of all perfections. However, these religions promote radically different world-views. This is why medieval French and Persian poetry are very different in content and aesthetic.[2]

Yet the poetry of these two cultures has some striking similarities. Both recognize the luxury of nature and its appeal to man. Both uphold order as a perfection, and recognize it within the natural world. In both cultures, men emulate natural order through skilled craftsmanship. Art is the refinement of the order created by man.

The garden is at once the operant symbol of man-imposed order, and an exclusive collection of natural luxury. In the garden space, order imposed by the mind weds luxury that appeals to the body. Gardens represent the harmonious union of mind and body. Denis Hue describes the garden as a "lieu étrange, à mi-chemin entre la nature et l'ouvrage de main d'homme."[3] But the ideal garden implies no moderation of either force; rather, it is "à la fois réalité corporelle et réalité spirituelle."[4] The luxurious, orderly garden represents a civilization that is at once refined and primal.

> And when we reach the true culture that is our aim, we attain to that perfection of which the saints have dreamed, the perfection of those to whom sin is impossible, not because they make the renunciations of the ascetic, but because they can do everything they wish without hurt to the soul, and can wish for nothing that can do the soul harm, the soul being an entity so divine that it is able to transform into elements of a richer experience, or a finer susceptibility, or a newer mode of thought, acts or passions that with the common would be commonplace, or with the uneducated ignoble, or with the shameful vile.[5]

The garden represents the soul's perfection, man's constant aspiration. It is not surprising that both examples of garden poetry I've examined are panegyric. They extol a figure of perfection: the ideal king in Manuchihri's *Divan*, and the Virgin Mary in the *Roman*.

Both Manuchihri and the poets of the *Roman* saw the same thing when they looked at real gardens: places of order and luxury. But, writing in different cultures, poets associated these concepts of luxury and order with very different moral values. Through his poetry, each poet emphasizes that quality of the garden which his culture upholds as morally desirable. Julie Meisami writes:

> Perhaps the medieval Christian, ill at ease with the sensuous aspects of the garden, had difficulty incorporating them into an equilibrium with its spiritual symbolism; whereas the medieval Muslim did not experience the same tension between natural and spiritual and could entertain both in harmony. Such differences in outlook may have led to the persistent allegorization of the garden in medieval western literature, where Paradise itself becomes a figure of the human soul; while in the literature of medieval Islam we find, by contrast, an emphasis on the sensual aspects of the earthly garden which is a reflection of a Paradise conceived of in terms more physical than spiritual.[6]

Different cultural associations with the garden result in two bodies of poetry with opposing messages and radically different aesthetics, despite their poets' similar perceptions of the physical garden.

In Islamic culture, the garden is associated with celestial paradise, pure and simple. Thus Manuchihri invokes the garden as a metaphor for the ideal kingdom to instruct his terrestrial monarch by its example. At the same time, he directly instructs his public audience to enjoy the world's luxuries, which are fragments of heaven on earth. However, in the Christian world, the garden setting is inextricably associated with Eden, where man first turned from God. It is a fearful place, full of things that tempt the senses and thereby lead to the sin of carnal love. Thus when Guillaume and Jean invoke the garden, it is to warn Christians away from natural luxury. But unlike Manuchihri, these poets instruct indirectly and by non-example; the behavior of the "hero" of the *Roman* represents exactly what an aspiring Christian must not replicate.

The two methods of instruction differ as much as the messages they convey. Jean and Guillaume emulate the process by which God conveys truth according to Christianity. On the surface, their poetry is as deceptive as the luxurious garden that is God's Creation, and its true meaning can only be discovered through rigorous intellectual application. Manuchihri, however, conveys his message literally in the text and sensually in the music of his poetry. This reflects the directness which for Muslims characterizes Allah's aesthetic.

It is my hope that this study will introduce further comparative exploration of the garden as setting and symbol in medieval French and Persian poetry. Poetic gardens are a rich inroad to fruitful comparative study of these two very different cultures, in which the present body of medieval scholarship is fiercely deficient. Medieval poetry holds a menagerie of unexplored gardens: the palace gardens of the *Chanson de Roland* and the epic *Shahname*; the fantastical gardens of Nezami's *Haft Paykar*, and Hafez's mystical garden spaces. "Garden poetry" represents man's search for paradise on earth. Let us continue this search.

ACKNOWLEDGEMENTS

First, I must thank the gardeners of Keukenhof, Versailles, and Prospect, the artificial spaces that have inspired me.

Next, sincere thanks to those who have listened to my garden talk: Erin B. Habelt, David J. Kaley, and all of my friends in the Mudd Library, the Tower Club, and the Triangle Club; and to Kessler.

I am indebted to Irene Nicastro and Gale Bartholf, and Roger Babb, Robert Hollander, and Stanley Corngold.

Sincerest thanks to my wonderful advisers, John Fleming and Jerome Clinton, for their accessibility, insight, and enthusiasm for this study.

Finally, loving gratitude and tender affection to my wonderful family: you make my world a garden.

APPENDIX A: GLOSSARY OF PERSIAN WORDS

I've used transliterations and definitions from Clinton's *Manuchihri*, 155–156:

Bayt: distich.

Divan/davavin: a poet's collected lyric poems.

Duca: concluding section of a Persian *qasidah*.

Chahar bagh: rectangular walled garden quartered by two streams intersecting at right angles.

Madh: the section of panegyric in a *qasidah*.

Misrac: half-*bayt*.

Musammat: a kind of strophe poem in which the first five *misracs* of each strophe rhyme, and the sixth *misrac* is the refrain.

Næzm: poetry; order.

Næsr: prose.

Nasib: the prelude or introductory section of a *qasidah*, also called *tashbib* or *taghazzul*.

Nowruz/naw ruz: New Year's Day, celebrated on the first day of spring.

Qasidah: an occasional poem of a dozen or more *bayts* which rhymes in the first two *misracs*, and by *bayt*; normally composed of a *nasib*, *madh*, and *duca*.

NOTES

Introduction and Chapter 1

[1] William L. Hanaway, Jr., "Paradise on Earth: The Terrestrial Garden in Persian Literature," *The Islamic Garden: Dumbarton Oaks Colloquium on the History of Landscape Architecture IV*, eds. Elisabeth B. Macdougall and Richard Ettinghausen (Washington: Dumbarton Oaks, 1976) 63.
[2] Jonas Lehrman, *Earthly Paradise: Garden & Courtyard in Islam* (Berkeley: University of California Press, 1980) 62.
[3] Lehrman 67.
[4] Lehrman 64.
[5] Ernst Robert Curtius, *European Literature and the Latin Middle Ages*, trans. Willard R. Trask (London: Routledge & Kegan Paul, 1965) 195.
[6] Oscar Wilde, "The Decay of Lying," *Oscar Wilde*, ed. Isobel Murray (New York: Oxford UP, 1989) 215.
[7] Wilde 223.
[8] Oscar Wilde, "The Critic as Artist, Part II," *Oscar Wilde*, ed. Isobel Murray (New York: Oxford UP, 1989) 286.
[9] "Ghazal, Persian," *Encyclopaedia Islamica* 1035.
[10] An interesting avenue of research remains in discovering how much influence antique Hellenistic literature had on the flowering of medieval Persian poetry. There is the possibility of some buried connection; however, it should be remembered that much of Greek literature itself, including Homer's epics, has Middle Eastern roots.
[11] Persian painting often depicts the monarch, surrounded by subjects and attendants, holding court at the center of a palace garden in full bloom. This illustrates the conflation of court with garden.
[12] Jerome W. Clinton, "Court and Poetry at the Beginning of the Classical Period," *Persian Literature: Columbia Lectures on Iranian Studies*, 3, ed. Ehsan Yarshater (New York: Persian Heritage Foundation, 1988) 76.
[13] Appendix A is a glossary of meanings for transliterated Persian words.
[14] Julie Scott Meisami, *Medieval Persian Court Poetry* (Princeton: Princeton UP, 1987) 11.
[15] Meisami 11–12.
[16] The most competent translation of Manuchihri's *Divan* into a Western tongue is A. de Biberstein-Kazimirski's anthology of French translations. See *Menoutchehri: Poète Persan du IIème Siècle de Notre Ère* (Paris: E. J. Brill à Leide, 1886). For ease of citation, and to make this study accessible to a wider audience, I shall cite French translations of Manuchihri's work in the body of the study.
[17] Translated elsewhere as "the gardens of perpetuity." The celestial garden that represents Islamic Paradise is sometimes translated as "Eden." This should not be confused with the Christian Eden, which is morally ambiguous and transient.
[18] Hanaway 41–63.
[19] Hanaway 50.
[20] Hanaway 62.
[21] A. Bartlett Giamatti, *The Earthly Paradise and the Renaissance Epic* (Princeton: Princeton UP, 1966) 11.
[22] Jerome W. Clinton, *The Divan of Manuchihri Damghani: A Critical Study* (Minneapolis: Bibliotheca Islamica, 1972) 44.

Chapter 2

[1] C. S. Lewis, *The Allegory of Love* (Oxford: At the Clarendon Press, 1936) 119.

[2] John V. Fleming, *The* Roman de la Rose: *A Study in Allegory and Iconography* (Princeton: Princeton UP, 1969) 58.

[3] Fleming 58.

[4] In light of the drama that is to take place in the garden, this explicit connection between the *Jardin* and the Islamic world demonstrates that "Europeans imagined that cultural/religious entity [Islam] to be characterized by moral licentiousness, permissiveness, self-gratification, and cultural decadence." See Maria Rosa Menocal's *The Arabic Role in Medieval Literary History: A Forgotten Heritage* (Philadelphia: University of Pennsylvania Press, 1987) 44.

[5] Alain Labbé, "Nature et artifice dans quelques jardins épiques," *Vergers et Jardins dans l'Univers Médiéval: Sénéfiance N. 28* (Aix-en-Provence: Publications de CUER MA, 1990) 180.

[6] Charles Dahlberg, "Love and the *Roman de la Rose*," *Speculum* 44 (1969): 575.

[7] Fleming 81.

[8] Pierre Grimal, ed. *Le Jardin Médiéval: Colloque à l'Abbaye de Saint-Arnoult Monastère des Arts les 3 et 4 Septembre 1988* (Saint-Arnoult: Éditions A.D.A.M.A., 1990) 11.

[9] Grimal 11.

[10] Note, however, that capitalization of allegorical figures is adopted for ease in translation; the original old French text did not capitalize words like *oiseuse, raison*, and *nature*, all of which represent personifications of these ideas.

[11] Winthrop Wetherbee, "Some Implications of Nature's Femininity in Medieval Poetry," *Approaches to Nature in the Middle Ages: Papers of the Tenth Annual Conference of the Center for Medieval & Early Rennaisance Studies.* Volume 16, ed. Lawrence D. Roberts (Binghampton: Medieval & Renaissance Texts & Studies, 1982) 59–60.

[12] Dahlberg 575.

[13] D. W. Robertson, Jr., "The Doctrine of Charity in Mediaeval Literary Gardens: A Topical Approach through Symbolism and Allegory," *Speculum* 26 (1951): 42.

[14] To see; to draw near in conversation; to touch; to kiss; to do the deed.

[15] Alain of Lille, *Summa de arte praedicatoria*, in Migne, *PL*, CCX, col. 122, qtd. in Lionel J. Friedman, "Gradus Amoris," *Romance Philology*, XIX, 2 (Los Angeles: University of California Press, November 1965) 169.

[16] Robertson 25.

[17] Fleming 101.

[18] Dahlberg 576, my emphasis.

[19] Lionel J. Friedman, "Gradus Amoris," *Romance Philology* XIX, 2 (Los Angeles: University of California Press, November 1965) 170.

[20] Fleming 75.

[21] His behavior is of course irrational to good Christian readers. Readers eager to cast the first stone, however, should remember that they cannot *see* the irresistible beauty that seduces Amant.

[22] Mary Ellen Faughnan, "Morality in the *Romance of the Rose*: The Misunderstood Figure of Nature," *Proceedings of the PMR Conference: Annual Publication of the Patristic, Medieval and Renaissance Conference, Vol. 4*, ed. Joseph C. Schnaubelt and John J. Hagen (Villanova: Augustinian Historical Institute, 1979) 139.

[23] Faughnan 139.

[24] Faughnan 140.

[25] Wetherbee 58–59.

Chapter 3

1 Clinton, "Court and Poetry," 94.
2 And vice versa: carpets were often woven with floral garden designs.
3 Lehrman 67.
4 In all of these images, the transforming agent of the garden is concretely characterized as the wind. In his *General Prologue*, Chaucer uses the term *Zephirus*, denoting the West Wind, "as a personification of the natural forces of springtime," referring to the "seasonal force of renewal and reanimation." See Paul B. Taylor's article, "The Alchemy of Spring in Chaucer's *General Prologue*," *The Chaucer Review* 17, 1 (1982): 1–4. I deal with alchemy as a metaphor for poetic creation in Chapter Four.
5 C. H. de Fouchécour, *La Description de la Nature dans la Poésie Lyrique Persane du XIe Siècle* (Paris: Librairie C. Klinchsieck, 1969) 176.
6 De Fouchécour 176.
7 De Fouchécour 247.
8 It may surprise Western readers to discover that in Islam, celibacy is considered a sin.
9 Clinton, *Manuchihri* 44.

Chapter 4

1 William Paley, "The Argument from Design," *Reason and Responsibility*, ed. Joel Feinberg, 9th ed. (New York: Wadsworth Publishing Co., 1996) 43.
2 Paley 41, 45.
3 Unlike Manuchihri's *qasidas*, the *Roman* was not written for live performance; its true message is not conveyed in its music, although its rhyme and meter are carefully crafted.
4 Curtius 83.
5 Susan Stakel, *False Roses* (Anma Libri: Stanford University, 1991) 115.
6 Stakel 116.
7 Curtius 84.
8 Cicero, *De Inventione*, I, 16, 22, qtd. in Curtius 84.
9 Stakel 5.
10 Mary Ann Caws, "In Search of the Crystal: The Poetics of Literary Alchemy." *Voices of Conscience*, ed. Raymond J. Cormier (Philadelphia: Temple UP, 1977) 246.
11 Caws 248.
12 Franz Kafka, *The Diaries of Franz Kafka, 1910–1913*, trans. Joseph Kresh (New York: Schocken, 1949) 173.
13 In this context, it is amusing that Augustine titles his book *The Literal Interpretation of Genesis*.
14 Augustine, "On the Literal Interpretation of Genesis: An Unfinished Book," *The Fathers of the Church*, Vol. 84, trans. Roland J. Teske (Washington: The Catholic University of America Press, 1991) 145.
15 Jean Leclerq, "The Love of Beauty as a Means and an Expression of the Love of Truth," *Mittellateinisches Jahrbuch, Band 16* (Stuttgart: Anton Hiersemann, 1981) 62.
16 Julie Scott Meisami, "The World's Pleasance: Hafiz's Allegorical Gardens," *Comparative Criticism: A Yearbook*, Vol. 5 (Cambridge: Crit., 1983) 154. Quotes: Terry Comito, *The Idea of the Garden in the Renaissance* (New Brunswick, 1978) 105 and S. H. Nasr, *Science and Civilization in Islam* (Cambridge, 1968) 24, respectively.

Conclusion

1 Wilde, "Critic as Artist, Part II," 274.
2 Undeniably, religion is not the only force that shapes art. Geography, cultural history, tradition, and government are a few of the many forces that influence a culture's art in addition to religious doctrine.
3 Denis Hue, "Reliure, Clôture, Culture: Le Contenu des Jardins," *Vergers et Jardins dans l'Univers Médiéval: Sénéfiance* N. 28 (Aix-en-Provence: Publications de CUER MA, 1990) 155.
4 James Dauphine, "Du Paradis Terrestre," *Vergers et Jardins dans l'Univers Médiéval: Sénéfiance* N. 28 (Aix-en-Provence: Publications de CUER MA, 1990) 91.
5 Wilde, "Critic as Artist, Part II," 276.
6 Meisami, "Hafiz's Allegorical Gardens," 156.

BIBLIOGRAPHY

Primary Sources

Augustine. "On the Literal Interpretation of Genesis: An Unfinished Book." *The Fathers of the Church, Vol. 84*. Trans. Roland J. Teske. Washington: The Catholic University of America Press, 1991, 143–188.

Guillaume de Lorris and Jean de Meun. *The Romance of the Rose*. Trans. Charles Dahlberg. Princeton: Princeton UP, 1995.

Guillaume de Lorris and Jean de Meun. *Le Roman de la Rose*. Chronology, preface and editing by Daniel Poirion. Paris: Garnier-Flammarion, 1974.

Kafka, Franz. *The Diaries of Franz Kafka, 1910–1913*. Trans. Joseph Kresh. New York: Schocken, 1949.

Manuchihri Damghani. Divan. *Menoutchehri: Poète Persan du 11ème Siècle de Notre Ère*. Trans. A. de Biberstein Kazimirski. Paris: E. J. Brill à Leide, 1886.

Nietzsche, Friedrich. *Beyond Good and Evil*. Trans. and commentary by Walter Kaufmann. New York: Vintage Books, 1989.

Wilde, Oscar. "The Decay of Lying." *Oscar Wilde*. Ed. Isobel Murray. New York: Oxford University Press, 1989, 215–239.

——. "The Critic as Artist, Part II." *Oscar Wilde*. Ed. Isobel Murray. New York: Oxford University Press, 1989, 267–297.

Secondary Sources

Badel, Pierre-Yves. "Alchemical Readings of the Romance of the Rose." *Rethinking the* Romance of the Rose. Eds. Kevin Brownlee and Sylvia Huot. Philadelphia: University of Philadelphia Press, 1992, 262–285.

Caws, Mary Ann. "In Search of the Crystal: The Poetics of Literary Alchemy." *Voices of Conscience*. Ed. Raymond J. Cormier. Philadelphia: Temple UP, 1977, 243–261.

Clinton, Jerome W. *The Divan of Manuchihri Damghani: A Critical Study*. Minneapolis: Bibliotheca Islamica, 1972.

——. "Court and Poetry at the Beginning of the Classical Period." *Persian Literature: Columbia Lectures on Iranian Studies*, 3. Ed. Ehsan Yarshater. New York: Persian Heritage Foundation, 1988.

Curtius, Ernst Robert. *European Literature and the Latin Middle Ages*. Trans. Willard R. Trask. New York: Pantheon Books Inc., 1953.

Dahlberg, Charles. "Love and the *Roman de la Rose*." *Speculum* 44 (1969) 568–584.

Dauphine, James. "Du Paradis Terrestre." *Vergers et Jardins dans l'Univers Médiéval: Sénéfiance* N. 28. Aix-en-Provence: Publications de CUER MA, 1990, 87–96.

Deluz, Christiane. "Le Jardin Médiéval, Lieu d'Intimité." *Vergers et Jardins dans l'Univers Médiéval: Sénéfiance* N. 28. Aix-en-Provence: Publications de CUER MA, 1990, 97–101.

Faughnan, Mary Ellen. "Morality in the *Romance of the Rose*: The Misunderstood Figure of Nature." *Proceedings of the PMR Conference: Annual Publication of the Patristic, Medieval and Renaissance Conference*, vol. 4. Ed. Joseph C. Schnaubelt and John J. Hagen. Villanova: Augustinian Historical Institute, 1979, 137–141.

Fleming, John V. *The* Roman de la Rose: *A Study in Allegory and Iconography*. Princeton: Princeton UP, 1969.

de Fouchécour, C. H. *La Déscription de la Nature dans la Poésie Lyrique Persane du XIe Siècle*. Paris: Librairie C. Klincksieck, 1969.

Friedman, Lionel J. "Gradus Amoris." *Romance Philology* XIX, 2 (1965): 167–177.

"Ghazal, Persian," Encyclopaedia Islamica.

Giamatti, A. Bartlett. *The Earthly Paradise and the Renaissance Epic*. Princeton: Princeton UP, 1966.

Grimal, Pierre, ed. *Le Jardin Médiéval: Colloque à l'Abbaye de Saint-Arnoult-Monastère des Arts les 3 et 4 septembre 1988*. Saint-Arnoult: Éditions A.D.A.M.A., 1990.

Hanaway, William L., Jr. "Paradise on Earth: The Terrestrial Garden in Persian Literature." *The Islamic Garden: Dumbarton Oaks Colloquium on the History of Landscape Architecture IV*. Eds. Elisabeth B. Macdougall and Richard Ettinghausen. Washington: Dumbarton Oaks, 1976, 41–67.

Herman, Gerald. "The 'Sin Against Nature' and Its Echoes in Medieval French Literature." *Annuale Mediaevale XVII*. Ed. Herbert H. Petit. Atlantic Highlands: Humanities Press, 1976, 70–87.

Hue, Denis. "Reliure, Clôture, Culture: Le Contenu des Jardins." *Vergers et Jardins dans l'Univers Médiéval: Sénéfiance* N. 28. Aix-en-Provence: Publications de CUER MA, 1990, 155–172.

Labbé, Alain. "Nature et Artifice dans quelques Jardins Épiques." *Vergers et Jardins dans l'Univers Médiéval: Sénéfiance* N. 28. Aix-en-Provence: Publications de CUER MA, 1990, 177–187.

Landsberg, Sylvia. *The Medieval Garden*. London: British Museum Press, 1996.

Leclerq, Jean. "The Love of Beauty as a Means and an Expression of the Love of Truth." *Mittellateinisches Jahrbuch, Band 16*. Stuttgart: Anton Hiersemann, 1981, 62–72.

Lehrman, Jonas. *Earthly Paradise: Garden & Courtyard in Islam*. Berkeley: UCP. Macpherson, Patricia Parker, and Julian Patrick. Toronto: Univ. of Toronton Press, 1983, 115–127.

TSUNG-I DOW

THE CHINESE ATTEMPT TO MINIATURIZE THE WORLD IN GARDENS

INTRODUCTION

What was the intent of the world-renowned landscape gardens in Suzhou, China, such as the garden of "The Master of the Fishing Net" (Wang Shi Yuan) from which the small courtyard of the "Jade Spring Abode" was reproduced in the Metropolitan Museum of Art (Astor Court) in New York City, or the garden of the "Unsuccessful Politician" (Zhou Zhen Yuam) that painter Weng, Zhen-Ming and poet Yuan Mei frequented, or the garden of "The Lingerer-On" (Liu Yuan), once owned by the family of I. M. Pei, or the "Blue Water Pavilion" (Can Lang Ting)?

These gardens serve as the original representational attempts of Chinese intellectuals, notably neo-Confucians or affluent, elite scholar-bureaucrats, to crystallize their Confucian-Taoists ontological views in miniaturized, realistic structures on the aesthetic level to evoke self-cultivation. It may seem like an extension of early Chinese endeavors to attain the same purpose through the creation of a virtual reality in abstract calligraphic landscape painting with brush and ink on paper. Had there been, at that time, in China a computer screen to simulate world phenomena it would have been a much more effective medium to reflect their naturalistic, if not materialistic, world outlook than landscape painting, which requires a certain degree of literary proficiency to decipher the essence of what the painter conveyed in the classical poem conspicuously above the painting. Classical landscape painting was a poem in a picture. Thus, the appreciation of this body of work was confined to a select few in China and hardly attracted the outside attention of westerners, whose landscape gardens avoided this literary requirement for appreciation. By evoking the sensual pleasures of visitors to these gardens through multi-sensory participation, they were more powerfully affecting than still, monochrome landscape paintings. However, these landscape gardens are bound to be associated with Chinese feng shui, geomancy, and its prevailing correlative view of the world. In light of modern physics, feng shui is considered by some to be mysterious, if not based on superstition. Therefore, the essential philosophical assumptions for the landscape garden layout face critical challenges today and its design risks extinction in the current world of architectural tradition.

A.-T. Tymieniecka (ed.), Analecta Husserliana LXXVIII, 139–150.

The influences of advancements in science also contribute to these challenges. In particular, the advancement of quantum mechanics, which puts limits on the relevancy of Newtonian physics, also deepens the understanding of the world at a particle and molecular level. Its fundamental principle, according to one of its founders, Niels Bohr, consists of complementarity in contradiction. This principle, amazingly, resonates with Confucian-Taoist ontological assumptions in the trilogy of Taiji Yin-Yang, which Bohr adopted as his coat of arms after his visit to China in 1937. With this in mind, the philosophical assumptions underlying Chinese landscape gardens deserve reexamination, even if their medium of expression remain debatable.

The levels of exploration are: (1) the ultimate unity of all existing phenomena and the appearance of the world and role of humans in such a setting; (2) the twofold structure and function of the complementarity and contradiction of all phenomena; (3) the process of attaining the harmonious balance of this twofold structure and function constitutes the existence of all things and events; and (4) the significance of a landscape garden in achieving multi-purpose life enhancement and possibly its therapeutic effect.

PART ONE

THE PHILOSOPHY OF THE GARDEN

1. Ultimate Unity of all Existing Phenomena: The Appearance of the World and the Role of Humans in Such a Setting

Undoubtedly, what we can describe about the world is what we know about it. In an effort to acquire reliable knowledge of the world, no greater advancement has been made than scientific investigation. What science can do that other branches of knowledge cannot is to penetrate the appearance of a phenomenon in search of its ultimate reality. Theories of quantum mechanics with the subsequent invention of electronic instruments help physicists today to visualize objects that are trillions of times smaller than the eye can see and that move millions of times faster than the eye can follow. Evidence has convincingly shown that there are only four fundamental interacting forces in nature: strong, weak, electromagnetic, and gravity. And also, all matter consists of only 12 elementary particles: six kinds of quarks and six kinds of leptons (Hort Breuker, et al., p. 58). At the present level of knowledge, no one is certain how life originated or how consciousness arises in our body and brain. Science asserts that all living cells consist of quarks and leptons. Quarks and leptons have no "life," let alone conscience. Yet, the discovery of DNA tells us that virtually every cell in most living organisms contains DNA and within DNA the totality of information determining the organism is encoded. This genetic code closely resembles the universal language, which means the same thing for all forms of life. DNA consists of only four nucleotides, the four letters of which (AGCT) account for the genetic sequences of all living things. It is obvious that relatedness exists between humans and nature through the commonality of DNA. The idea of singularity indicated in the prevailing theory of the "Big Bang" provides a rational explanation for the ultimate unity of the world.

From this premise of unity, a correlative explanation of the phenomena of the world and its myriad objects may have intrinsic value despite its inadequate formulation. Either "A" causes "B" or is changed into "B" or vice versa. "A" and "B" must have something in common. If "A" or "B" occurs separately, no relationship can occur between them. In terms of particle physics, nothing happens in the material world except the displacement of electrons. Therefore, there must be some kind of harmony and conformity between the way we think and the way the universe operates, allowing us to know the world to some extent and predict the course of development of some events to some degree. Better stated, the state of the human mind bears a natural causal and teleological relationship to the objects around us in order to attain the process of cognition (W. C. Lycan). Furthermore, the correlative relationship suggests that the conditions for human existence preexist in nature, that the process of human evolution accounts for how humans best adapt and utilize their environment for the benefit of life. Humans constitute part of the whole universe. Physicists today probe the subatomic world and show the inseparability of the observer and the observed.

In less than a second, physicists rely on an electronic system to sort about 25 billions bits of data collected from approximately 50,000 electron-positron collisions and a trail of ionized atoms in clouds, bubbles and chambers. They have succeeded in translating the data into images of elementary particles. The most important device in this endeavor is the human brain. Computers cannot match the ability and capacity of the human brain to recognize complicated patterns collected by LEP (Large Electron Positron). By pointing to an object on the screen, physicists can call up specific information about the object to determine its meaning. The image on the screen acts as an index to all the information gathered. The brain can translate spatial information on a flat computer display into three dimensions. Still, the event is presented from only one point of view. Optical illusion is inevitable. Misinterpretation of the event can occur. Facts about atoms were known long before the scintillation screen and mass spectroscopy helped to define quantum mechanics. In fact, the term quark according to is originator, Murray Gellman, was merely an imposition. In the same line of argument, sunshine and air, indispensable for life, are abundant and available for all through eternity. We are accustomed to and often unaware of the ultimate unity with nature making it difficult to perceive that the world exists in a twofold contradictory structure in complementarity, which functions in a continuum. This may be the reason why Nobel laureate Linus Pauling was unable to reconcile the contradictory nature of DNA, while the young James Watson succeeded in deciphering it.

Confucius also restricted his explorations and perceptions. He insisted on exploring and knowing ourselves first. He deferred discussions of supernatural phenomena. This heritage has obscured issues, such as revelations, in current Chinese thinking. However, the concept of oneness in myriad diverse manifestations with a correlative relationship evolved to become the central contention of the Confucian-Taoist worldview. In efforts to illuminate the philosophical principle of the *Book of Changes (I-Ching)* the neo-Confucian Zhou, Dong-Yi presented the *Tai-Ji Yin-Yang Diagram* while Shao, Yung arranged the sequence of 64 hexagrams in a mathematical table from one to 64 to reveal world unity in diverse independent origination. In the *Tao-Te Ching* with which Confucians share their philosophical base, the concept of oneness is categorically stated: that the Tao evolved into one, from one to two and so on.

Landscape painters put this idea of oneness into a picture on paper similar to the virtual reality on a computer screen, using large-scale, monumental mountains to depict the durability of nature, and on a considerably smaller scale, humans traveling a winding road among the peaks symbolizing a contemplative mood and implying that humans are not the measure of all things but part of the whole.

Naturally, landscape gardens are much more realistic than paintings. Stylized stones, either natural or sculpted, are used in special arrangements to represent all mountains rather than any particular one. The visitors occupy center stage, and interpret the garden's meaning through their sensual pleasure. At the same time, the basic idea of oneness remains, conceived in these gardens in a sophisticated manner.

In the *Prelude to the Orchid Pavilion* (Lan Ting) in the Hangzhou area, calligrapher Wang Xi-Zhi wrote, "Heaven and earth provide hostelries for all things in the world; only time reveals the to-and-fro of those transient guests. The fresh breezes above the river from which the ears gain sound and the full moon on top of the peaks from which the eyes acquire color, are free to all and never exhausted." The couplet describing the significance of the Blue Water Pavilion said that fresh air and the bright moon are free to all and only have the value humans attach to them, as well as the lake in the forefront and the hill afar.

In Chinese, landscape painting is described as "mountain and water," and in the landscape garden, which is tied to feng shui, it is described in terms of wind and water. Thus, water is paired with the mountain to constitute the basic design principle for both media. In both venues, the rivers symbolize the arteries of the earth's body, while the mountain is its skeleton. In a sense, the mountain and water are contradictory in characteristics, but they complement each other to form a whole. The ontological assumption that ultimate unity is manifested in a twofold structure and function permeates Chinese artistic expression.

2. The Manifestation of Unity as a Twofold Contradictory-Complementary Structure in Landscape Gardens

Long ago Confucius implied that Taoists sneered at those who claimed to have known something about the world, comparing them metaphorically to a frog looking at the sky from the bottom of a well, thereby having only a one-sided view. This reminds us of the difficulties of a reductionist approach, which accounts for the primary method of scientific investigation. While Einstein was perplexed about the idea of togetherness in separation, Bohr clarified the point by explaining that electrons operate in what he called a vacuum state. Taoists contended that a twofold contradictory structure functioning in complementarity can best be conveyed by the concept of Yin-Yang, which constitutes every thing or event in the world. Confucians later absorbed this assumption in their correlative views.

Whether there are two, co-existing worlds as Taoists portrayed them, a visible, light and known world, "yang jian," and an invisible, dark and unknown world, "yin jian," may be too complex to resolve. The existence of dark matter seems very convincing and could occupy up to 90 percent of space. In 1996, the Hubble Space Telescope unveiled that the estimated number of galaxies was quintuple what was originally thought, now numbering 50 billion. Our sun is only one of 50 billion stars and now appears to have a life span. The laws of Newton were no longer more universal than the Confucian-Taoist assertion that there is only one sun in the sky that would shine forever (John N. Wilford).

Could dark matter be the cosmic glue that holds the Milky Way together? Does this new discovery affirm a twofold structure of the universe? Opinions vary. However, despite the existence of this invisible dark matter, it is apparent that matter, which constitutes the known world, exists in a twofold structure, quarks and leptons, so that all particles are coupled to each other, contradictory yet complementary. For every kind of particle there is an anti-particle that is identical in mass but opposite in other properties, such as electric charge. The basic constituents of DNA are coupled in a similar manner.

In a scientific understanding of the world, neither classical physics nor quantum mechanics can be excluded. They co-exist, reminding us that human cognition proceeds in two stages from the primary level of yes-no, much like causal, linear, predictable Newtonian physics to the higher level of non-linear, non-deterministic, uncertain quantum mechanics. On the primary level, a wave is a wave and a particle is a particle. On the higher level of quantum mechanics one can see that light can be both wave and particle depending on the point of view. Scientists today are resolved that Newtonian physics applies as long as quantum mechanics can be ignored.

How does the transition from the Newtonian level to the quantum level proceed? No definite rules exist to resolve this difficulty. Murray Gellmann's view is that complexity resides in a kind of transition phase between order and randomness. Water in its frozen or vapor state is not very complex. Between these two states water is liquid and can move in complex patterns that are mesmerizing (Murray Gellmann).

It is in this many-sided approach to analyzing things that the cognitive process follows a logical, sequential process of yes-no and separate-together, which reflects a twofold process. "Yes" affirms while "no" rejects in synchrony when the study of a subject is underway. The subject exists unchanged no matter what the disposition is. "No" in this context does not mean that the subject no longer exists. Once a subject is "tagged" with features the cognitive process is simpler. The twofold structure of the cognitive process is based on the phenomena of the known and the unknown in the process of acquiring knowledge.

Cognition is but one example of the twofold structure of the world. The human body is another example. Human beings are structured in male and female physiology, identities and relationships. An individual consists of body and mind. The body is structured in two major components: axial and appendicular, two sides of the heart, two hands, and so on. The mind consists of reason and emotion. Research can be identified as logical-illogical, inductive-deductive, digital-analog, among others. Emotion displays love-hate, happiness-sadness, anger-joy, among others. This dual structure can be extended to all human phenomena and to all that human beings touch. This is reflected in the design of landscape gardens.

Classical Chinese landscape designers based their forms on the precepts of Yin-Yang. In order for the landscape garden to symbolize the universe, it must capture its relational duality in miniature. The designer used the materials at hand and relied on the finely tuned patterns of paired opposites. Rocks and water reveal the essences of nature as a void in a solid and vice versa. Even the seasons are counterbalanced in the design. In every landscape garden, there is a lake embellished by island pavilions, causeways and bridges in which all the delicate artistry of irregular shorelines, rocky banks and background hills are reflected on the surface of the water (Maggie Keswick).

3. Harmonious Balance Reflected in Gardens Constitutes the Essence of the Manifestations of World Phenomena

The concept of balance here implies that one element is balanced against the other to achieve the formation of a whole in all things and events. For example, the minimum requirement of two legs for a table indicates, in a sense, the relationship of two opposite elements complementing each other horizontally in supporting a tabletop, and vertically through reciprocal duality. From such a point of view, harmonious balance means that the primordial pair, the constituents of the twofold structure, interact in just the right proportions rather than in precisely equal or perfect symmetry. Precise equivalence implies immobility that is without interaction. Perfect symmetry implies sameness that is without differentiation or diversity. In short, contradictory structures can only become complementary in a condition of harmonious balance and complementary actions cannot take place except in an interactive process. Therefore, harmonious balance as the ultimate principle of existence of things and events can best be exemplified in nature. For example, it can be seen as the neutralization of the positive and negative charges needed to generate an electric current. In human events, it can be seen as homeostasis in sustaining life, harmonizing emotions and the role of reason in balancing the mind. Even the principle of checks and balances in the two-party political system and the role of the balance sheet in business fit this example.

In philosophical terms, the relationship of interdependent and intertwining but mutually constraining and/or mutually facilitating elements in the spiraling progression of the primordial pair is dialectical. The primordial pair remains intact throughout the process of change, while organisms pass the

DNA polymerase process in the cycle of life. Further illustration of this interactive relationship in essential aspects of world phenomena may be necessary, although it may be very brief and approximate.

The standard model of particle physics discloses that all interactions of the primordial pair, quarks and leptons, can be understood as a result of four fundamental forces exerting influences on each other by exchanging a third particle, the carrier. These forces in nature are: strong, weak, electromagnetism and gravity. According to Steven Weinberg, weak force and electromagnetic force can be merged into one. Except in gravity, the dual nature of the strong and weak forces is apparent. The strong force binds and attracts while the weak force contributes to decay and repulses. In most cases there is a negative charge to every positive charge and a south magnetic pole for every north magnetic pole. Whether or not there is a "gravion" carrier, which effects gravity, is not a certainty. Common sense dictates that we are not pulled down into the earth by gravity, therefore there must be some counter force to balance the gravitational pull. The fact is that atoms do not change in size as the size of our body increases. Since our body is not self-compressed or increasing without limit, no materials would have sufficient strength to withstand the gravitational pull. Particle physics revealed that interactions in the twofold primordial pair not only allowed one to generate the other through the carrier or by other processes, but also to put a limit on their reciprocal development. When one rises, the other falls, or one waxes and the other wanes to sustain the harmonious balance in a continuum. In addition to mass and charge, particles such as protons and electrons can be visualized as tiny spheres spinning clockwise and ending up in a counter-clockwise orientation, continuing to cycle for their existence with no other orientation.

Philosophically, each and every identifiable entity in the world must have a boundary so that its structure can be established. The boundary itself is a self-limit. Life ends in death. One can also say that the reverse is true, death ends in life, sustaining the harmonious balance of continuous existence. The body's cells are constantly dying and renewing and the integrity of the DNA information is maintained. Every particle has an anti-particle with which it can be annihilated in order to synthesize into a new or different structure, yet the twofold structure remains. Some feel this cycle applies to the life-sustaining food chain. Research indicates that the transition from prokaryote to eukaryote in the development of an organism's cells is based upon the function and process of eating (C. de Duve, p. 102).

Mathematics provides the foundation for science. As far as the relationship between mathematics and reality is concerned, mathematics is a symbolic representation of the statistical average of events. Its major approach in digital representation must ignore a certain amount of potential information in segmenting a dynamic process for prediction. Thus, the mathematical representation of reality is no more accurate that the computer's simulation of a virtual reality. For example, the shortest straight line portrayed in Euclidean geometry can only be possible on a flat surface on Earth. Such an ideal condition of smoothness is almost impossible to locate. Fractal formation reflects more accurate or natural contours than the Euclidean precepts. S. Ramanujan demonstrated mathematically that the square and the circle are the same and that one can equal two. K. Godel and G. J. Chaitin discovered incompleteness in mathematics. In general, the more symmetry in a system, the greater its predictive power in a mathematical equation. Yet, under some circumstances symmetrical laws can give rise to asymmetrical phenomena. At a microscopic level, symmetry is broken in a renormalization process (Schrodinger's cat). The question of whether the Big Bang was a symmetrical event has been raised. At a distance of 10 to the negative 29th power a centimeter, is impossible to distinguish quarks, neutrinos and electrons. They appear one and the same in ultimate unity.

Complementary interactions in the twofold contradictory structure, achieving a harmonious balance, are manifested not only in the patterns of one generating or limiting the other but also in interdependent relationships that one initiates and another must complete via some action. Take the male and female relationship as an example. While the male initiates life, the female completes it. Nothing is more perfectly self-contained or self-complete. The simple DNA iteration evolves into heterogeneity, thereby providing the seeds for evolution.

Whether or not the principle of harmonious balance constitutes the essence of existence for all phenomena in the world and whether or not its process consists of the oscillation between polar opposites within the primordial pair, will remain a challenge. There is no question that the elite, scholar-bureaucrats were deeply ingrained in this Confucian-Taoist, ontological conviction. The landscape garden's design reflects this. The exterior walls of the garden were typically square, enclosing the garden and delineating its exterior boundaries. Walls within divided the space into compartments to depict diverse units, but provided a unity within the whole. A labryrinth was rendered in which the space was layered by paths through gateways that wind among trees, shrubbery with rocks and pavilions, terraces, water and hills, towers on the ground in fine-tuned patterns of opposites.

A visitor to a landscape garden might be led through twisting galleries over a bridge with a glimpse of a distant tower or vista, which is intended to evoke the feeling of infinity, before being led on. In this type of garden design, the linear, geometric style did not occupy a dominant role, as it did in places, cities and formal residences. It was balanced with non-linear, zigzag patterns to reflect the realities of nature. Undeniably, the twisting style was intentional. Rather than being concerned with architectural efficiency or self-indulgent, fanciful projects, these gardens were an attempt to create an artificial environment that would best facilitate self-cultivation. The goal was to assist the visitor in becoming the ideal Confucian (junzi) rather than to establish a monumental and imposing presence that evoked awe or power, such as a church or the Supreme Harmony Hall in Beijing. The goal was to have the visitor look inward as a method for addressing a world outlook. The designers' aesthetic expression and its impact call for further examination.

PART TWO

THE LANDSCAPE GARDEN AS A MULTI-PURPOSE CENTER FOR SELF-CULTIVATION AND ITS THERAPEUTIC CLAIM

In a sense, attaining the lofty goals that these landscape and garden designers set may have been unrealistic. Recent criticism has likened these gardens to amusement parks for the privileged landlords at leisure. Indeed, they were the handiwork of a few affluent neo-Confucian scholars. The basic doctrine of neo-Confucian self-cultivation stemmed from a new synthesis of the Confucian insistence on the innate goodness of human nature with the Great Vehicle School of Buddhism approach, intent on disciplining the mind. It was set in a Taoist, naturalistic, if not materialistic, outlook. In the view of neo-Confucians, if humans constitute part of the cosmos, the mediator of heaven and earth, and are part of nature, they cannot be evil. Evil was seen to arise from inadequate livelihood and lack of proper education. Life on earth was not thought of as repentance for disobeying the ultimate command or pre-destined suffering to search for final liberation. The ability of humans to be educated was seen as an indication that everyone possessed the capacity to carry out their obligations in society. Self-reliance was the key and the theme was "you reap what you sow."

The neo-Confucian designers entertained such an optimistic view of human amenability. They thought sensual pleasure in miniaturized, cosmic gardens could induce a few gifted visitors to attain a more lofty personality and, thereby, enlighten others. They expected multi-compartments clustered in a single garden to accomplish their goals. In the meditation section, one could discipline the mind by recharging or strengthening its power; in the study section, one could improve one's literary attainment or compose poems for greater self-expression; in the entertaining section, one could socialize with friends to preserve harmonious social relationships; in the family section, one could consolidate and refresh the bonds with blood relatives. Occasionally sections for fertility and retirement were added.

The optimistic goals of the landscape garden in its unique correlative setting to catalyze and internalize a greater sense of morality in a visitor were based on the conviction by the designers of a direct connection between humans and nature in the medium of "qi" through breathing. What is "qi?" It defies scientific explanation at present. Nominally "qi" is air or gas. In its essence, it may mean vital energy or a life force. Confucians and Taoists contend that it arises from the ultimate oneness, in what is known as the "great qi of oneness," yuan qi and evolved into a twofold primordial structure, with the positive (zhengqi) and the negative (xieqi). An enlightened person (jungzi) is expected to uphold and enhance the positive aspects of life to modify and restrain the negative in order to maintain a harmonious balance and sustain life. Death is described as the stoppage of "qi" (duanqi). The concept of "qi" is almost incomprehensible in terms of modern science. However, in acupuncture and the recent scientific confirmation of its analgesia, it is hard to dismiss what the practitioners profess as a method to restore the balance and flow of "qi" to promote healing. This appears to be much more than a placebo effect.

Everyone experiences things differently. Yet, everyone has positive and negative sensations. For example, the smell of roses can arouse pleasant feelings. On a microscopic level, quantum mechanics discloses that every molecule is linked to all others in an endless chain of connections, on which life depends. The discovery of pheromones and their effect on attraction and repulsion of males and females serves as an example in humans. Landscape design, in classical Chinese gardens, represented the cycle of life in miniature. Whatever is needed for sustaining life exists in the universe already. For example, the healing properties of plants are undeniable. The composition of the landscape garden, flowers for four seasons, three friends of winter as well as rock and water settings, correlate with the cosmos and appear to facilitate life-sustaining "qi." No wonder "qi" exercisers prefer a park, if not a garden.

CONCLUSION

Obviously, the symbolic, miniaturized expression of the world, according to the concept of Taiji Yin-Yang, as reflected in landscape garden design, was a hindrance to architecture meeting commercial demands in terms of efficiency and economy. It is now considered outmoded. Take, for example, its negative impact on the financial success of I. M. Pei's Fragrant Hill Hotel in Beijing. Some think Pei was thwarted because he used the traditional landscape garden as the foundation for exploring a synthesis for the future.

What impact on aesthetic expression is foreseen in view of the apparent affinities between the concept of the Yin-Yang contradiction in unity, or Taiji, and the principle of complementarity in contradiction in quantum mechanics as well as the double helices in DNA, the binary structure in computation, systems in symmetry and equilibrium and recent ecological perspectives? In the past, the view of art for art's sake in the freedom of expression has been subordinated to the role of evocation and attention to moral persuasion for harmony, balance and the "golden rule" in life. Can aesthetic expression proceed without a purpose, free from the cultural environment on which art depends for existence? A Yin-Yang dialectical, monistic view may provide a framework for reference when contemplating a new synthesis, a theory that everything is included and interdependent. Can the classical, Chinese landscape garden be seen as a prelude?

Florida Atlantic University, Boca Raton, FL

REFERENCES

Hort Breuker, et al., "Tracking and Imaging Elementary Particles," *Scientific American*, August 1991, p. 58.

C. deDuve, "The Birth of Complex Cells," *Scientific American*, October 1996, p. 102.

Murray Gellman, "Researchers on Complexity Ponder what Exactly They Study," *The New York Times*, May 6, 1977.

Maggie Keswick, *The Chinese Garden*, St. Martin Press, New York, N. Y., 1986.

W. C. Lycan, *Consciousness*, M. I. T. Press, Cambridge, MA, 1987.

John N. Wilford, "Suddenly, the Universe Gains 40 Billion More Galaxies," *The New York Times*, January 17, 1996.

SITANSU RAY

AESTHETICS OF ANCIENT INDIAN SYLVAN COLONIES AND GARDENS: TAGORE'S REFLEXIONS

I. INTRODUCTION: MEMORY AND NOSTALGIA

"*Dao phire se aranya, lao e nagar*" (give us back the sylvan past, take away today's cities), is Rabindranath Tagore's (1861–1941) earnest prayer as expressed in the poem "*Sabhyatar Prati*" (to modern civilization) of his poetical work *Chaitali.*[1]

Actually, the seats of ancient Indian culture and civilization were in the forests, far from capitals, cities and villages. Populated places were not convenient for learning and meditation. The wise sages and hermits lived in *tapobanas* or forest colonies. Aryan civilization along with Vedic wisdom flourished amidst sylvan solitude. The truth-seeking human spirit obtained full freedom there by virtue of sylvan beauty along with the vast terrain, majestic mountains and flowing rivers. The sights of sunrise and sunset and of the noctural sky full of stars and planets added to their communion with nature and infinity.

Pupils were sent to such *tapobanas* for training and tutelage under the Gurus. Learning the *Vedas*, associated with a sense of wonder, creative imagination and spiritual traits found easy flow and articulation in the forest. Plain living and high thinking were the motto of the sylvan colonies. But those colonies were never like monasteries. The pupils enjoyed the company and affection of their master's family. They tended the cattle, milked the cows, collected firewood and fruits. On the whole, they looked after their master and his family and guests. They did not have to pay money for their studies.

One-sixth of the revenue was spent by the kings for the well-being of the forest colonies. The kings were sincerely devoted to the great masters. Though the colonies were outside of society, they were to society what the sun is to the planets.

Apart from "*Sabhyatar Prati*," there is quite a cluster of poems depicting the nostalgic memory of ancient Indian sylvan colonies and gardens, materially simple but spiritually rich. Two are especially noteworthy: "*Bana*"[2] (forest) and "*Tapobana*"[3] (forest colony for meditation), the first one portraying the beauty and sublimity of the green in contrast to artificial urbanization; and the second one depicting the imagery of the hermit seated in meditation, disciples engaged in studies under the shade of trees, young

A.-T. Tymieniecka (ed.), Analecta Husserliana LXXVIII, 151–160.

daughters of the sages sprinkling water on the plants of the garden and, last but not least, the king getting down from the horse-drawn cart and keeping his crown off to bow down to the hermit.

"*Prachin Bharat*"[4] (ancient India) is another beautiful poem, describing the glory of Brahmins residing in the forest, in contrast to the materialistic grandeur of the ruling *Kshatriya* kings of Ayodhya, Panchal, Vidarbha, Virat, Kanchi and other such kingdoms.

II. SYLVAN GLORY IN VEDIC CHANTS

All-encompassing monistic fullness is metaphorized as a great tree in the Vedic sentence, "*Briksha iva stabdho divi tisthatyekastenedam purnam purushena sarvam.*" In the essay "*Prachin Bharate Ekah*"[5] (the One in ancient India), Tagore explains it: That absolute and sovereign "One" is like a huge tree. He is all-abounding and complete in Himself. The tree is the appropriate metaphor for apprehension of His unity and variety. A verse is worth quoting:

> "*Yatha saumya bayamsi basabriksham sampratisthante.*
> *Evam ha vai tat sarvam para atmani sampratisthante.*"[6]

(O beautiful, as the birds find their resort in a tree, so also each and every thing has its rest within the Absolute Soul.)

The Aryan sage addressed his pupil as *saumya*, i.e., the beautiful. Through various similes the pupil is taught about the *Brahman*, who is to be realized only and cannot be explained in words. The motto of ancient Indian teachings was to see the world in God and to realize oneself and all in Him. The toiling and tired wayfarers can rest in the shade of a tree. The tree is a comprehensible metaphor for the Supreme Soul.

Literally too, the first vibration of life came to this earth in plants and weeds prior to the animal kingdom. "*Kena pranah prathamah praitiyuktah*"[7] (wherefrom does the first life-force come to this world)? The *Kathopanishad* says: "*Yadidam kincha sarvam prana ejati nihsritam*"[8] (everything in this world is full of life, emitted by the Supreme Life).

A chant from *Shwetashwatara Upanishad* goes like this:

> "*Yo devoagnau yo apsu yo vishwam bhuvanamavivesha*
> *Ya oshadhishu yo vanaspatishu tasmai devaya namo namah.*"[9]

(We bow down to the God, who is all-pervasive in fire, water, plants, crops and throughout the whole universe.)

For the present context, we should give attention to "*oshadhishu*" (God, present in the seasonal crops) and "*vanaspatishu*" (in the big plants too).

It will not be out of place to note that the above-mentioned verse was introduced by Tagore for chanting during the evening prayer session at Santiniketan.

III. THE SYLVAN EPISODE IN THE *RAMAYANA*

"*Aranyakanda*" (the sylvan episode) of the great epic *Ramayana* depicts Rama's exile in the forest. Rama accepted it as an enchanting adventure rather than as punishment. His dear wife Sita and his brother Lakshmana accompanied him. Banished from royal luxury, they lived in a hut. They had to fight against ferocious monsters and demons. Otherwise they were happy in every respect. The heroic fight also added to their pleasurable adventure. They felt close kinship with the beautiful forests, rivers and hills. They derived joy from their association with beautiful birds and animals. They created a plain and beautiful lifestyle in the natural sylvan environment. Sita was joyfully inquisitive about wild creepers and weeds and the flowers she had never seen before. Her brother-in-law Lakshman collected them for her. The picturesque scenery of cranes and swans on the sandy river beach added to Sita's joy and happiness.

The romantic poet Valmiki most aesthetically depicts various other scenes:

Suramyamasadya tu Chitrakutam
Nadimcha tam Malyavatim sutirtham
Nananda hristo mrigapakshijushtam
Jahau cha duhkham puravipravasat.[10]

(The sight of the beautiful mountain Chitrakut, the sacred river Malavati and the forest land haunted by deer and birds added to Rama's pleasure. He completely forgot that he was exiled from the capital.)

Na rajyabhranshanam bhadre
Na suhridbhirbinabhavah
Mano me badhate dristwa
Ramaniyamimam girim.[11]

(The scenic beauty of this mountain does not allow me to grieve for losing the kingdom and my royal companions.)

On the whole, we see that sylvan beauty compensated for all loss of royal status. Moreover, the forest also gained some novel riches, the riches of human affection and conjugal love. This rhetoric of transferred epithets enriches Valmiki's depiction and Tagore's evaluation thereof.

In later events, when Sita was kidnapped by Ravana, Rama's grief touched the whole sylvan environment including all the flora and fauna, which were part and parcel of Rama's expedition for Sita's deliverance and emancipation.

IV. SYLVAN COLONIES AND GARDENS IN KALIDASA'S WORKS

During the reign of Vikramaditya, India was highly prosperous, so this was called the "golden age" by the historians. But the age of *tapovana* had already passed by then. Yet we see that the famous poet and playwright Kalidasa, one of the nine jewels of Vikramaditya's royal court, held in high esteem the aesthetic environment of the ancient Indian *tapovanas*. This legacy of sylvan aesthetics still remained in the nostalgic heart of aristocratic Indians.

The very beginning of the poetical work *Raghuvamsham*[12] is full of charming descriptions of ancient Indian sylvan colonies and gardens: The young ascetics collect *kusha* (a kind of grass needed for religious ceremonies) and firewood. The young girls pour water into the ridges around the plants. Flocks of birds gather there to drink water. A number of deer move freely about eating *nivara* (a kind of wild paddy). Nature, animals and man are harmonious in such an environment. Furthermore, aromatic smoke pervades the *tapovana*. The guests cosily feel welcome.

Kanwamuni's *ashrama* (hermitage) in the drama *Abhijnanashakuntalam*[13] is beautifully portrayed, along with the feminine playfulness of the three damsels Shakuntala, Anasuya and Priyamvada, and Shakuntala's love with the king Dushyanta, all in the setting of a sylvan colony.

After the ordained spell of troubles, Shakuntala, along with her son, is reunited with her husband. The venue of this last scene, extremely moving, is in the sylvan colony of the hermit Maricha.

Six seasons of ancient sylvan India are variedly depicted by Kalidasa in his lyrical work *Ritusamhara*[14] and everywhere nature is closely associated with man-woman love relations, aesthetic and erotic. As an ardent reader of Kalidasa's works, Tagore refers to *Ritusamhara*[15] in *Tapovana*. Furthermore, in the poem *Ritusamhar*,[16] Tagore metaphorizes six seasons as six pretty damsels serving drinks to the ancient poet.

In Kalidasa's *Kumarasambhavam*,[17] the sylvan seats are located in the Himalayas. Vernal beauty around Shiva's seat is most aesthetically depicted in the third canto. The love-god Madana's spell, Shiva's wrath and ultimate bliss in his union with Uma are excellently depicted in Kalidasa's work.

Meghadutam[18] is a lively personification of the cloud as Yaksha's messenger carrying his message to his lover. Kalidasa's grasp of Indian geography, from the Ramagiri to Kailasa, is amazing. Offering the fragrant *Kutaja* (*kurchi* in Bengali dialect), Yaksha makes friends with the monsoon cloud and spreads his painful estranged heart over the hills and dales, rivers and forests. The pangs of the human heart are not confined to the utter estrangement of lovers alone, for each and every object of nature is surcharged with affectionate feeling.

V. FURTHER SYLVAN REFLECTIONS IN TAGORE'S WORKS

Vanabhatta's *Kadambari* and Bhavabhuti's *Uttararama-charitam*[19] are also discussed by Tagore in connection with vital sylvan aesthetics of ancient India. Most of Tagore's rain songs bear his assimilation of the essence of Kalidasa's *Meghadutam*.

Tagore's poem "*Varshamangal*"[20] is a nostalgic reverberation of the blissful rains. As part of a rain festivity, young women, beautifully dressed and clad in natural flower ornaments, are singing with the tunes and rhythms of various musical instruments. They are dancing too. On the whole, the atmosphere is an aesthetic depiction of mankind and nature. A few stanzas of the poem are set to music.

"*Swapna*"[21] (dream) is a poem of imaginative journeying in the land of beauty and intimate love of ancient India.

The poetical work *Bithika*[22] (avenue) suggests a sylvan atmosphere. The poem "*Aditama*"[23] (the most pristine) connotes the first vibration of life, that of an *ashattha* tree, enlivening the whole forest as it were. The poet's sensitive heart feels the natural *Omkara*, prior to man's voice. In the poem "*Vanaspati*,"[24] we can hear "*sristir pratham vani*"[25] (the pristine message of creation) in a *vanaspati* (old large tree). Its life vibration continues throughout the ages. The tree has been reciting, as it were, the age-old *sama* chants.

Another poem of the same title, *Vanaspati*,[26] takes place in the poetical work *Puravi*. The phenomena of original life-vibrations as manifested in a big tree move the poet's consciousness and imagination in a unique way.

The *kadamba* flower, *kadamba* tree and *kadamba* forest are profusely mentioned in the Vaishnava lyrics as well as in Tagore's poems and songs. The *kadamba* has become a symbol of love between Krishna and Radha along with her *sakhis* (girlfriends). Furthermore, the *kadamba* flower has become a symbol of the monsoon in Tagore's songs. *Kadamba* trees, full of ball-like yellowish flowers, and Krishna with her damsel mates, have become a popular theme of Indian painters. It is also called *neepa* in a Tagore song:

Eso neepavane chhayabithi tale
Eso karo snan nabadhara jale.

(Let us gather together under the neepa trees. Let us take our merry bath in the new rains.)

The title of Tagore's poetical work *Mahuya*[27] denotes and connotes youth, love and non-urban zeal for country life. The *mahuya* is a tall tree bearing fragrant flowers in the spring. Their thick, soft and juicy petals are sweet to taste. Eaten profusely, *mahuya* petals may cause pleasant intoxication. The Santals of Birbhum and Bihar consume them as their meals. *Mahuya* forests are found in these regions, especially in Chotanagpur in Bihar.

Even in Tagore's earlier works, forests take a very important place. For example, his opera-like music-dramas *Valmiki Pratibha*[28] and *Kalmrigaya*[29] may be cited. The settings of both are in the forest. The dreadful forest, the stormy night, the aweful Kali temple and the merciless dacoits along with suitable music and acting in *Valmiki Pratibha* keep the audience in utmost suspense. Saraswati, the Goddess of Learning in the disguise of a little girl, changes Valmiki's heart altogether. The leader of the dacoits turns into the poet Valmiki. Each and every scene is in the forest.

Kalmrigaya means an ominous hunt. The king Dasharatha is hunting in the forest. Mistaking him for a young elephant, he actually hits Andhamuni's (the blind sage) son in the water of the Sarayu River. The venue of this drama is also in the forest.

The enactment of *Vasanta*[30] (spring) is a unique form of applied aesthetics incorporating the forest terrain, the mango grove, *karabi* (oleander), the bamboo forest, the myrtle grove, the *sal* avenue, the *bakul* tree, the river, the southern spring breeze, the sylvan path, the jhumko creeper, the *akanda*, *dhutura* and last but not least *jaba* – all are personified as sonorously singing and colourfully dancing characters on the stage. The king and his court poet are just in the background to introduce the characters. The forest aesthetics are in their superb form in this nature drama.

Some other plays and works on the theme of nature are *Phalguni*[31] (adjective of the month of Phalgun, February–March, i.e., early spring), *Shesh Varshan*[32] (the last rains), *Nataraja Riturangashala*,[33] etc. In the last one, Nataraja is the Dancing Lord, the creator of Music and Dance. He employs six seasons as actors and actresses. The earth is the theatre house. The scenes change with the change of seasons.

VI. *VANAVANI*: THE MESSAGE OF THE TREES

A special mention of Tagore's *Vanavani*[34] must be made since it is akin to the theme of the present topic. It is a cluster of poems and songs on trees, creepers, flowers, and the tree-planting ceremony as well.

The poem on "*Devadaru*"[35] (a kind of Indian cedar) was written while Tagore was in Shillong, the beautiful hill station of north-east India. Nandalal's letter with a painting of hills full of *devadaru* forests inspired the poet to write this poem. The straight, upward and tall *devadaru* is aesthetically metaphorized as music of the worldly flute making harmony with the music of the sun: "*Suryer sangite meshe mrittikar muralir sur.*"[36]

Apart from poems on the *sal* avenue, the mango grove, the coconut tree, etc., Tagore also writes on some unknown yet beautiful plants and creepers grown at Santiniketan. Some of them are *nilmani lata* (producing blue flowers), *kurchi* (wild but having sweet fragrance) and *madhumanjari* (honey bud) etc. The *Chameli* (a kind of jasmine) with peacocks inspired Tagore to write the poem "*Chamelibitan.*"[37]

Vanavani ends with some songs and poems meant for *Vriksharopana Utsava*,[38] or the tree-planting ceremony. Tagore had introduced the tree-planting ceremony as well as other seasonal festivals with a view to reviving the close relationship between man and nature.

In the song *Maru vijayer ketan udao*, Tagore uses a nice metaphor. The leafy trees are like flags for conquering the desert. The dumb earth becomes redolent with the trees, the first and foremost offsprings of the earth.

The song *Ay amader angane* is an address to the young plants: "Do come to our countyard as our affectionate companions. The sunshine and the rain will give you vitality."

The five basic elements (*Pancha bhuta*), including *Kshiti* (earth), *Op* (water), *Teja* (sun), *Marut* (air), and *Byom* (space) are personified and evoked with five beautiful pieces of poetry as part of the tree-planting ceremony.

Such other portions of Tagore's works and creative activities serve as a boon to the awareness of environmental, ecological and aesthetic interaction between man and the forest.

Tagore's short speech *Aranyadevata*, delivered at Sriniketan on the occasion of a land-tilling and tree-planting ceremony, explains the utility of the greenery. The whole area of Santiniketan and Sriniketan was at that time just a vast barren land. Over the decades, these campuses have been thoroughly changed. After the fashion of the Tagorean *Vriksharopan* ceremony, the *Vanamahotsava* or *Aranyotsava* (forest festival) is being held in other parts of the country and even at the government level.

VII. SANTINIKETAN: TAGORE'S ENDEAVOUR TO REVIVE THE ANCIENT INDIAN SEAT OF LEARNING IN THE TAPOVANA STYLE

It is known to all that the four-walled classroom system of education did not suit the child Tagore's temperament. In later years Tagore retained disdainful memories of his school days in Calcutta, in at least four schools in short succession.

With a view to reviving the ancient *tapovana* system of education, Tagore set up his school at Santiniketan with only a few boys and teachers. He says:

> I consider it as a part of education for my boys to let them fully realize that they are in a scheme of existence where trees are a substantial fact, not merely as generating chlorophyll and taking carbon from the air, but as living trees.[39]

He adds:

> In India we still cherish in our memory the tradition of the forest colonies of great teachers.[40]

On this model, he started his school at Santiniketan, which is now Visva-Bharati, i.e., the World University. Classes are held here under the trees. Morning classes and evening studies are still preceded by the chanting of Vedic hymns. Weekly congregations are held with Vedic chants and suitable devotional songs of Tagore. The Vedic and Upanishadic songs are part of the annual convocation. The Upacharya (Vice-Chancellor) blesses the *Snatakas* (graduates) in the fashion of ancient Aryan masters of *tapovanas*.

"Santiniketan" means "the abode of peace." Let us observe how Tagore feels the spiritual-aesthetic essence of the sylvan environment. In his own words:

> I sat alone on the upper terrace of the Shanti-Niketan house and gazed upon the treetops of the *sal* avenue before me. I withdrew my heart from my own schemes and calculations, from my daily struggles, and held it up in silence before the peace and presence that permeated the sky; and gradually my heart was filled. I began to see the world around me through the eyes of my soul. The trees seemed to me like silent hymns rising from the mute heart of the earth, and the shouts and laughter of the boys mingling in the evening sky came before me like trees of living sounds rising up from the depth of human life.[41]

Tagore dreamt of Visva-Bharati Santiniketan as an *ashrama* or a *tapovana* type of world university where scholars and students from all over the world would gather "for the highest end of life in the peace of nature."[42] He felt sure that what was most necessary was the breath of culture in the midst of nature, with no severe formal method of lesson-giving and lesson-taking. Music, fine arts and histrionic art should be part and parcel of this forest colony. At the same time, the young and the old, the students and teachers must not ignore the daily sunrise and sunset and the silent glory of stars. Nature's festivities of flowers and fruit, of seasonal beauties, the bliss of the moonlit night, must find their joyous recognition in us.

VIII. CONCLUSION

Just the material resources of forests cannot be the be-all and end-all. Science, technology and industrial revolution must not be allowed to mar the aesthetic, spiritual, ecological and ethical value bases of the forests. Interaction between man and the forest from the perspectives of these value-bases will enrich humanity as a whole. Ancient Indian glimpses, newly reflected in Tagore's works, will serve as a boon to global awareness of the extensive and intensive aspects of sylvan aesthetics.

Visva-Bharati (World University), Santiniketan, India

NOTES

1 Rabindranath Tagore, "*Sabhyatar-Prati*" (to modern civilization), *Chaitali, Rabindra Rachanabali* (collected works of Tagore), henceforth abbreviated as *R.R.*, Birth Centenary Edition, Govt. of West Bengal, 1961, Vol. I, 550.

Chaitali is the adjectival noun form of *Chaitra*, the last month of the Bengali year. *Chaitra* spreads from the middle of March to the middle of April. *Chaitali* also signifies the harvest-rich end of spring. A bunch of *Chaitali* poems was composed in *Chaitra* of the Bengali era 1302 (i.e. 1896).

2 *Ibid.*, p. 551.

3 *Ibid.*, p. 551.

All of the above three poems were written by Tagore on the same day, i.e., the 19th of *Chaitra* 1302 B.S. (Bengal *sal* or Era).

4 *Ibid.*, p. 552.

5 "Prachin Bharate Ekah," *Dharma* (religion), *R.R.* 12, pp. 28–36.

6 *Ibid.*, p. 28.

7 *Vanavani*, *R.R.* 2, p. 837.

8 *Kathopanishad* 2.3.2.

9 *Shwetashwatara Upanishad* 2.17.

[10] Quoted and explained by Tagore in the essay "Tapovana," *Siksha* (education), *R.R.* 11, p. 598.
[11] *Ibid.*, p. 599.
[12] There are various editions in the libraries.
[13] Translated into English and Bengali by renowned professors and published by various publishers from time to time.
[14] Several editions are in the libraries.
[15] Rabindranath Tagore, "Tapovana," *op. cit.*, *R.R.* 11, pp. 592–3.
[16] *R.R.* 1, p. 552.
[17] Along with the original Sanskrit, there are various translations in English as well as Bengali and other Indian languages.
Tagore refers to Kalidasa's *Kumarasambhavam* in "Tapovana," *R.R.* 11, p. 593.
[18] Kalidasa's *Meghadutam* is extremely popular among educated Indians. Tagore's elder brother Dwijendranath Tagore translated it into Bengali verse.
Tagore refers to *Meghadutam* in "Tapovana," *R.R.* 11, p. 600. Kalidasa and his *Meghadutam* are eulogized in Tagore's poem "Meghdut" in *Chaitali, R.R.* 1, p. 552.
[19] *R.R.* 11, p. 592 and *R.R.* 11, p. 600.
[20] "Varshamangal," *Kalpana, R.R.* 1, pp. 696–7.
[21] "Swapna," *ibid.*, pp. 699–701.
[22] *Bithika, R.R.* 3, pp. 241–346.
[23] *Ibid.*, pp. 254–5.
[24] *Ibid.*, pp. 300–301.
[25] *Ibid.*, p. 301.
[26] "Vanaspati," *Puravi, R.R.* 2, pp. 719–20. This poem, like some others in *Puravi*, was composed by Tagore during his voyage and visit to Argentina in 1924–25.
[27] *Mahuya, R.R.* 2, pp. 755–834.
[28] Even before reaching the age of twenty, Tagore had composed and staged *Valmiki Pratibha* (February 26, 1881).
[29] *Kalmrigaya* (1882).
[30] *Vasanta* (1329 B.S., 1923) *R.R.* 5, pp. 585–601.
[31] *Phalguni* (1916), *R.R.* 6, pp. 441–488.
[32] *Sheshvarshan* (1922), *R.R.* 5, pp. 603–617.
[33] *Netaraj Riturangashala* (1926), *R.R.* 5, pp. 619–663.
[34] *Vanavani* (1926), *R.R.* 2, pp. 835–867. The "Bhumika" (preface) of the book was written on October 23, 1926 while Tagore was in the Hotel Imperial in Vienna. The book was dedicated to the famous scientist Jagadishchandra Basu, who worked intensively on the life of trees.
[35] *Ibid.*, pp. 842–3.
[36] *Ibid.*, p. 843.
[37] *Ibid.*, pp. 855–8.
[38] *Ibid.*, pp. 864–867, dated Santiniketan, July 13, 1928.
[39] Rabindranath Tagore, *Personality* (lectures delivered in America) (first edition 1917), Macmillan Indian edition 1985, pp. 118–9.
[40] *Ibid.*, p. 127.
[41] *Ibid.*, p. 133.
[42] *Ibid.*, p. 147–8.

SECTION III

GEORGE R. TIBBETTS

OPERA AS A MIRROR OF THE INFINITE: THE TRIUMPH OF THE HUMAN SPIRIT OVER NATURAL FORCES IN *RIDERS TO THE SEA*

Nature can be a source of great joy and pleasure to us, or a source of great sorrow and pain. In the latter case, we either triumph over it or are crushed by it. Whether we triumph or are crushed does not depend upon physical reality, but upon our perception of that reality. Thus, what can be a devastating event or series of events can also, when filtered through our perception, become something positive.

This is certainly true of Maurya, the chief character in *Riders to the Sea*, a play by the Irish playwright John Millington Synge that later became an opera at the hands of the contemporary British composer Ralph Vaughan Williams. Maurya is an old woman who loses her husband, father-in-law, and six sons to the sea. Initially her loss has an understandably devastating effect upon her. However, she is eventually able to come to terms with it and achieve peace within herself. She comes to the realization that the sea can take no more from her and that, considered in cosmic terms, there is, in God, a power that can enable not only her, her sons, and her family, but every human being, to transcend, at least within their own lives, the malevolent power of nature and, indeed, death itself.

Both Synge in his play and Vaughan Williams in his opera are concerned with the ability of the human spirit to draw on the power that comes from God and to use this power to triumph over the malevolent power of nature. Taken together, the play and the opera are, in effect, two looking glasses or mirrors that enable us to observe the struggle within Maurya's life between the infinite power of God and the still infinite but not insurmountable power of nature.

Vaughan Williams's setting of *Riders to the Sea* was one of several settings or attempted settings of this work. Ethel Smyth and Ivor Gurney had made unsuccessful attempts prior to the First World War. Fritz Hart set the work, as did Henri Rabaud in his *L'Appel de la Mer. Riders* foreshadowed Vaughan Williams's later attempt in his music for the film *Scott of the Antarctic* to deal with the theme of human endurance in the face of an unfriendly nature. He later incorporated this music into his symphonic poem *Sinfonia Antarctica*.

A.-T. Tymieniecka (ed.), Analecta Husserliana LXXVIII, 163–169.

Vaughan Williams uses an operatic style similar to that used by many British composers. Unlike much Italian opera, with its numbers interspersed with recitative, *Riders* is, in effect, a sung play. This places great demands upon the acting abilities of the singers. Maurya, in particular, must not only be an accomplished singer, but also an accomplished actress. It is for this unfortunate reason that the opera has so often been performed by student or essentially amateur groups. Its first performance was in London in November 1937 by students of the Royal College of Music conducted by Sir Malcolm Sargent. In spite of the essentially student nature of the performance, the *London Times* critic wrote at the time that *Riders* "revealed the extraordinary qualities of an exquisite opera . . ., [that it was] powerful . . . interesting, but above all . . . purely beautiful"[1] (Michael Kennedy, p. 256). Its next performance was in 1938, by the Cambridge University Musical Society. It received a somewhat more professional performance at the coronation of Queen Elizabeth II when it was performed along with another of Vaughan Williams's operas, *Hugh the Drover*, at Sadlers Wells. One of the more interesting performances was at the University of California at Santa Barbara during a Vaughan Williams tour of America when both Synge's original play and the opera were performed on the same program by students in the theater and music departments of the university.

Vaughan Williams set Synge's play virtually as is. Moreover, he set Synge's words in such a way that they follow the natural inflections of human speech. This can be traced in part to the influence of Debussy's *Pelleas et Mélisande* which exhibits a similar fusion of words and music. To a far greater extent, however, it grew out of a long process of development in vocal style. This probably began with Vaughan Williams's experiences as a young man with folk song, and with a specific experience of listening to the changing inflections in the speech of an itinerant preacher, as that preacher became more and more eloquent. By the time he wrote *Riders*, from 1925 to 1932, Vaughan Williams's approach to setting words had reached its fullest maturity.

To understand *Riders to the Sea*, one must first understand the role played by the dead in early twentieth-century Irish culture. The distinction between the living and the dead was a blurred one. The Irish believed that the dead could intervene and, indeed, control the lives of the living. This extended all the way from the ability of the dead to communicate with those who were alive, to their ability to return from their realm to that of the living, and to take those who were alive back with them to the "other side." Declan Kliberd, in speaking of two prominent politicians of the day, says, "In the national politics of the period a similar mentality prevailed: figures from the Home Ruler Parnell to the hunger-striker MacSwinney seemed to exercise even more influence in death than they had ever managed to achieve in life. In an Irish democracy the dead are always recognized to have rights, if not actual votes"[2] (Declan Kliberd, p. 6). In light of all this one needed, therefore, to be careful not to offend the dead. If one did, the consequences could be severe.

This power of the dead over the living requires the latter to follow certain rules if they wish to avoid incurring the wrath of the dead and the misfortune that can ensue. Thus Maurya, who is the chief character in *Riders*, is in many ways the source of her own pain, both in her own eyes and in the eyes of those around her. For example, she fails to bless her last remaining son, Bartley, as he starts out on his journey. Bartley himself also contributes to his downfall, however. He dares to wear the shirt of his dead brother, Michael, and this breach of the rules governing the relationship between the living and the dead convinces Maurya that she has seen Michael riding on the grey pony that is trailing along behind Bartley. This is a clear indication to her that Michael has returned from the dead and is about to take Bartley with him to that realm. She is convinced of this even though one of her daughters, Cathleen, tells her that she has just looked out the window and seen Bartley with only a grey pony trailing along behind him. Moreover, so deeply intrenched are these beliefs in the Irish culture of the time that Cathleen, in spite of what she has just seen with her own eyes, immediately shares her mother's fears.

Vaughan Williams uses his orchestra to provide an emotional backdrop for what is happening on stage and to underline aspects of the drama that are not expressed in the text, in particular God and the sea. He uses a Wagnerian system of continuous melody and leitmotifs associated with specific characters or events (the sea; Maurya; Michael, Maurya's son; prayer). The sea, for example, takes on an almost human character and becomes, in fact, a player in the drama.

The opera proper is preceded by a brief orchestral prelude in which the motif associated with the sea is overwhelming. This symbolizes the malevolent nature of the sea and its dominance over the lives of the people in the opera either as an instrument of death, as in the case of the male members of Maurya's family, or as a source of sorrow and pain to the women and, at least throughout most of the opera, to Maurya herself.

The opera is set in a fishing village on the west coast of Ireland. Maurya has lost all the male members of her family to the sea except for two sons, Michael and Bartley. There is considerable concern for Michael because he has not been heard from. A young priest has retrieved clothing that was taken from a drowned man and has given it to Maurya's two daughters, Nora and Cathleen. They fear that the clothes may belong to Michael. They are also concerned with the effect that this development may have on their mother.

The discussion between Nora and Cathleen turns to Bartley, Maurya's other remaining son, who plans to travel by sea to the Galway Fair to show some horses. The young priest has told Nora that he will not attempt to stop Bartley from making his journey, but, to allay their fears, he calls attention to the fact that Maurya is engaged in constant prayer. God will heed Maurya's prayers, the priest says, and will not take away her one remaining son.

Nora and Cathleen express concern over the way in which the wind is rising from the south and west, and the danger this may pose for Bartley should he choose this time to begin his journey. As if to underline their concerns, the sound of the wind and the sea resumes full force in the orchestra.

At this point, accompanied in the orchestra by the motif associated with Maurya, Nora goes to the door of Maurya's room and listens. When she hears her mother stirring, she tells her sister that they better hide the clothing that the young priest has brought because of the fear that Maurya may see the bundle before they have determined whether the clothes are Michael's or not.

Maurya is convinced that Bartley won't start his journey that day because he will be deterred both by the weather (the wind rising from the south and west – always a bad omen) and the persuasive powers of the young priest, who, she feels, will try to dissuade him from making the trip. However, her daughters tell her that Bartley is determined to go and that the young priest has said he will not try to stop him.

At this point, Bartley enters. Maurya tries once again to persuade him not to go, but Bartley remains adamant. Maurya is concerned about losing her one remaining son to the sea. Bartley may hear the words she is speaking, but he does not hear what she is saying. He has a completely different point of view. His sisters side with him rather than with their mother. Cathleen, for example, says: "It's the life of a young man to be going on the sea, and who would listen to an old woman with one thing and she saying it over"[3] (Ralph Vaughan Williams, pp. 30–31).

Bartley invokes the blessing of God on them as he leaves. Cathleen is shocked that her mother will not give Bartley her blessing. There is enough misfortune in their household, she contends, without her mother failing to perform such a basic act as giving her blessing to her son as he sets out on his journey. Both women blame their mother for not having given Bartley a cake that has been baking in the oven. Not only will Bartley go hungry, but they will be utterly destroyed because of Maurya's preoccupation with the fate of the male members of her family and her lack of attention to the practical actions that could ameliorate the severity of that fate. She does agree, however, to take him the cake that her daughters have baked. That way, as they remind her, she can give Bartley her blessing at the same time that she gives him the cake, and thereby undo the unfortunate results of her neglect. As she goes out, she comments on the curious way in which customary circumstances are reversed in her case: "In the big world the old people do be leaving things after them for their sons and children, but in this place it is the young men do be leaving things behind for them that do be old"[4] (Ralph Vaughan Williams, pp. 40–42). The roaring of the sea in the orchestra serves to remind us of the reason for this state of affairs.

After Maurya leaves, Nora and Cathleen retrieve the bundle of clothes and discover that the clothes are indeed Michael's. They are concerned both with Michael's death and with the effect that his death will have on Maurya. As the sea continues to make its presence felt in the orchestra, Nora laments the fact that its power can reduce a once robust man to part of an old shirt and a stocking.

Maurya returns, to the accompaniment of the motif associated with her. Nora and Cathleen notice that she still has the cake that she planned to give to Bartley. The only response her daughters can get to the questions they put to her is a keening sound. When Maurya is able to speak, she asks her daughters if they have seen Bartley. Cathleen looks out the window and tells her mother that yes, she sees Bartley riding on the mare with the grey pony following behind him. She cannot see, however, that there is anything unusual about this. Then Maurya tells her daughters that she has seen Michael. Cathleen replies that that is not possible, because they have found Michael's dead body. Maurya persists and says that she saw Bartley come first on the red mare. When she tried to give him her blessing, she found that she could not do so. As he passed quickly by, however, he blessed her. And then, on the grey pony that was following the mare, she saw an elegantly dressed Michael. Cathleen, understanding the significance of what Maurya believes she has seen, begins to keen and says: "It's destroyed we are from this day. It's destroyed, surely"[5] (Ralph Vaughan Williams, pp. 65–66). Maurya agrees and resigns herself to the fact that she will lose Bartley to the sea, as she has lost all the other male members of her family. She is so distraught that she asks Nora and Cathleen to call in Eamon, a local woodworker, to make her a coffin, because she believes that she won't survive their deaths.

Maurya, who for all practical purposes has ceased to be aware of what is happening around her, launches into a soliloquy in which she relives the experience of losing her husband, father-in-law and four of her six sons to the sea. In the midst of this soliloquy, a chorus of women begins to keen, the sea motif reappears in the orchestra, and Nora and Cathleen hear the voice of someone crying out in the vicinity of the seashore. They realize the significance of what they are hearing – that the body of someone who has been lost to the sea has been found. Then Cathleen hands Maurya Michael's clothes, that the young priest had earlier given to her and her sister. Maurya now has physical evidence that Michael is dead.

This is followed by the appearance of a group of women who enter the cottage crossing themselves. When Cathleen asks what has happened, one of the women tells her that they have found Bartley's body. The grey pony that was trailing along behind him knocked him into the sea. At this point, the orchestra plays the sea motif in counterpoint with the motif associated with Maurya.

The men then carry in Bartley's body and lay it on a table. Maurya kneels at the head of the table and asks Nora to give her holy water. She takes Michael's clothes and drapes them across Bartley's feet, sprinkles holy water on him, and in this way blesses him. Then she stands up and spreads Michael's clothes beside Bartley's body and sprinkles the remaining holy water on them. Cathleen and Nora cannot understand Maurya's attitude and her behavior. They expect her to be more distraught by the death of Bartley. Finally, Maurya takes the cup that contained the holy water, turns it downwards on the table, and places her hands together on Bartley's feet. The action with the cup is necessary because if one does not dispose of a utensil associated with the dead, the dead person will surely return and claim the life of whoever uses the utensil.

The opera closes with Maurya's affirmation of the peace she has found and her consequent triumph over the efforts of nature to destroy her: "They're all together this time, and the end is come. May the Almighty God have mercy on Bartley's soul, and on Michael's soul, and on the souls of Sheamus and Patch, and Stephen and Shawn . . . and may He have mercy on my soul, and on the soul of everyone is left living in the world. Michael has a clean burial in the far north, by the grace of Almighty God. Bartley will have a fine coffin out of the white boards, and a deep grave surely. . . . What more can we want than that? . . . No man at all can be living for ever, and we must be satisfied"[6] (Ralph Vaughan Williams, pp. 89–97). The sound of the women keening accompanies Maurya's arioso. Gradually, this sound diminishes to the point where there is a single solo voice in the chorus of women. The sea is also silent. Musically, this symbolizes the fact that the sea, which has for so long dominated the life of Maurya and the other women in the village, no longer dominates Maurya's life. Her triumph is likewise symbolized by the fact that the key, which has alternated between a somber C# minor and an affirmative E major, modulates into E major as Maurya begins her blessing, and remains in that key to the end of the opera.

Institute for the Psychological and Pedagogical Study of Voice Performance
P. O. Box 541
Philadelphia, Pennsylvania 19105–0541, U.S.A.

NOTES

[1] Michael Kennedy, *The Works of Ralph Vaughan Williams* (2nd ed., London: Oxford University Press, 1992), p. 256.
[2] Declan Kliberd, "Introduction – The Living and the Dead: Synge and Yeats," in *Shadows: A Trinity of Plays by J. M. Synge and W. B. Yeats* (London: Oberon Books, 1998), p. 6.
[3] Ralph Vaughan Williams, *Riders to the Sea* [by] *J. M. Synge, Music by R. Vaughan Williams* (London: Oxford University Press, 1972), pp. 30–31.
[4] Ibid. pp. 40–42.
[5] Ibid. pp. 65–66.
[6] Ibid. pp. 89–97.

TRICIA MEEHAN

LATE MODERNITY AND *LA VILLETTE*: "UNSETTLING" THE OBJECT/EVENT DIALECTIC

I. LATE MODERN CULTURE AND SYMBOLIC COMMUNICATIONS

Johan Fornas, in *Cultural Theory of Late Modernity*, places symbolic communication at the heart of cultural production.[1] He presents the inherited dialectical understanding of communicative structures and uses the limitations of this construct to investigate potential communicative processes that operate closer to the intersubjective and polyphonic nature of late modern culture.

The time-space considerations found within his "culture considered as symbolic communication" demonstrate the importance of the constructed world in late modern culture. This also introduces a notion of place as a potential force capable of diverting dialectical forces and introducing polyphonic frameworks. Fornas defines culture in this way:

> Culture is everywhere in human life and society. We are human by understanding and interpreting what we perceive, that is, by constructing symbols where something stands for something else. Symbols make it possible to think of what is not present, and thus to reflect upon the past and to plan the future, to explore the other(s) and speculate about the unknown. By collectively shaping such symbolic patterns we construct a world and give ourselves specific positions in it.[2]

Fornas's emphasis on symbolic communication also requires consideration of memory and reminiscence, which he does not directly address. His concerns echo Aristotle's. Both authors search for the experience of something not present – the presence of absence. They both regard symbols, figures, and traces as objects, and memory is what moves and becomes the event rooted in them. Any understanding of place, in relation to the object/event dialectic, must also consider memory.

The two sides of the object/event dialectic can be identified within Fornas's discussion of culture. This passage introduces objects determining events. He states:

> The temporal dimension may be contrasted with aspects of culture that are usually conceived in spatialized terms: the social spheres which simultaneously frame and enable cultural activities. The time-flows of modernization run into spatial spheres formed by order structures in the physical and social spaces of modernity; relatively fixed structures and institutions which, like sluices, delimit which human actions are possible.[3]

A.-T. Tymieniecka (ed.), Analecta Husserliana LXXVIII, 171–180.

He also describes events determining objects. He states:

> No one descends twice into exactly the same cultural river: culture is not a static structure, but an everchanging flux. No one escapes being moved by it: subjects are not fixed entities but evolving constructions. No one emerges from this flow without leaving some small trace of his or her movements: human actions have effects.[4]

Fornas understands the constructed world both as directing our actions and as their result. The concept of "trace" is therefore critical as both an action leading to the formation of objects and objects leading to the formation of actions. It occupies the tension "in between" the object/*/event dialectic. It might, therefore, prove to be a useful analogue for examining the mediating potential of late modern place.

Paul Ricouer defines the concept of "trace" as the result of the human desire for the fixation of the "course of events." He states:

> Could we not say that what we call the course of events plays the role of the material thing which "rescues" the vanishing discourse when it is written? . . . An action leaves a "trace," it makes a "mark" when it contributes to the emergence of such patterns which become the documents of human action.[5]

Fornas understands that this concept carries with it meaning produced as "more a process than a fact."[6] Ricoeur names this process of an object freeing itself from its context, "materialized alienation."[7] This means that the work escapes its time, author, context, and audience. It is free to form a world of meanings around itself.

Trace, then, as the result of meaningful action, inhabits the in-between of the object/*/event dialectic by confronting the phenomenology of the present within Ricouer's understanding of materialized alienation. Things experienced out of time and out of location are present in a given moment, a given experience. If one inhabits a world where actions leave traces or marks, one potentially moves through the world searching for the acknowledgment of their own existence. Understanding patterns in relation to one's life, finding objects that allow interpretations to be formed which make sense to the subject, require that experiences of memory be part of that process. Thus reflexivity, or mirroring,[8] is a condition of the spatial presence of absence as well as the temporal one.

II. TRACE AND THE PRODUCTION OF PLACE AND MEMORY

This implicates both place and memory (space and time) as considerations for meaningful action. The production of place is a question of doing in the moment, of trying to mark our presence in the world. The production of memory is the event of recognizing ourselves reflected in the world in some way. In the first, events result in objects. In the second, objects result in events. Thus place and memory form a dialectical opposition in this context, with an analogue to trace potentially occupying the "in-between."

Distinguishing differences in their production further clarifies the oppositional nature of memory and place. The production of the built world, versus our seeing ourselves in that world, is facilitated with the use of the words "make" and "construct." "Make" here means the allowing for bringing something new into existence. This contrasts with "construct" which denotes forming by fitting parts together, or the act or result of construing, interpreting or explaining.[9] In this sense, places (events determining objects) are made. Memories (objects determining events) are constructed.

Each trace, then, manifests the object/event dialectic both through the "making" of the not yet known or something new, and the "constructing" of the interpretation or understanding based on what is given. On one side, events leave their imprint in time – interpretive memories are constructed – while on the other, objects leave their imprint in space – new places are made.

Now the question of the term "in-between" the making place/*/constructing memory dialectic becomes central. This requires further consideration of the "presence of absence" question inherent in Johan Fornas's three characteristics of modernization. He states:

> The processes of change which combine to form modernization have three things in common: they tend towards *an irreversible dynamization*, an *ambivalent rationalization* and a *differentiating universalization*. This means they are intense and cannot be easily reversed, even if they might take different forms; they promote a two-sided rationality with both positive (creative and emancipatory) and negative (destructive and oppressional) potentials; and they are increasingly general and global in their capacity to create new pluralities by separating social and individual life-spheres.[10]

These traits frame a series of important dialectical pairings. Irreversible dynamization operates within the change/continuity dialectic. Ambivalent rationalization presents a truth/false or norm/deviation dialectic. Differentiating universalization delineates the local/universal dialectic.[11]

These pairings acknowledge place and memory not only as dialectical, but also as necessarily mediated. *Le Parc de La Villette*, by Bernard Tschumi, is an example of a project operating from similar assumptions, and contributes to the exploration of the middle term. Tschumi states: "Space, movement and event are inevitably part of a minimal definition of architecture. . . ."[12] This definition introduces movement as a characteristic of that central term and reflects generative processes at play in late modern culture.

III. THE *PARC DE LA VILLETTE* – INHABITING THE OBJECT/EVENT DIALECTIC

The *Parc de la Villette* was built in Paris during the presidency of François Mitterrand. The master plan was awarded to Tschumi in March 1983. The competition brief described the project as an "Urban Park for the 21st Century."[13] The polemical nature and explicit call to address both cultural issues and questions of modernity allowed Tschumi to investigate the idea of culture as symbolic communication. He envisioned the park as "trace or play of language"[14] and challenged architectural norms with "an interchangeable relation between object, movement, and action" (event).[15]

This acknowledgment of the dialectic, and the introduction of a central term, "in-between," as essential, reinforce Tschumi's goal of diverting dialectical forces and "unsettling both memory and context" (place). The generative nature of this middle term emerges through investigating the conditions of the *Parc* in relation to the dialectical implication of Fornas's three characteristics of modernization.

Differentiating Universalization

The third characteristic identified by Fornas, the first considered here, is differentiating universalization, which delineates a local/universal dialectic. Modernism's universalizing tendencies result from systems with global reach, but places, cultures, and societies continue to be recognizably different as well. The processes of universalization are not merely one-way, from center to periphery, but also move from periphery to center. Fornas states:

> Modernity does not imply a universal standardization, however, but rather a growing universal differentiation in society. . . . The concepts for stratifying modernity are themselves historically created by processes of modernization which make it possible to think of culture, society, and individual as different categories.[16]

Fornas argues that once the components of a totality have been differentiated, that differentiation is an always understood possibility. The resulting need for recombination requires an intentionality for the new reunion to potentially have meaning beyond the pre-differentiated unity. The resulting interplay requires that the modern project be reflexive by nature.[17]

This understanding of differentiation assumes that movement is, for both Fornas and Tschumi, a characteristic of the central term, "in between" local and universal. It is rational, systemic, and autonomous, allowing frameworks of reflexivity to be established. This movement cannot be grounded in one or the other extreme of the opposition if the cause and effect limitations of the polarities can be effectively challenged and dismantled.

The design strategies employed by Tschumi clearly demonstrate this trait of modernization. He challenges two of architecture's universals, both the synthetic Vitruvian trilogy of firmness, commodity, and delight, and the time-honored form-function correspondence. He utilizes a "strategy of difference" not constrained by dialectical oppositions.[18] This allows the composite nature inherent in mediating dialectical oppositions to be challenged. He states: "The Park became architecture against itself: a dis-integration."[19] Thus, he challenges the status quo aspects of the discipline with this project.

The differentiated formal and systemic elements at *La Villette* are: the system of points, the system of lines, and the system of surfaces – or objects, movements, spaces.[20] Program itself becomes a diverting factor within all three systems. Tschumi's interest in this sort of differentiation builds upon the idea of system or structure as self-referential in origin. This internal logic[21] establishes the autonomy of each component, thus creating differentiated systems with universal extension.

Architecture severs its traditional relationship to tradition in order to generate this directed and evolutionary movement. These differentiated, abstracted, autonomous elements are now ripe for processes of juxtaposition and recombination that open up a multiplicity of potential meanings and outcomes. Differentiation has destabilized the synthetic tradition, allowing the "infinite horizon of interpretation," or the "free play of signs" to be set into motion.

Irreversible Dynamization

This introduces the next characteristic, irreversible dynamization, with its change/continuity dialectic. Fornas identifies the goal of processes of modernization as directional and evolutionary. He states:

> All historical changes are in one sense irreversible, since nothing can ever be undone or repeated in exactly the same way – at least the fact that a similar thing happened before will lend its repetition a slightly different meaning.[22]

Fornas links this with the modern "urge to historicize." As processes of change intensify, history discovers new relationships between past and future. He introduces Reinhart Koeselleck's three observations of modern history. First, history is a human construct – it is made. This "makeability" "indicates a new space of experience and horizon of expectation." Second, history is incomplete and opens towards the future. A plan of action is separated, through time, from its realization. Third, there is an increasing distance between expectations for the future and lessons learned from past experiences.[23] The resulting "time-space compression," identified by David Harvey, extends the perception of the present, problematizing "the reproduction of intersubjective patterns."

Tradition now, whether conscious ritual or semi-conscious habit, has "not a necessary but [a] desired continuity, actively selecting among received and recovered elements from the past."[24] Continuity is determined instead by four historical process-types that Fornas identifies at work in modernity: stubborn structures, accidental events, periodic cycles, and directional processes.[25] The juxtaposition of these systemic continuities now creates that time-space compression characteristic of the third term found "in between" the change/*/continuity dialectic.

This range of process types is visible at *La Villette* and does support the creation of this time-space compression:

- Differentiation and the deployment of three autonomous systems (point, line, plane) previously discussed are stubborn structure processes. These systems reach out across space and time heeding only their own internal logic.
- Accidental events result from the superimposition of these three systems on each other. The collisions that result create fragments, seen now in relation to other fragments, rather than in relation to their systemic origins.
- Periodic cycles issue from the frame and sequence structure of the cinematic promenade. The frames are discrete, ordered, and recognizable in their repetition. This allows the contents to vary freely, without disrupting the flow of the experience. Memory is both required for, and altered[26] in, such cyclical processes.
- Directional processes appear in two ways. First, the *folies* are generated through substitution strategies based on mechanical operations that create both recognizable repetition and constant change.[27] Second, substitution is also a pragmatic strategy for the larger consideration of the *Parc*. The park master plan had to incorporate the contributions of many participants, both in the initial design and over time.[28]

Time-space compression results from the juxtaposition of these process types on each other and demonstrates how Tschumi addresses the change/continuity dialectic at *La Villette*. Past experience, present encounters, and expectations for the future collide in acts of recombination, thus creating Tschumi's challenge to "architecture's obsession with presence." He states: "Presence is postponed and closure deferred as each permutation or combination shifts the image one step ahead."[29]

Ambivalent Rationalization

The remaining characteristic identified by Fornas is ambivalent rationalization, and it introduces the truth/norm dialectic. His definition of the term "ambivalent" is important:

> While equi-valence refers to two things of equal value, the prefix ambi- means "both", so that ambivalent has two (opposite) values simultaneously. It does not denote a general vagueness, but rather that one singular force has two precise but contrary implications.[30]

Fornas identifies rationalization processes of modernization as essentially dialectical and ambivalent. On one side there is knowledge based on "an analytic truth," "freed from tradition." Criticism yields true and false judgments. On the other is knowledge based on the "ontology of the present," "freed from gods." Criticism establishes judgments of norm and deviation.[31] This ambivalence establishes reflexivity as essential to the advancement of the modern project. The presence of one outcome allows for both the understanding and the critique of the other.[32] In fact, modernity recognizes itself through reflexivity. Both identity and roles, in relation to the larger systems – whether goal-oriented rationality or communicative rationality – are defined in this way.[33]

Judgments are made by something that exists "in between" the true/*/false or norm/*/deviation dialectics. This entity is autonomous from the modernization processes and can therefore make judgments concerning them. It is a subject critically constructing its identity-making valuations based on whether or not it finds itself mirrored in these processes.

Ambivalent rationalization manifests itself at *La Villette* as the necessary and desired outcome of differentiation and superimposition strategies. Both the rationality of systemic response and the irrationality of incidental accidents are central to this. The disjunction and recombination that result are not programmed according to true and false, or norm and deviation valuations. The recombinations produced contain the traces of the restless dynamic of processes of modernization at work and are open to the self-reflexive, mirroring processes of identity in construction.

Acknowledging ambivalence allows the recognition of the other. Tschumi enables both norm and deviation as necessary for developing identities based on a critical understanding of modern conditions. The psychological urgency driving the *La Villette* rhetoric reflects the desire to acknowledge the "Other" and give it a rightful place within the system. Tschumi observes that deviant behavior (madness) is often "negated in order to preserve a fragile cultural and social order."[34]

Tschumi, like Fornas, understands criticism as the life-blood of modernity. Tschumi states: "Deviation is both the excess of rationality and irrationality. As a norm, it contains the components of its own explosion. As a deviation, it frees them."[35] Tschumi's interest in psychological concerns at *La Villette* extends this understanding. The something "in between" norm and deviation is the subject critically producing its identity.

Making Place/Theory/Constructing Memory

Fornas's three characteristics of modernization, as well as Tschumi's exploration of space/movement/event at *La Villette*, clarify the nature of the central term under consideration. Its characteristics include movement, time-space compression, and critical identity formation. This central term demonstrates the intentionality of meaningful action and clearly differs in nature from the bracketing terms rooted in the object/event exploration. It is not a resultant composite of the dialectic, but rather a generative force capable of uncoupling the bounded bracketing terms.

Mediation, a foundation for the design of *La Villette*, does just this. It diverts the closure of polar forces allowing recombination to occur through juxtaposition and contiguous relationships. This "in-between" thus serves to "unsettle" the making place/*/constructing memory dialectic. Tschumi's mediation parallels Fornas's identifying the "in-between" as intersubjective, dialogic, and textual. It is more than the trace alternating in its cause-effect implications.

Fornas's discussion of theory positions it as such a mediating force.[36] It is autonomous and capable of operating according to its own internal and systemic logic. It creates time-space compression in its ability to separate and reconnect traditions.[37] It generates and directs evolutionary movement. Its deployment requires both intentionality and critical depth if the "problematized reproduction of intersubjective patterns" is to be acknowledged and addressed.

Theory is precisely what mediates production at *La Villette* by inhabiting the "in-between" of place/*/memory, "unsettling" their established generative roles in the production of architecture. Theory prevents either the construction of memory or the making of place from dominating production. Because of this condition, "presence is denied" and theory must become productive both in terms of constructing and making. Fornas states: "To elaborate the contradictory relations between different theoretical systems can be very creative work, resulting in new concepts and new meanings. . . ."[38]

La Villette demonstrates that this notion of theory must insist on an experiential dimension. The sensual, emotional and intellectual, clearly differentiated in late modern life, are simultaneously called into play and recombined through theoretical mediation. Continuity and change, understood as processes rather than traditions, intensify as movement increases and memory and recollection processes based in habit weaken. Theory mediates both the marking of paths and recollecting of selves along the way. It is theory that persists as place and memory are unsettled. What is more important, it is theory that salvages the necessity of both place and memory for acts of recombination which counter the unsettling effects of late modern processes.

University of Arkansas
Fayetteville, Arkansas

NOTES

1 Johan Fornas, *Cultural Theory and Late Modernity* (London; Thousand Oaks, CA: Sage Publications, 1995), pp. 1–2.
2 Ibid., p. 1.
3 Ibid., p. 49.
4 Ibid., op. cit., p. 1.
5 Paul Ricoeur, *Hermeneutics and the Human Sciences* (Cambridge; New York: Cambridge University Press, 1985), p. 206.
6 Fornas, op. cit., p. 180.
7 Ricoeur, op. cit., p. 199.
8 Fornas, op. cit., pp. 258–278.
9 *Webster's New Collegiate Dictionary* (Springfield, MA: G. & C. Merriam Co., 1979), pp. 241, 688.
10 Fornas, op. cit., p. 20.
11 Ibid., pp. 20–31.
12 Bernard Tschumi, *Cinégramme Folie: Le Parc de la Villette* (Princeton: Princeton Architectural Press, 1987), p. 27.
13 Ibid., p. ii.
14 Ibid., p. VIII.
15 Ibid., p. 27.

16 Fornas, op. cit., p. 31.
17 Ibid., p. 31.
18 Tschumi, op. cit., p. V.
19 Ibid., p. VII.
20 Ibid., p. 6.
21 Ibid., pp. 4–5.
22 Fornas, op. cit., p. 27.
23 Ibid., p. 21.
24 Ibid., p. 22.
25 Ibid., p. 25.
26 Tschumi, op. cit., p. 12. He quotes here a passage from his unpublished text, "Etude de Définition."
27 Ibid., pp. 24–27.
28 Ibid., p. IV.
29 Ibid., p. VII.
30 Fornas, op. cit., p. 29.
31 Ibid., p. 27.
32 Ibid., p. 28. He states: "The goal of this reformulation of modern rationality is to let the critique of reason . . . radicalize the modern project of enlightenment, rather than liquidate it."
33 Ibid., p. 30.
34 Tschumi, op. cit., p. 17.
35 Ibid., p. 27.
36 Fornas, op. cit., p. 6.
37 Ibid., p. 11.
38 Ibid., p. 12.

GARY BACKHAUS

THE LOOKING-GLASS SELF: SELF-OBJECTIVATION THROUGH THE GARDEN

INTRODUCTION

The Initial Experience of the Artifact under Investigation and Various Modalities of Intentionality

This study concerns the estate garden located in Annapolis, Maryland that was built and owned by William Paca. Among his many outstanding achievements, Paca was an eighteenth-century lawyer, revolutionist, signer of the Declaration, Governor of Maryland, Judge of the Federal District Court of Maryland, and anti-federalist. I visited the estate garden for the first time in the summer of 1999 without any intention of formulating a research project; but as a phenomenologist, I became motivated to engage in the phenomenological epoche.

By remaining within the natural attitude, the visitor to the garden is unaware of shifts in noetic acts through which noematic correlates of the garden and associated objects undergo presentational modifications within the context of the experience. One can shift into the aesthetic order of reality, which brackets the pragmatic motive of the paramount province of the everyday in order to savor, for the sake of *aisthetikos*, the perceptual/sensuous presentation of the object. Within the natural attitude this aesthetic province is easily conflated with the cognitive province of knowledge. In the cognitive order, the object is analyzed by applying concepts that serve to increase understanding and to acquire new data in a circumscribed context. William Paca's garden can be experienced both aesthetically and as an object of cognitive analysis. A cognitive analysis may either enhance or stifle the aesthetic experience, and vice versa.

From within the natural attitude, other orders of reality are also naively intended, and the observer usually shifts from one attitudinal mode to another without achieving excellence in any one province of meaning-intention. The perceiver at times relates the artifact to himself personally. It then becomes subjugated to the role of psychological trigger, prompting the observer's thoughts about his own life. The Paca garden merely functions as the occasion for introspection in which the self-reflective content only obliquely concerns the garden. The artifact also can serve sociality, whereby consociates are

A.-T. Tymieniecka (ed.), Analecta Husserliana LXXVIII, 181–217.

escorted to the Paca grounds for their enjoyment. The value of the artifact in this instance is in its function of providing for a favorable social impression. One may focus on the object for its moral edification or as a vehicle for political propaganda. One may attend to how the object had been produced, or perhaps to the stature of the producer/owner or his historical milieu.

During the first visit to the garden estate, I shifted between the natural attitude and the phenomenological attitude, which allowed me to make these preliminary observations concerning attitudinal modifications. A few days after my visit, I formulated a problematic in social phenomenology: Is it possible to experience the phenomenon of the looking-glass self through an artifact that had been owned by a predecessor?

The Explication of the Phenomenological Problematic of this Study

This phenomenological exploration leads to the investigation of subjective meaning-contexts. A subjective meaning-context is comprised of what an objectivation (the manifestation of human expressivity in products of activity) means in the mind of the agent whose objectivating acts have resulted in a productive concretization (the perceptual manifestation of an action or artifact). An artifact, in this case the estate garden of William Paca, first must be interpreted in terms of an objective meaning-context, which requires that the observer attend to the garden's status as an object. This focus consists in determining characteristics of the object with no reference to the constituting acts of the person who made or used it. In terms of a subjective meaning-context, all rational action consists of planned intermediate goals and an interpretive reconstruction of those steps can be carried out by direct or indirect observation. In this project I am not so much interested in the building and maintenance as in the subjective meaning-context of the cultural significance that Paca inscribed in the landscape and in the use of the garden. I intend to learn how Paca constituted in his own mind the social and politico-ideological functions of his garden. The problematic requires the construction of a personal ideal type based upon this course-of-action typology. The personal ideal type is a description of the kind of person, e.g. gentry or commoner, who has acted in such and such a way (course of action). Typologies are less problematic when considering a contemporary who shares one's culture, but type construction becomes a more difficult hermeneutic problem with personages of other cultures or historical times. Typologies are always more deeply founded when courses of action are connected to inquiry involving the motivational context. Motivation concerns why the person acts as he does – the adequate subjective ground for the objectivation in question. Motivational understanding requires a diachronic investigation of Paca's past and future, before and after the building of his garden estate. Evidence for the *in-order-to motive* explicates the future-oriented purpose that would be served by building the garden estate. The *because-motive* consists of evidence from the past that had led Paca to formulate the project of building the garden estate.

More specifically, the study concerns Paca's self-presentation, interpreted through his garden. Whereas a personal ideal type can be highly general and anonymous (the habitual type, e.g., lawyer, father), the concrete personal ideal type concerns a specific personage (the characterological type). Biographical information is collected as a source for constructing his characterological type. In some sense all of our objectivations are self-presentations; we project an image of ourselves to others, sometimes more consciously than at other times. We must be careful to distinguish how Paca understood his self-presentation, the persona that he expressed, from our image of him. The meaning-interpretation of self-presentation requires a historical context, a hermeneutic of the historical milieu of William Paca, the context out of which he expressed himself. Paca's self-presentation is compared with our own image of him, which is an essentially hermeneutic history that brings the past alive in the present.

Self-presentation admits of degree, and in a specific objectivation, how much of someone's personality manifests is philosophically problematic. Obviously, a deeper interpretation will involve evidence through which we gain greater insight into personality and not merely glimpses of a person's trivial affects, or more opaquely the sublimation of personality behind a function or role. Specifically, in order to understand Paca's self-presentation through the garden, it is necessary to clarify the relation of being and having, i.e. personality and ownership.

The final stage of this study concerns a specific class of subjective meaning-contexts called the looking-glass self. The looking-glass self is an inter-subjective, socially formed self by which a person projects a self-presentation, which elicits an evaluative response from the social partner(s), and from the attitudinal response exhibited by the social partner(s), a self-feeling or attitude toward oneself is experienced. The particular phenomenological problematic is to investigate how the looking-glass self, which is based in face-to-face social interaction, undergoes modifications due to the fact that the garden estate is the artifact and William Paca is the predecessor with whom we have an indirect mediated social experience.

Procedure for the Solution of the Phenomenological Problematic[1]

The first section describes the structures concerning self-presentation, which manifest through social and communicative processes of objectivation. Next, the relation between being and having is explicated, for it must be shown how the personality of an owner can be apprehended through the investigation of possessions. Thirdly, the indirect social relationship is explicated as it is modified in the relation with a predecessor. Lastly, the looking-glass self is explicated in terms of its structural modification through the experience of an artifact produced by a predecessor.

The second section begins with an exploration of Paca's biography (courses of action) so that a characterology can be constructed. This context is synchronically widened through an examination of the material culture of Paca's historical milieu, which makes it possible to apprehend the *Weltanschauung* of the emerging class of American gentry to which Paca belonged.

The third section addresses the objective meaning-context of the aesthetic artifact, the garden. The physical outlay of the garden is described – its spatial configurations and its contents. Also of importance is the context of the objective meaning, which involves placing the garden into the history and philosophy of landscape architecture. Since a course-of-action ideal type can be regarded as an anonymous meaning-context, some discussion concerning planning, construction and maintenance of the garden is germane. The symbolic meaning inscribed in the garden is discussed in light of its ideotechnical function.

The fourth section of this study interweaves the inter-subjective context of the looking-glass self with the actual garden experience, which involves a description based on a perambulation through the garden conducted within the phenomenological attitude. The conscious experiences of William Paca are grasped in the mode of quasi-simultaneity with that of the interpreter. The researcher imagines what Paca wanted to express about himself and responds by evaluating this self-presentation, and then imagines what Paca's self-feeling would be as the result of the researcher's own impression. The researcher presents his own self to Paca and imagines what attitude toward that presentation Paca would have, then the researcher inspects his own self-feeling. This research raises this looking-glass phenomenon to the level of scientia, which in the natural attitude is only intended in a primitive and naive form. Of course, such an imaginative investigation cannot change the past, but its horizon has included the future that it projected. One's contemplation of the subjective contexts of history create a living-relational valuing hermeneutic with the past, and not merely the distancing and passively apprehended factologies of objective history.

The concluding section of this paper will be devoted to an evaluation of the value of this form of study. The discussion involves themes of socialization and axiology and reflects on the influence that the world of predecessors has on our own being.

SECTION ONE: THE SOCIAL PHENOMENOLOGY OF THE SELF

Self-Presentation in Objectivating Acts of Consciousness

After graduating from the College of Philadelphia in 1759 with a law degree, William Paca moved to Annapolis to start his practice. In 1763, he married Mary Chew, a wealthy Annapolis heiress and they immediately purchased lots on Prince George Street and began construction of a five-section Georgian-style mansion and garden. The Pacas moved into the mansion in 1765.

With this paltry amount of historical evidence it is already possible to begin the theoretical exegesis of subjective meaning-contexts by examining the significance of the following type of putative statements: "I have planned the construction of my home and garden," "I am a lawyer who owns a mansion and garden," or simply, "I think that I would like to maintain a garden estate as my residence in the city of Annapolis." These are statements of a typical form – with the employment of the shifter "I" as the grammatical subject. It must be elucidated how these statements, by which a person indicates that she is the one performing the action or the one who predicates something about herself, relate to the problem of self-presentation.

Charles Horton Cooley describes the self of everyday experience that is designated by the pronouns – I, me, my, mine, myself.[2] These pronouns engender a social component, for social partners also utilize the very same words to denote themselves. The "I" is a social phenomenon, because the ideas expressed by it are formed through communicative life. The shifter, "I", is unambiguous in meaning, but it is an abstract universal in terms of its referent. It does not denote a concrete personage; it rather indicates an assumed social role. The person who utilizes "I" concretizes its meaning by implicating himself as the person who has articulated a judgment. The significance of statements that employ "I" as the subject implies a self-presentation in connection to the judgment-content.

An experiential unity, e.g. person-who-built-the-house-and-garden-on-Prince-George-Street, is an inter-subjectively formed conception. When Paca makes the claim that, "It is I who owns the house and garden on Prince

George Street," he identifies himself as the object of thought. The predication dissects the experiential unity in order to form a judgment concerning the p-quality of S. The object-identity/self-predication is a socially recognized meaning-category. So, one identifies oneself with an objective meaning that is socially mediated. "The person who built the house and garden on Prince George Street is me," and "I am that person," are forms of the socially objectivated self. There are two senses by which the (S)/I and (p)/me distinction is then a socially mediated phenomenon: the assumed role as speaker, and the social mediation (the imputing of socially formed categories to oneself) that forms the I-me distinction.

"I" involves a certain feeling component of ideas, because it is accompanied by the my-attitude. Self-feeling emerges within social life resulting from those ideas that one stakes out as one's own. This appropriation to one's self involves the activity of distinguishing those meanings that belong to the individual consciousness from the rest of the world. The act of judging results in its objectivation in a socially communicable sign-system (language) and the "judge" assumes responsibility as the dative, "to me," for the manifestation of a fact. All claims, even those that do not utilize the first person pronoun, can be preceded by "I think that X." When the "I think" is expressly thematic, the speaker intends that the idea not only point to a state of affairs, an alleged fact, but also to herself, the speaker, as the person who has appropriated the judgment as her own. The act of appropriating ideas also entails the act of establishing one's own individuality. In this way the "I" carves out of the general social current its individual position. Paca would have formulated judgments similar to the following, at least to himself: "I want to construct a complex waterway with a pond, springhouse, bathhouse, and water canals." It is not our intent to put words in Paca's mouth, but to reconstruct from his objectivations *the individual* who built an elaborate waterway based on what we would call "the technology of eighteenth-century landscape design." In other words, statements need not be expressly recorded in a written document since it is through the act of doing, that the reflective judgment, "I am the doer," is attributable. So, even if Paca never said, "I own the estate on George Street," the owning itself implies self-presentation, for we generally think in words, and we especially like to explicate to ourselves the meaning of our actions in reflective acts. It is through reflective acts that one apprehends oneself, which occurs after the lived experience has "run-off." *Knowledge about oneself* is always judicative. All objectivations (activities and artifacts that are the medium for subjective expressivity) indicate a self-presentation, and some are conscious projections of self. There remains a question of transparency, depth, and breadth: that is how visibly, how deeply, and how much of the self becomes manifest in an objectivation.

Being and Having: Self-Objectivation through Ownership

The relationship between being and having is necessary for addressing the problem concerning the levels of transparency, depth, and breadth of the personality that can manifest through an owned artifact. The significance for self-presentation as an objectivation of personality in the concrete relationship between Paca and his possession requires elucidation. Georg Simmel's treatment of the self/property relationship expands the doctrine of self-presentation. In the self's power to manipulate objects (dealing with the resistance that material reality presents), one secures one's own development and enlarges one's place in the general social current. Simmel maintains that possession is an activity of the self, an act of having that is at once an act of doing, and vice-versa. Simmel establishes the thesis that the passive concept of property is a fiction and that an inactive ownership is a mere abstraction. He states, "Static possession exists only in the imaginary aftermath of the processes that precede it, and in the imaginary anticipation of future enjoyment or use."[3]

Subjective acts of ownership are limited by the type of object. Subjective activities are determined according to the possession's qualities, which are objectively imposed. The type and condition of the possession impose both a qualitative and a quantitative temporal structuring of activities. The possession presupposes that the owner has certain capacities not only for adequate maintenance of the possession, but also for the self to develop through the possession. A talented owner can shape the possession as an objectivation of his personality. Ownership entails an interaction between the nature of the owner and the nature of the possession, which establishes the mutual dependence of having and being.

> Just as the possession of specific objects is all the more genuine and active the better and more certainly the human subject is suited for it, so the reverse is just as true: the more fundamentally and intensively the possession is really owned, that is made useful and enjoyed, then the more distinct and determining will be the effects upon the internal and external nature of the subjects.[4]

To own something commits self-development to a certain direction. For Paca to live within the parameters of the estate means that his own self-development is shaped by its demands. Investment of energies into the possessing of an object is also the investment of one's capacities into certain actualities, to concretize the self in a certain way. There is a stock of knowledge that is requisite in order for Paca to be successful in the endeavor. With the construction of the garden, landscape architecture must be studied; the nature of plants, seeds, soils, and water flow has to be understood. Time must be invested in the maintenance of the garden, which must be ordered according to rationally constructed strategies. Paca could have left everything up to his gardener so that he could completely ignore this aspect of managing his estate, but this notion does not square with historical facts about the significance of gardens for those times, and with what can be established about the personage of Paca. At the College of Philadelphia, Paca studied Euclidean and plane geometry, Palladio's treatises on architecture, mechanics, and natural philosophy.[5] Paca invested his being into the cultural requisites for this form of having, which created the possibility of this specific path for self-development, a growth of personal being through a concrete project of ownership.

An important component for the objectivation of personality in the being/having relation is the level of independence that is achieved from the limitations imposed by ownership. Ownership can tie down the self in a confining manner. Ownership, paradoxically, can be both self-developing and self-thwarting. Negative freedom or "freedom from" can be translated into the discussion of possession. "Freedom indeed means the ability to bring about the independence of being from owning."[6] Let us illustrate the measure of this freedom through an example in which very little independence can be achieved. Suppose someone has a very small plot of ground, a shack, and a few tools as their only possessions. Their sole source of food has to come from the plot of ground, which only through extremely hard and attentive work would yield enough food for them to survive. In this case, the having controls to an extremely great degree the being of this individual. There exists little independence from that which is owned in this instance. Without one's being transcending the limitations imposed by the having, the personality as the subjective totality is not developing and thus does not inform the objectivation.

Ownership provides the opportunity for the will of the self to find expression in possessions. To own possessions entails the ideal that the objects do not resist the intentions of the owner. Possessions form the sphere whereby the tendencies and character traits of the owner become visible reality. Simmel states that the self would collapse and lose its dimensions if external objects were not available for the expression of the self's own tendencies. Human freedom, in this sense, ideally consists of a person developing each of her capacities exclusively for its own purpose, unfettered by imposed external obstacles and internal interference from other capacities. Each individual consists of a bundle of various aptitudes, qualities, and interests, and the ideal would be for each energy direction to evolve according to its own excellence. But no personage is a mere collection of capacities, for these are parts of a whole. Developing one aspect of the personality colors the whole. The personality is the interrelation of traits, the organization of self as an open dynamic system. Thus, ideal freedom, in the sense that each of one's traits/talents is to develop unfettered by any imposition placed on it by the other traits/talents, is impossible. Nevertheless, the more aspects of the personality that can be developed without at the same time thwarting other aspects, the more a person can achieve human freedom, which concretizes the personality.

Paca developed his talents in law, politics, liberal education, and as the owner of an estate. Thus, a certain amount of freedom for personality development in these areas had to be achieved, which means that a certain independence, that is, a relative freedom from the interference of each of these capacities on each other, would have had to be achieved. This means that Paca had to have achieved relative independence in the relation of being and having. This involves an important process: being has to gain separation from having but in a way that the personality does not withdraw from the possession. Rather, the personality is even more so reflected in the object, because the object no longer imposes itself on the freedom of the owner. The question then is, how is this possible?

Simmel maintains that not only is this possible, but that property can also involve growth of the personality beyond the individual. As with procreation, whereby the sphere of the individual extends beyond its original limits into another self, with possession the sphere of the individual can extend beyond the original limits into the property. The being of the object is molded through the activity of ownership, but that development can extend the personality of the owner beyond the person. The dynamism inherent to the being of some types of possessions continues the objectivation. This dynamism extends the boundaries of the owner's personality in these instances by allowing for the being-development of the object itself. The possession becomes the owner's advocate. "This concept of property as a mere enlargement of the personality is not refuted but rather is strongly confirmed by cases in which the self-awareness of the personality has been transposed from the core of the Ego to its surrounding layers, to property."[7] The enlargement of personality through the transmutation of the person's being into the person's possession is very important for investigating the case of

Paca's garden. In order for this possession to bear his personality, he would have to have achieved some independence from the imposed responsibilities of the garden estate. Paca extended his personality into the possession while maintaining separation between being and having.

It must be possible, then, for this mutual relation of being and having to be separated and yet to be brought back together through a higher level of synthesis.

> That self-awareness has transcended its immediate boundaries, and has become rooted in objects that only indirectly concern it, really proves to what extent property as such means nothing other than the extension of the personality into the objects and, through its domination of them, the gaining of its sphere of influence.[8]

Without the process that brings being and having back together at a higher relational form of systems organization after a relative independence has been achieved, ownership would become vacuous. The synthesis must involve the transmission of the being of the owner into the being of the possession.

> On the one hand, the whole significance of property lies in the fact that it releases certain emotions and impulses of the soul, while on the other hand the sphere of the Ego extends both over and into the "external" objects just as the process in the mind of the violinist or the painter is continually transferred to the movement of the violinist's bow or the brush.[9]

What is important is that the processes of the personality be maintained through the external objects, autonomously, so to speak. The personality must be objectivated in the object, which means that personal being permeates the object. The possession reflects the significance of the owner's personality.

The objectivation of personality through possessions can further lead to symbolization, an "authentic status symbol," by which the possession stands in for the being of the person. The spirit of the person is transferred to the object. We point to a painting and say, "This is a Renoir." Marilyn Monroe's wardrobe bears the stamp of her personality; it is not merely a collection of garments. In an inauthentic symbolization of status, e.g., the person who points to an expensive dress and states, "This is really me," the higher synthesis of being and having is missing. This higher synthesis is a process of degrees. The personality is not wholly missing in cases whereby little independence is achieved. But the concretization of personality reaches a phase (at exactly what point that occurs cannot be ascertained) in which the self-organization does shift to a higher level whereby the objectivated personality in the possession becomes thematic.

In the case of William Paca, the evidence of his life and achievements points to the possibility of inquiring into his personality, as presented through his possessions. An investigation of Paca's garden should at the same time reveal aspects of this personality. It is then possible to construct a concrete and rich personal ideal type (characterological) from the examination of the artifact, Paca's garden. If the garden does not reflect the personality of William Paca to a high degree, then the personal ideal type would be only a typifying scheme exhibiting a high degree of anonymity and generality. A course-of-action ideal type must be constructed concerning the garden, including the ways in which Paca was active in his ownership.

Interpretation of an Artifact Produced by a Predecessor

The two problems that need to be explicated concern the interpretation of the subjective meaning-context through the social relation with a predecessor, and the interpretation of the subjective meaning-context from the observation of an artifact. The interpretation of objectivations, which indicate another person's subjective experiences, is carried out through an investigation either of an act-objectification in the movements and gestures of the body, or of an artifact that the other person has produced. Artifacts are either signs or manufactured external objects. Our evidence is an artifact in the form of a manufactured external object, Paca's garden. As a product of action (behavior oriented toward the future on the basis of a previously formulated project), the garden manifests as evidence of the mental processes in the mind of its producer. "To know the subjective meaning of the product means that we are able to run over in our own minds in simultaneity or quasi-simultaneity the polythetic Acts which constituted the experience of the producer."[10] What is established is the step-by-step process that had been built up in the experience of the producer. How these experiences occur and are formed requires further knowledge of the producer who is to be interpreted.

In the genuine simultaneity of two streams of consciousness, the action of the producer is witnessed and the polythetic Acts take place in real time in the we-relation. In the case of Paca, an indirect social relationship is formed. The productive actions are not witnessed but can be imagined, and thus quasi-simultaneity is achieved. Quasi-simultaneity is a more problematic probability, for one cannot check one's interpretation through direct contact with the agent. Paca's intentions, which concern the functional use of his garden, are presented quasi-simultaneously in our experience of the garden. As a gentry member, his concern was to inscribe a socio-cultural expression of himself in the landscape, and what we run through in our minds in quasi-simultaneity is how Paca envisioned the affect of the garden estate on his contemporaries and successors.

The Looking-Glass Self

The sociological fact that the presentation of personality in an artifact is interpretable implicates our own attitude toward, or evaluation of, that self-objectivation. Cooley describes the looking-glass self:

> In a very large and interesting class of cases the social reference takes the form of a somewhat definite imagination of how one's self – that is any idea he appropriates – appears in a particular mind and the kind of self-feeling one has is determined by the attitude toward this attributed to that other mind.[11]

The structure of the looking-glass self begins with presenting oneself to the other. Secondly, one apprehends the other's judgment or attitude toward that presentation. Thirdly, the self-feeling that one experiences is imputed from the judgment or attitude of the other(s) – examples are pride and mortification. Alfred Schutz makes mention of the looking-glass in describing the face-to-face situation.

> A further essential component of this experience is that I also grasp his attitude to me. . . . I experience myself through my consociate, and he experiences himself through me. The mirroring of self in the experience of the stranger (more exactly, in my grasp of the Other's experience of me) is a constitutive element of the we-relation. As Charles H. Cooley has already shown in a penetrating fashion, the reciprocal mirroring is of fundamental import for the process of socialization.[12]

The problem is to describe how the looking-glass self is operative in social stratifications other than the face-to-face and through the mediated experience of an artifact. The key involves acts of imagination – one imagines the attitude of another toward a projected self-presentation from which one modifies one's behavior accordingly.

Any attitude directed toward an "other self," Schutz calls "Other-orientation." "Other-orientation" can be one-sided but when two people become reciprocally oriented, there exists a social relationship. The looking-glass self involves a social relationship. Social action, which involves behaving according to an antecedent project, can have as its in-order-to motive the intention to bring about certain conscious experiences in the other person. Schutz calls this "affecting-the-Other." If during the project one has imaginatively anticipated the conscious experience of the Other whereby a presentation of one's self is the thematic content, then this is a "consciously-projected-self-presentation." In the direct face-to-face reciprocal social relationship, the Other witnesses the project and enters into a motivational context. The social actor projects a self-presentation to the Other in order to elicit a response. This projected in-order-to motive elicits a because-motivated response. The self-presenter witnesses how the Other responds, which mirrors the self back from the Other, the socially reflected self.

With contemporaries that are not consociates, the social relationship is indirect. They are kept in mind while planning one's activities. These partners are anonymous and so one experiences them as "ideal types." Ideal types constitute one's knowledge of the social world and so they come into play even in direct face-to-face relationships. But types are continually modified in the face-to-face as the interaction progresses and the actions of the Other reinforce or interfere with the ideal-types that one uses to interpret the Other. In the looking-glass phenomenon, the self is presented as a certain type to be evaluated by the Other, who responds by typing the self-presentation in question.

The world of predecessors is already completely fixed. One can only know it through ideal types. Schutz is correct when he states that in taking up an orientation toward predecessors, one can never influence them.

> To say that an action of mine is oriented toward the action of one of my predecessors is to say that my action is influenced by his. Or, to put it another way, his action conceived in the pluperfect tense is a genuine because-motive of my own. I never influence my predecessors, they only influence me.[13]

Schutz states that there is no distinction between social observation and social relationship concerning predecessors, because it will always turn out to be a case of one-sided Other-orientation on the part of the living person. But this structure turns out to be quite complex. Schutz states,

> There is only one kind of situation in which I can meaningfully speak of a reciprocal interaction between myself and one of my predecessors. This is the situation in which he acts upon me and I respond by behaving in such a way that my conduct can only be explained as oriented to his act, having the latter as its because-motive. This would be the case, for instance, if he bequeathed some property to me.[14]

Did not Paca leave his garden estate for his successors? Would there not then be a reciprocal interaction?

Predecessors are known through records and monuments that are interpreted as signs, whether or not they intended them for contemporaries only or to include posterity. These signs are detached from the predecessors' streams of consciousness, but every sign is expressed by someone.

> It is therefore perfectly proper for me to ask myself what a given predecessor meant by expressing himself in such and such a way. Of course, in order to do this, I must project myself backward in time and imagine myself present while he spoke or wrote. Now, historical research does not take as its primary object the subjective experiences of the authors of source materials.[15]

One knows predecessors, like one knows contemporaries, through ideal types. However, the schemes one uses to interpret the predecessor are different from the expressive schemes of the predecessor. Historical interpretations remain tentative, but history as the genuine experiences of Others can be understood as a continuous we-relationship in subjective time. Schutz's social phenomenology grounds the possibility for a genuine subjective history.

Paca's stream of consciousness can be imagined in quasi-simultaneity with one's own. Because his garden is a sign that he intended as a monument to his being, there exists an indirect social relationship with him. The garden was meant as a self-presentation, which, indeed, objectivates his personality, and thus through quasi-simultaneity, it is possible to engage in the looking-glass phenomenon with Paca. Paca's self-presentation inscribed in the garden is reflected in my experiences of it, which reveal my attitude toward Paca's personality.

The merits of this hermeneutic history (subjective meanings across the boundaries of time in a living relation with the Other) are that one engages the continuous we-relation of humankind in a vital non-detached form of scientia and lived experience. I am attuned to historical self-presentations, which informs my attitude (the looking-glass), and thus alters the manner in which I engage the world. It is not mere objective historical facts, but informed imaginative acts that create a consciously lived relation with predecessors.

SECTION II: THE SOCIO-HISTORICAL CONTEXT[16]

Biographical Sketch of William Paca: Characterological Typology

The few quotes about Paca from contemporaries remark that he was tall, corpulent, and handsome. It is said that he was graceful in gesture and generous toward others. He is said to have been of noble actions, sound understanding, and respected in deliberation. Historians Gregory Stiverson and Phebe Jacobsen write,

> Although William Paca was not particularly gregarious or outgoing, he recognized the necessity of establishing social and professional contacts. Perhaps because he found relating to others on an individual basis somewhat difficult, he was attracted to clubs, where he could cultivate social relationships within a formalized and structured environment.[17]

Paca coveted his privacy and in his very active public life preferred not to be in the center of attention. An important clue to Paca's personality is found in his course of actions in terms of the relationship between the public and private spheres of life.

Paca was born in Northern Baltimore County in 1740. At the age of twelve, along with his older brother, he was sent to attend the Academy and Charity School in Philadelphia. His brother did not further his education, for he was to take over the plantation from their father. William entered the newly chartered College of Philadelphia and completed his degree in 1759. Having acquired prominent friends, he went to Annapolis to begin a career in law. Paca helped to establish the Forensic Club, which expanded his social connections and provided an opportunity to practice his skills for argument. On one occasion Paca argued for aristocracy against democracy and won, and on another occasion he argued that slavery was a legal practice and lost. It is likely that he was representing his own inclinations in these debates, but his sensibilities were much more complex than a simple pro or con on these issues.

In late 1760 or early 1761 Paca went to England to continue his legal education, and remained for at least six months. With his name on the roll of the Inner Temple, he obtained a unique status among his fellow lawyers. Upon his return he began taking cases in which the rhetorical battles with other attorneys (such as his friend Samuel Chase) in the courtroom became notorious. Paca married Ann Mary Chew in 1763. "Molly" was related by bloodlines to nearly all of the most prominent families in Maryland. This marriage was not merely fortuitous, for each of his lineage, from great-grandfather to himself, married wisely in terms of social prestige and economic gain.

> Mary could have chosen a wealthier and more socially prominent man than William Paca, but what he lacked in these attributes he compensated for with his education, graceful manners, and physical charm. . . . The success of Paca's own law practice was assured by his education and skill and augmented by his marriage.[18]

Paca now had established the wealth, status, and influence that afforded him the opportunity to assert political power.

The Pacas set out in their first year of marriage to establish their social prominence by building their garden estate, which conspicuously exhibited their genteel social status in this transformational period for the Chesapeake region.

Molly's family was a large one, and most of her brothers and sisters resided in substantial dwellings nearby, but the Pacas embarked on the most ambitious building plan yet seen in Annapolis, constructing the first of the great brick townhouses that adorned the city in the decade before the Revolution. . . . Once the house was finished, however, it established a standard that influenced the building plans of other wealthy merchants, lawyers, and government officials in the provincial capital.[19]

Paca would have been content to establish himself as an important attorney in a maritime, mercantile economy and as a member of the emerging American "aristocracy," but major political events swept him into other concerns.

One other course of action sheds light on how Paca carefully calculated and orchestrated important social events. A sickly niece of Molly's moved in with them after an elderly relation died. When the sickly girl died, Paca meticulously attended to every detail of an extremely elaborate funeral. Funerals were a way in which families established social rank, and Paca respectfully followed what was expected but nevertheless seized the opportunity to create the most favorable social impression. There was nothing about him to suggest that he was merely an opportunist or a cynic; this was definitely not the case. Yet he prudently calculated social effects in order to present and maintain his social standing.

Paca recognized the objective significance and utility of creating social connections for establishing political influence and power. The Forensic Club and the funeral arrangements support the thesis that he recognized that certain socially visible, carefully constructed self-presentations were necessary for establishing "appropriate" Other-recognition. Even though Paca was politically a great believer in individual freedom and rights and democratic political ideals, his courses of action connote a philosophy of class hierarchy and social distance from which he very much derived power and privilege as a member of the gentry. The building of the garden estate involved a well conceived plan to inscribe social status and political control in the material culture, which was typical for the cultural transformation of the 1700s.

Eighteenth-century Chesapeake gentry and artisans alike enjoyed viewing gardens on their journeys and in their local neighborhoods and believed that one could tell a lot about people by the gardens they kept. Serious gardeners even recorded their observations. William Farris, strolling the streets of Annapolis, jotted notes in his journal about his neighbor's gardens. John Adams was an inveterate garden watcher and often judged the status of his contemporaries by what he saw.[20]

Nevertheless, Paca guarded his privacy and politically maintained the notion that private life (the site of freedom) must be defended against the intrusion of public life. He realized that his private affairs served as a symbol of his patriarchal stature, which was beneficially projected into the public sphere through his genteel form of sociality.

With these bits of evidence we can better construct Paca's motives for building the garden estate: in order to establish the social status of his private life in a way that would promote his position of prominence in the public sphere. The brick wall of his garden serves as a metonym for this point.

William Paca (1740–99) had enclosed his garden with a brick wall. . . . Brick walls were expensive and available to only a few in early America, but they served useful purposes. . . . They *announced* that the owners were persons of means [italics mine].[21]

The because-motive for formulating the project in the first place was to establish residence for himself and his wife in the town in which her immediate relatives lived (both land and urban cosmopolitan gentry) and where he was practicing his profession as a lawyer in the growing mercantile economy. The perceptual effect of the garden estate was to represent in the landscape his social, economic, and political power. What may appear less obvious is the level of consciousness about this plan that we must impute to Paca's subjective meaning context. As a lawyer in an unstable maritime economy, he would have been well aware of the economic instability of the colonies. The building of his garden estate in 1763 coincided with the culmination of the French and Indian War. England was in deep trouble financially and pressed the colonies with proclamations and taxes. The colonial economy was becoming less agrarian, more diversified, and quite delicate. In order to secure stability, the illusion of security and power had to at least be visibly manifest. In this context, Paca would not have taken socio-political organization for granted. His efforts were not merely to enter into an already established socio-politico-economic structure, but to create and maintain one. The garden estate served the function of propaganda, which was a meaning-component in Paca's intentions.

To better understand Paca, then, it is necessary to turn to his political life. The Stamp Act of 1765 brought much alarm in the English Colonies. The first public protest in Annapolis occurred in August with Samuel Chase leading the meeting and with Paca very active, but not with the same "rabble rousing" conspicuity. In 1766 Paca and Chase formed the Annapolis chapter of the Sons of Liberty.

In Maryland the resistance to the Stamp Act gave public exposure to William Paca, Samuel Chase, and other leaders and enhanced their reputations. The popularity Paca attained from his efforts to achieve the repeal of the Stamp Act enabled him to gain his first public office when the citizens of Annapolis elected him to the city's Common Council in May 1766.[22]

Paca was elected to the Lower House of Maryland's General Assembly in 1767. Paca was a member of the antiproprietary party (the colony of Maryland had been established as privately owned by George Calvert, Lord Baltimore) and he helped to assert its power against the Upper House and the governor. Parliament had passed the Townshend Duties and Maryland took up opposition to this in 1768. Massachusetts had circulated a grievance letter that Governor Sharpe of Maryland had thought best to ignore, but Paca and the Lower House responded in favor of the Massachusetts redress.

Slipped in between the furious political events was the death in 1770 of Mary Paca's brother, who left Wye Island to his two sisters.

Thirty years old, successfully established as an attorney, and with a reputation as an influential and popular political leader, William Paca now found himself master of a lucrative plantation and a member of the colony's planting gentry.[23]

With his town estate just receiving its finishing touches, Paca now owned a portion of the most fertile land in Maryland. However, the Pacas remained in Annapolis, for they enjoyed the social life that they had established. Paca was selected to become a member of the Homony Club, an elite social club based on the pleasurable entertainment of sociality. Some of the typical motives involved in gardening suggest the ulterior motives that derive from the relationship between gardening and sociality.

The gentry painstakingly planned their gardens and grounds to provoke genteel conversations with guests. Spirited discussions of garden plants and techniques mingled with social and political gossip. . . . Colonial gentlemen walked admiring visitors around garden paths to impress then with their skill and knowledge and control. . . . Chesapeake gentlemen wanted to make sure their gardens were seen and that the layout paid tribute to them.[24]

Gardens were a medium to promote their owners. Not only does the garden promote prestige for a member of the gentry, but it also symbolizes the individuality of the owner. In Charles Wilson Peale's portrait of Paca in 1772,

Paca is shown standing in a room on the second story of his mansion with a large open window to his right. The back portion of the garden is depicted, including a classical temple with the figure of Mercury on its dome. This area is in the English Garden style, which was rare at the time for an American garden. All of this symbolizes much about Paca himself and would have been legible semiotics to his elite contemporaries. Thus, the garden inscribed in its geography the status of the particular personage for the benefit of social promotion.

The next political issue that involved Paca concerned the renewal of the Tobacco Inspection Act, which had to do with setting the salaries of clergymen by pounds of tobacco. Due to the two houses of the assembly being locked in debate, no compromise had been managed before the act expired. Since the original law of 1747 was a bill that was passed only after King William's death by a legislative body that lost the validity of its power with the death of the king, Paca argued that there were no obligations to the original poll act as law. When Joseph Harrison, a member of the Lower House, did not pay his poll tax, Paca, Chase, and Thomas Johnson defended him and they won their case. A new law passed, substituting money payment for tobacco, which had unfavorable financial implications for the clergy. The shrewd Paca "had achieved wide recognition as a mature and accomplished politician with few equals in Maryland."[25]

> Paca preferred fighting injustice and oppression by constructing finely argued newspaper essays that traced constitutional precedents and appealed to man's natural rights, but he never forgot the lessons learned during the successful protests against the Stamp Act and Townshend Duties. Strong public support was crucial if the antiproprietary leaders in the Lower House were to prevail, so occasionally Paca joined Chase and other politicians whose personalities were more suited to mingling with crowds and appealing directly to the people.[26]

"To appreciate Paca's contribution to Maryland politics in the years preceding the War of Independence, one must realize that to a large extent he was the intellect behind Chase's rhetoric."[27] Although Chase and Paca teamed up on the same political issues, Paca assumed the more genteel posture, while Chase, who was of humbler background and situation, acted in accordance with his lesser social station.

Mary Chew Paca died in January of 1774, and although devastated, William Paca had little time to grieve. After the Boston Tea Party of 1773, Paca was organizing local committees in Maryland to oppose the Boston Port Act, and he joined the first Continental Congress in Philadelphia in September of 1774. Paca left his children for months at a time as he served on many committees, drafting many documents for the Congress. When in Maryland, Paca did everything that he could to help raise troops and arms. He supported independence and signed the Declaration.

In 1777 Paca married Ann Harrison of Philadelphia. As her father was a wealthy Philadelphia merchant, Paca again married wisely. He moved her onto the plantation on Wye Island. He finally sold the garden estate in Annapolis in 1780, the same year that Ann Harrison died. On the Eastern Shore of Maryland, Paca assembled a militia for defense against the British. Tories were routed out and made prisoners. But an incident in 1778 is extremely insightful in understanding Paca's politics and personality. The Continental Army, invariably and drastically short of supplies, at times had to resort to impressment (confiscation of private property). An American officer once attempted to seize some of Paca's horses. In a letter, Paca describes the incident:

> A Mr. Rudolph, an officer of the light Horse with a Sergeant paid me a visit a few Days ago. They were in Pursuit of Horses and demanded some of me. Mr. Rudolph shewed me an Authority from Major Lee to press all Horses between Philadelphia and Delaware that would suit for light Horse. After reading his authority I told him if he attempted to seize any of my Horses I would blow his Brains out.[28]

It can be argued that Paca's participation in the fight for independence was based on his firm belief in the right of the individual. He was not interested in forming a federation or even a strong state government except in the sense of a government that would guarantee the private rights of individuals. In 1782 Paca was elected Governor of Maryland and served three consecutive terms. Afterwards, Paca refused to attend the Constitutional Convention in 1787 as a delegate for he backed the Articles of Confederation and believed that individual rights would be threatened by a stronger central government. He became a leading Antifederalist, but finally compromised concerning the Constitution after he was granted permission to introduce amendments.

> Paca's main quarrel with the Constitution was that he felt it did not adequately safeguard individual rights, and many of the guarantees he sought – particularly freedom of religion, freedom of press, and legal protection for those accused of crimes – were later incorporated into the Bill of Rights.[29]

The last ten years of his life were spent on his Wye Island Estate. In 1790 he decided to build a new mansion, Wye Hall, which was "the most splendid house in America when completed."[30] Paca was one of the largest landowners in the state; he also owned over 100 slaves, and had collected a vast quantity of expensive furniture and silver plate. "Paca was a cultured man who had long cultivated a taste for the finest in English architecture and landscape gardening,"[31] the consummate gentry-man.

There are what appear to be blatant inconsistencies in Paca's beliefs, common to most Revolutionary leaders. He defended individual rights yet was a slave-owner, he fought for democratic ideals, yet was aristocratic. Mark P. Leone remarks that "the major contradiction we see in Paca's life, and which arose at the time of the Revolution, was between a slave-holding society and one proclaiming independence in order to promote personal freedom and individual liberty."[32] Leone well describes how these contradictions were disguised in material culture. But I would put forth that for Paca, it was his strong partitioning of the public and private spheres that led to viewing these spheres as two orders of reality in which the ethics and social organization of the one did not apply to the other. This seems to be the best way to understand what appear as contradictions in Paca's beliefs, which were not viewed as such in his subjective meaning-context. Historian Joyce Appleby recognizes the tension between classical republicanism and Lockean liberalism in the belief system of the revolutionists. "During the course of the eighteenth century, upper-class colonists became more independent as well as more English. . . . The classic republican ideas they had picked up from the English country party taught them to beware of political corruption and to fear designs upon their liberties."[33] The public sphere was to provide *freedom from* interference while the private sphere provided *freedom for* the development of the person, which meant that those persons with the highest qualities would rise to their "natural" privilege of power.

> Liberty in the classical republican paradigm and liberty in the historic rights tradition are distinct and potentially contradictory concepts. The classical liberty of freemen to participate in political decisions celebrates the public arena and the disinterestedness of civic virtues. The liberty of secure possessions protects private, personal enjoyments – liberties that become vested interests.[34]

Paca's political beliefs included republican and individual rights strains. Further research would better construct his subjective view of what we see as the tension in his politics.

Historical Context: Symbolism in the Material Culture

Historians of material culture and historical archaeologists have established symbolic analyses of Chesapeake society and specifically Paca's garden. The historico-cultural milieu must first be examined before distinguishing the specific symbolism that Paca inscribed in his garden.

James Deetz reports that until circa 1660, Anglo-American culture was based on the English tradition of the Middle Ages.[35] The colonial settlements, however, were much more of a homogeneous peasant culture than that of class-stratified England. After the restoration of the Crown in 1660, a greater interest in the colonies brought a re-Anglicization. After circa 1760, however, cultural transformations were instigated by cosmopolitan city dwellers. Paca would have been one of these cultural innovators. The colonies, in this transformation, were following the English, who had already begun to transform their material culture based on Renaissance ideals. The medieval worldview was being replaced by the new Georgian worldview (the Age of Reason, the Enlightenment). In this worldview mechanics replaces the organic, balance replaces asymmetry, and individuality replaces the corporate. Artifacts expressed this worldview, which was becoming transcribed into the cultural geography of the young country.

Richard L. Bushman remarks on the different pattern of living whereby houses and gardens were "outward signs of what the inhabitants hoped would be an inward grace. They wished to transform themselves along with their environments."[36] The material culture was meant to express "the ideal of a cultivated and refined inward life."[37] Gentility was exclusively the culture of the upper class in the eighteenth century, and along with their Georgian houses and gardens came "new modes of speech, dress, body carriage, and manners."[38] The Georgian order created a conspicuous social distance between gentry and commoner. The emergence of gentility brought about a heightened self-consciousness, as one was performing for the eyes of others. "Genteel society created beautiful stage sets on which people performed in public view."[39] Paca's garden was the site of one of these stage sets. The looking-glass self became an important social phenomenon. Bushman remarks, "Not only was criticism directed outward to others, but people had to watch themselves through the eyes of others."[40] The visual expression of gentility created the perceptual standard for inclusion in, and exclusion from, genteel society. The material culture of gentility "actually had the comforting effect of reinforcing the established social order."[41]

Gentility as an aristocratic form of life is obviously incongruous with American democratic ideals. "Gentility was worldly not godly, it was hierarchical not egalitarian, and it favored leisure and consumption over work and thrift. These values ran at cross-purposes with religion, republicanism, and the work ethic, powerful complexes of values subscribed to by the same people who wanted to become genteel."[42] Although the values of gentility are incongruous with the values of capitalism – self-denial, work ethic, maximized production versus genteel leisure and aesthetic refinement – the gentry helped usher in the modern economy, as the first consumer class.

The gentry used their social power, education and wealth, to maintain their social status politically. Rhys Isaac, in *The Transformation of Virginia 1740–1790*, discusses two phases of the landed gentry's control of the region: They first established a tightly controlled social hierarchy in order to isolate their position of power. When this position began to erode, the social hierarchy was expressed in the Georgian order, which was inscribed in architecture, behavior, and other forms of material culture.

> The new style of building conveyed a whole set of social values and assumptions through attention to mathematical proportion and through the invariable usage of a three-part design. A strong sense of gradations of dominance and submission was expressed in the elevation of a central unit by means of balanced, subordinated lateral elements.[43]

Paca adopted this Georgian-style architecture which symbolized a political form. "The prominence of an elevated center, or 'head,' to which all other parts, or 'members,' were subordinate silently reinforced the dignity and claims to obedience of the gentleman who was styled 'the head' of the household."[44]

Material culture involves a dialectic through which behavior is modified by symbolism and symbolism is intensified by behavior. Thus, the symbolism is lived, in the sense that it is acted out. The acted-out symbolism constitutes a prescription for a political establishment of power.

> The construction of the great houses was part of the consolidation of this gentry dominance – a process that by the fourth decade of the eighteenth century in turn inaugurated a stable political authority in Virginia to a degree that was exceptional among the British colonies in America.[45]

Maryland, however, outdid Virginia in the drafting of its state constitution. Ronald Hoffman notes of Maryland that "in some respects the very ideology of self-determination that they had employed to justify their opposition to Great Britain now worked against them. The ideas that the people should govern had been widely expressed. . . . Chase, Carroll, Johnson, Paca, and the party's other leaders had not meant for the message to be taken literally."[46] Thus, "government by the people" can be interpreted as an ideological ploy that masked the elite's self-interest. "Of all the state constitutions written during the period, Maryland's was by far the most conservative in character.

The document was designed specifically to insure the continuance of elitist rule."[47] However, the elite also enacted fiscal legislation by which the aristocrats bore the burden of the revolution. It can be argued that the elite bought their power with that political maneuver.

Lewis Binford examines cultural artifacts in terms of three functions: technomic function (strictly utilitarian), sociotechnic (things used not merely for need but rather to enhance the sociality of occasions), and ideotechnic function (artifacts that form ideological contexts).[48] Ideology serves to mask the real conditions of existence, and instead expresses one's axiological ideals or imagined relation to the real conditions. The ideotechnic function of artifacts is the expression of ideology through material culture.

> By the mid-eighteenth century an important series of changes had begun and was still continuing. First, distinct gentry families emerged and came to be more and more set apart by an increasingly refined way of life. The refinement was expressed in architecture, since the great houses had passageways with rooms opening off of them to create segregated spaces for special persons and functions. Such movement away from communality was most dramatically signaled in the custom of "dining," with its reserved space, its linen-covered table, and its fashionably styled matching set of plates, knives, forks, and other eating accoutrements. At regular times select companies gathered at this ritual center, and each member would be seated on a chair in a carefully defined social space.[49]

These new forms of material culture functioned ideologically to mask the contradictions in American society, by legitimating segmentation, control, and hierarchy in a free democratic society. The elite gentry were able to mask their politics by inscribing and sublimating their worldview in the material forms of everyday life.

Mark P. Leone shows that this ideological symbolization is taken to another level with gardens. The inscribed segmentation into space that divided life according to a functional hierarchy of isolated elements becomes naturalized in the garden setting.

> But when it was imposed on nature, as in gardens, . . . the segmentation quickly became confused with nature itself. And the system of segmentation, ordering, and even grading toward hierarchy was mistaken as itself being natural and discovered within nature. This segmentation and its confusion with nature, as in gardens and astronomical observations, had the impact of making the social world, which was similarly arrayed, appear to be unquestionable.[50]

The garden, then, took on a very important ideological role, for, more than any other element of material culture, it masked the ideology by its inscription in the natural world, by fusing with it.

> The scientific equipment, books, observations, the ancient history, the classical and Biblical allusions are not the product of the idle time of the rich, or of arcane interests, or of cultivated and rare intellect; rather these activities placed hierarchy and control in nature and history, making hierarchy and distance appear inevitable. When this logic, which is the ideology of class in Virginia and Maryland, was believed, society was reproduced intact.[51]

The new scientific practices involved regular observation of nature, which resulted in compendiums of past occurrences, which were used as precedent for action in the present and as a guide for the future. Order (a rational basis for action) was placed into nature and history. Such books as *New Principles of Gardening*, by Batty Langley, advocated precedent in the classical ideals for formal gardens, which inscribed the ideology of universal history in nature. In other words, a specific attitude toward the past became naturalized. In the worldview/material culture dialectic Leone has shown how socialization emerged and became identified as genteel. From this, political power was institutionalized and earned legitimization through an ideology of naturalization.

SECTION III: DESCRIPTION OF THE GARDEN[52]

An Objective Description: Landscape Architecture (See Figure 1)

From the first terrace that runs along the back of the Georgian mansion to the lowest area of the garden there is a slope of sixteen and one-half feet. The

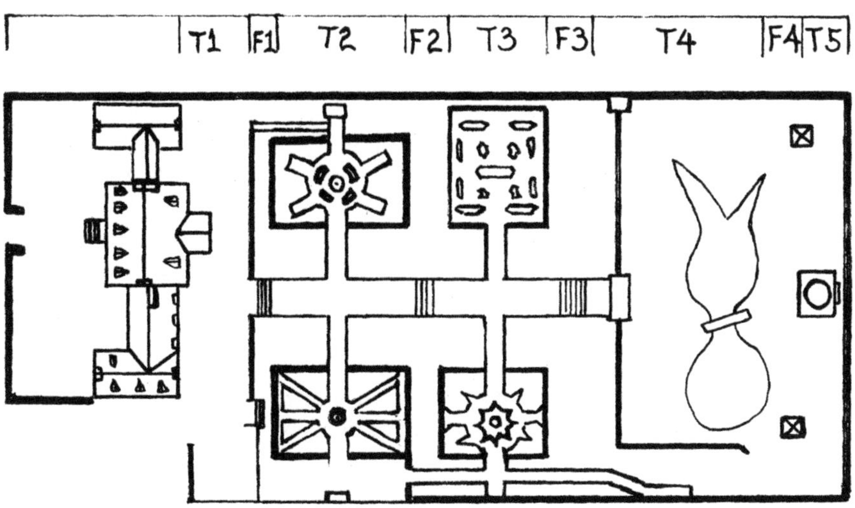

Fig. 1. William Paca Garden.

garden is fully enclosed by a brick wall, which contains thin rectangular slits to allow breezes off the bay to refresh the garden air. The garden consists of five terraces and four falls. The fourth terrace from the house, "the bottomland," is the lowest terrace. The terrace along the back wall is raised above the bottomland. A classical pavilion straddles the back terrace and the back fall, which slopes down to the bottomland. At the opposite end of the garden alongside the backside of the house is the first terrace, which is an uncovered piazza (patio) enclosed by shrubbery. This subsection really belongs as much to the house as to the garden.

The garden can be divided into two main areas. The first main area consists of three falls and two terraces and is landscaped in the geometric patterns of the formal French garden designs. Most of the bottomland, fourth fall and last terrace, is built in the contrived natural English garden style, the other main area.

A wide gravel path functions as the central axis and inscribes bilateral symmetry in the landscape. In the formal area, the central path manages each fall by a set of steps. On both the second and third terrace a perpendicular path leads to oppositely paired parterres, which establish the traditional four block symmetrical pattern. The third fall leads down to the bottomland. On the formal side of the bottomland is a lawn extending on both sides of the central path. Just on the other side of an open, stone water canal is the area of the English garden.

Standing on the first terrace patio and looking along the central axis created by the main pathway, the eye is immediately drawn to a classic-style pavilion (summerhouse) located between the back terrace and the last fall. Even though the central path ends at the water canal, the pavilion completes the central axis. The pavilion draws the eye along the vertical axis up to its domed roof and the figure of Mercury on top of the dome pointing skyward. Visual perception inscribes a triangle in the space of the garden. The eye is drawn down the central path to create one side of the triangle. The eye is drawn up vertically by the pavilion, and by resting on the statue at the top, a line of sight completes the triangle. The side enclosures of the garden participate in the visual gestalt by being gathered in by the pavilion (a landmark, which visually functions as the figure in the figure/ground structure) creating a volume. Unlike a flat stretch of land forming a volume through a hill or the tops of trees, the downward slope creates a fishbowl effect (with the pond literally in the shape of a fish). The sloping of the garden against the dramatic vertical "eye catcher" (the pavilion) makes the garden appear larger than it is. But paradoxically and ingeniously the pavilion is drawn in closer as if something distant was being looked at through a telescopic lens. This effect is facilitated by a "foreshortening" created in the dimensions of the parterres of 80 feet, 55 feet, and 40 feet respectively, which suggests a greater distance away than is objectively the case, so that the eye expects the pavilion to appear smaller than it is and actually appears.

The central path is conspicuously wide enough to allow several people to walk abreast. The light color of the path complements the rich green of the lawn in which it is laid. Much space is devoted to this path, which because of the slope and the terraces creates a hierarchical axial progression. This heightened attention given to the open-view path creates an aggrandizement of the individuals on the path, which functions like the wide open staircases in the mansions of the gentry.

Each of the parterres presents variations of symmetrical patterning about a point. Geometric patterns are formed by the juxtaposition of planting beds and paths. Three of the parterres contain an inner circle as its central thematic. The left parterre on the third terrace presents a quincunx grid. The long tube-like bed in the center is the only bed of that shape in the parterre and the perambulator circumambulates this shape, which means that fundamentally its design creates an experience similar to the others. Pathways in the parterre form geometric patterns that lead to and from the center planting and connect to a pathway that circumscribes it.

Borders of hedge enclose the parterres so that one has only two possible exit routes. The pathway into the parterre circumnavigates the center planting and continues along its axis, exiting the parterre and placing the perambulator onto a thin path that runs parallel with the brick wall enclosing the sides of the garden. Perambulation is oriented to the central planting in each of the parterres. Its thin geometrically ordered paths dictate a highly controlled locomotion. Perception is drawn to focus upon the inner horizon of the parterre. The spatial ordering, which includes carefully pruned specimens, heightens the experience of detail by pulling the contents of the garden in closer to the senses. The regular temporal rhythm of the experience is inundated with concentrated perceptual minutiae in refined spatial orderings. One spends an inordinate amount of time in a small area performing highly restrained bodily movements. One reason for the expansion of time is that each plant is individualized, demanding specific attention. The heightened sensory enrichment and delight of the mind in the geometric order coerce participation in the highly controlled setting.

Lawns and falls separate the parterres, and each parterre is organized around a theme. Spatial segmentation is by function and contents. On the second terrace to the left is the Rose Parterre, and to the right is the Flower Parterre. On the third terrace to the left is the Box Parterre, and to the right is the Holly Parterre.

Gently curving along the bottomland in the area of the English garden is the fish-shaped pond. The head is rounded, which is on your right when looking towards the back of the garden, and on the left are the two tail fins. A Chinese ornamental bridge allows the perambulator to cross over the pond at its thinnest place between the head and tail fin shapes. The plantings mimic natural asymmetrical patterning, but nevertheless have been planned for a pleasing perambulation. The springhouse is located to the left of the pavilion and off the back tip of the tail fin. Here, water is obtained and food items naturally refrigerated. On the right near the fish's head is the bathhouse. These three buildings are not symmetrically positioned and the two smaller buildings are partially hidden by plantings.

On the right side of the garden beyond the parterres is the kitchen garden. It contains three areas: the "physick garden" of medicinal plants, the herbs and vegetable garden, and the fruit garden. Fruit trees are trained along espaliers and the garden walls. Meticulous control of nature brings a very high bounty. On the left beyond the Rose Parterre is where the privies would have been, but now is the place for garden sheds.

Paca's garden is a typical garden of the Chesapeake gentry in several ways. Most importantly, it is a terraced garden with the residence placed at the highest level. It is both a pleasure and kitchen garden and is made up of formal geometric patterns in the French style, but resists French opulence and grandeur. Atypical are its brick wall, which signifies someone of very high means, and its English style area, which did not appeal to Americans who were still carving "civilization" out of the wilderness. The steps and gravel walkways are also upscale. Also, atypical for colonial gardens of the time is the elaborate engineering of the water system. The classical urns, pavilion, and statue were typical, and it was general practice not to overdo such classical motifs as found in the Italian garden.

The Symbolic Narrative Inscribed in the Garden (See Figure 2)

The central axis of the garden inscribes an ambiguous symbolism by which one set of meanings is linked to classical antiquity through the pavilion, and the other is linked to England. The simultaneous optical illusion of greater distance and the foreshortening of distance in the perception of the classical pavilion create a connection-at-a-distance, symbolizing a historical continuity

Fig. 2. The Georgian Order.

with antiquity through perspective. The universalism of the Enlightenment was becoming coupled with the idea of linear historical progress, which later reached its apex in the philosophy of Hegel. The central pathway with its axial progression ascending and culminating in the Georgian mansion, which stands for the genteel patriarch of the estate, symbolizes the link between universal ideas and historical perfectibility. The genteel society who literally walked on the pathway symbolizes the human agency of that progress. This historical return to antiquity symbolizes legitimatization through historical precedent. Autonomous political power needed a firm foundation in order not to appear arbitrary. Mercury symbolizes the relation with the cosmos for the apprehension of these universal laws in time.

Renaissance perspective, employed in the construction of the garden, creates the distancing based on partitioned observation. The first terrace, the patio connected to the house, is separated from the rest of the garden as the site of observation. This is the proper distance of the genteel from the rest of society, which is naturalized by the garden optics. The gentry observe the rest of the world (one perfected under their control) from a higher level. From their exalted position, the gentry may walk the paths while distinguishing themselves from the operations in the world. Since the gentry are the leisure class, their observation of the rationally ordered world is for their pleasure, entertainment, and consumption. The cultivation of the natural world mirrors their refined culture. The axial progression of the garden allows the gentry to view the hierarchy of levels, which from their position on the path identifies them as a special group separated from common concerns due to their socio-economic location.

Perspective also represents the position of the rational detached knower of the new modern science paradigm. To see from the standpoint of perspective is to take up an immobile position and to fix the eye on a vanishing point. This highly controlled embodiment establishes a static relation between here and there. Both become separated static positions in space. This orientation both individualizes and universalizes one's place in the world. Kant's ethics of autonomous rational agency follows from this form of embodiment. If one universalizes one's subjective maxim, or by analogy, if one stands still and fixes the eye on a vanishing point, anyone who also took up that same situation/position would do/see exactly the same thing. Renaissance perspective establishes as the paradigm a partitioning and distancing from the phenomenal world. Each standpoint would show a different perspective, but one that is universalizable. This creates a detached individualism, based on non-involvement.

The pond symbolizes the ocean that one crosses from England to the colonies. Paca would have been one of the first colonists to have an English-style garden, which merited much prestige. Since he had traveled to England, he had direct acquaintance with the new aesthetic sense. The English-style garden, then, represents an event in his life, but also the fact that the colonies were connected to, but distant from, England. Mercury symbolizes the earthly horizontal relation as well.

The English garden has a Chinese ornamental bridge over the pond. As a particular item, the style of the bridge was very fashionable and symbolic of someone of the highest refinement. What Paca also brought back from England was a heightened class-consciousness. The upper classes in England were the ones who embraced the imported ideals of the Renaissance and had the means to inscribe those ideals in their material culture. The "exotic" bridge symbolizes the conspicuity of genteel consumption. The strikingly white bridge diverts the eye toward it, away from the central axis, which shows the attention paid by the gentry to material curiosities. This was the first class of people to collect things for display – the personal and exclusive collection. As people of lesser wealth could not afford these items of refinement, they were barred from refinement. To be a person without refinement meant to be inferior. Class exclusivity was to be read from certain objects in the garden, the bridge, the classical urns, and the statue.

The parterres represent the Georgian ideal of partitioning functions and objects in a controlled setting. Life was to be organized around activities that were to be segmented and privatized. The geometric patterns inscribe a mechanized aspect into nature, supposedly showing its true reality and its perfectibility to those who understood the new mechanistic interpretation of Newtonian cosmology. But the ordered landscape meant ordered and controlled postures, which ambiguously symbolize the perfection of genteel manners and the new urban forms of work – regular, precise, efficient, diversified, and segmented forms of relations for the new city wage-earners. Not that the gentry would participate in common work, but they were the ones who had the refinement to organize it according to the mechanical laws already inscribed in the inner workings of nature.

The segregation of the parterres implies privatization and diversification of human endeavors. This precise ordering of space implies the need for coordination through universal time, the mechanization of time through clocks, which became the manner for measuring work. The individual occupied a unique place in this space/time but at the same time would be a function of the mechanical division of parts outside parts: causal reality – a Hobbesian attribute.

The collection of rare specimens and the creation of new varieties through grafting exhibited the owner's knowledge and skill at perfecting nature. This meant that he understood the inner workings of nature, and his skills translated into the demonstration of natural abilities to be in political control. The garden symbolized his natural abilities and therefore natural right to control and manipulate, like the factory or plantation owner. All relations are now externally arranged in a static ordering. No organic transformation is allowed, which means that the garden requires constant supervision and attention. So, every planting is given individualized attention. The garden is maintained the way one practices law: when human affairs are all brought under the control of law, then they are handled in this prescribed fashion.

SECTION IV: THE PHENOMENOLOGICAL EXPERIENCE OF THE LOOKING-GLASS SELF IN THE GARDEN

Paca was a clever man and this is gleaned from his understanding of the principles of optics, evidenced through his ability to manipulate perspective. He demonstrated his mastery of geometry through his ability to employ it in a geographical reorganization of the natural world. The management of a garden relies on a sound natural philosophy with special skills in botany, etc.

This was Paca's vehicle to publicly exhibit evidence of his mastery of scientific principles, his education through his sense of classical history and current events (the English-style garden), his refinement through his aesthetics, and obviously his material wealth. All of this I realized about Paca while standing on the terrace connected to his house. Paca intended that his social partners grasp this significance of him as the progenitor of the garden. If we could stand with Paca and view the garden from the first terrace, we would sense his dignity and guarded personal pride. Regardless of any contentions that I may have, I cannot help but give him his due. From my standpoint, his objectivated self-presentation is an ostentatious display. But I must recognize that in my life, there are ways that I display myself through whatever talents I objectivate. So my contentions do not concern the garden per se, even if the display of consumption is a little too self-consciously promoting. I am rather concerned about the ulterior significance of the self-presentation in the display.

More deeply, Paca gives to observation a scientistic attitude. The sense is that observation unlocks the mysteries of the world, and once these are understood according to universal laws, the world can be controlled, manipulated, and predicted. This enlightenment paradigm promotes detachment in science, and in the social sphere it promotes distance. This translates into the genteel form of social life, where the patriarch takes up a posture separate from the rest of the society that he has arranged according to these naturalized ideals. Paca presents a self that he deems civilized. The expression of civility is hierarchical, and he views social place as presenting an index of the level of humanity that has been reached. His detachment and distance lead to alienation. As a consequence, Paca would present himself as a more civilized human being, above the aboriginal and traditional communities from which his social world was distancing itself. Paca would have promoted the significance of his garden estate (self-presentation as civilized) in terms of his valuing the civilized form of life. As a mirroring of his self-presentation, this elicits my consternation.

I would see racism, inequality, injustice, and violence in the imposition of these values on Others. Paca would see the new science and its technological advancements as the progression of mankind and the perfection of nature, which would justify politics that promote this civilized advancement. He would see the non-civilized natures of others as their own self-exclusion from this perfection of humanity, and any appropriations that appear unjust as expedients necessary for progress to continue. He believed that through progress all can attain freedom, but I would see that the organization inscribed in his plan of the garden exhibits a non-freedom or a mitigated, manipulated freedom for the rest of the world. The resulting social stratifications introduced by his worldview create unnecessary forms of competition based on a questionable ideal for humanity.

When I walk on the central path, I feel the exalted position that it provides me. Paca intended for me to feel this and he expected a favorable reaction. If I am on this path, it is because I am his guest. I can only be his guest if I belong to the social elite. Even though I may choose the class of people with whom I want to associate, I feel uncomfortable with the exalted and privileged position. Just as I react to Paca's self-presentation, he would view me as a political opponent, one who favors a more egalitarian, democratic posture in the social arena, an opposing faction in elite politics. But he would probably view my reaction to him enigmatically, wondering why I would not want to take my natural place as a genteel personage. Would he view my mirroring of him as jealousy? Would he see my unfavorable posture as an inability to separate the freedoms that should be guaranteed for the private sphere from what is responsibility in the public sphere? Would he understand my unmasking of his political contradictions?

I enter a parterre and view domesticated nature in its highly manipulated, dominated, and controlled state. I experience the prescription of my own postures and movements in the geometric plan, and I am displeased. I can admire its significance only as a very limited form of life. However, the philosophical plan of the parterre has played out its dangerous implications from the standpoint of my times. The non-containment of the domination over nature has led to an exploitation of nature in a way that destroys its regenerative processes. The philosophical idea of nature as a machine is a metaphysical notion that has produced ecological disaster. The idea that human beings can penetrate nature's secrets is based on hubris. Nature is far too complex for us to master, and it is an interdependent whole. Thus, I would call into question the very basis of what Paca considers knowledge. I have the advantage of recognizing scientific paradigms, whereas he would not only believe in the mechanical model, but also in the linear progress of knowledge. Such assessments take on the character of a clash of religious beliefs, and my polemical attitude toward him would be that of a foe.

The controlled postures of the lived-body he would celebrate as refined, and the segmentation of functions and control of contents as the excellence of inscribing rationality into forms of life. I would see this segmentation as an abstraction, making life forms thing-like, especially in terms of work: not perfecting human embodiment at all, but rather making it another part in a machine, an automation. There is merit in the rational ability to conceive this modality, but it does not dignify human life, as the attempt to order life on the machine model actually destroys human life. Mechanical work in wage labor may lead to the possibility of higher consumption, but this is not a higher form of life. It is only because of Paca's social location that he has the leisure to really enjoy a richer life based on consumption – otherwise consumption of things is a substitute for a fuller human life. So the so-called civilizing of life is highly suspect. But I realize that I say this as someone who lives within and enjoys the benefits of this civilization.

By rejecting much of Paca's worldview, I mirror him in a way that rejects his self-presentation. I live in the consequences of "enlightened" forms of thinking and the influence of gentility on American life. The excitement that Paca would feel concerning what he considered economic and technological progress can only be followed by my concern for the evils those bring as well. As I am mirrored back to myself through Paca's self-presentation in the garden, I realize how much the parameters of my socio-historical situation are embedded in and generated from his ideals. At the same time I see that my ideals socially, economically, and politically belong to another continually emerging paradigm that may or may not be realized. Much of his paradigm, especially the social, remains entrenched in North American society and western culture.

CONCLUSION

There is a great difference between my feelings about the Paca garden at the time of my initial experience and now, after conducting my research. Without a meditation upon subjective history and the history of material culture, as a phenomenologist I would have described lived-body experiences in the garden. Such a phenomenology, while being extremely valuable in its own right, would have been "ahistorical" and would not have established the social relation with Paca as thematic. This subjective history involves an ongoing hermeneutic relation, a bridge to a certain past predecessor. Diachronic and synchronic difference shapes my relation with William Paca. By establishing a social relation with this predecessor, I have become aware of those differences in a lived way. I am not distanced from the predecessor through an objective set of facts; I have revitalized his being in quasi-simultaneity with my own, through which I have reconstructed his being. This involves conjecture of course, but the "problem of other minds" exists in all social relationships. With a predecessor, however, I must call into question my assumptions much more carefully. But by doing so, I become more aware of them as I proceed to see the world more and more from the eyes of someone who lived before me. Hermeneutics is a science of an ever greater deepening of understanding of relations between one's own context and that of some other entity or entities. This study has established a relation through typing William Paca as Chesapeake gentry. Further specific study of his character is needed in order to create a more personal relation. Hermeneutic inquiry is always an open project. The proper excellence of subjective history is that it establishes an existential relation with the past. The phenomenon of the looking-glass self establishes the modality of evaluation in subjective history. I experience a social relation with the past personage, because I make choices for the future through an evaluative assessment of my attitude toward their self-project. Otherwise, objective knowledge of history is my project, which does not entail a social relationship with my predecessors. Social phenomenology provides the methodological and epistemological basis for the establishment of this relation.

Morgan State University

NOTES

1 The solution to this problematic in social phenomenology involves us in what Schutz calls "subjective meaning in history." Even though this study is long for an article, it only presents a caricature of the kind of rigor necessary to accomplish this task in the most scientifically responsible manner. Nevertheless, the treatise does provide sufficient examples and schematics for this type of inquiry. I know of no rigorous examples of such a phenomenological history. See Alfred Schutz, *The Phenomenology of the Social World*, trans. George Walsh and Frederick Lehnert (Evanston: Northwestern University Press, 1967), 207–214.

2 See Charles Horton Cooley, "Chapter V. The Social Self – 1. The Meaning of 'I,'" in *Human Nature and the Social Order* (New Brunswick: Transatlantic Publishers, 1992), 168–210.

3 Georg Simmel, *The Philosophy of Money*, trans. Tom Bottomore and David Frisby (London: Routledge, 1990), p. 306.

4 Simmel, op. cit., p. 307.

5 Barbara Paca-Steele.

6 Simmel, op. cit., p. 321.

7 Simmel, op. cit., p. 323.

8 Simmel, op. cit., p. 324.

9 Simmel, op. cit., p. 322.

10 Schutz, op. cit., p. 133.

11 Cooley, op. cit., pp. 183–184.

12 Alfred Schutz and Thomas Luckmann, *The Structures of the Life-World*, trans. Richard M. Zaner and H. Tristram Engelhardt, Jr. (Evanston: Northwestern University Press, 1973), p. 67.

13 Schutz, op. cit., p. 208.

[14] Schutz, op. cit., p. 208.
[15] Schutz, op. cit., p. 209.
[16] I would like to thank historian Lawrence Peskin for reading a draft of my paper and for his research advice concerning Revolutionary and Maryland history. I would like to thank historian Brett Berliner for his help in the area of historical archaeology.
[17] Gregory A. Stiverson and Phebe R. Jacobsen, *William Paca: A Biography* (Baltimore: Maryland Historical Society, 1976), p. 36.
[18] Stiverson and Jacobsen, op. cit., p. 42.
[19] Stiverson and Jacobsen, op. cit., pp. 43–44.
[20] Barbara Wells Sarudy, *Gardens and Gardening in the Chesapeake 1700–1805* (Baltimore: Johns Hopkins University Press, 1998), p. 20.
[21] Sarudy, op. cit., p. 22.
[22] Stiverson and Jacobsen, op. cit., p. 48.
[23] Stiverson and Jacobsen, op. cit. p. 53.
[24] Sarudy, op. cit., pp. 103–104.
[25] Stiverson and Jacobsen, op. cit. p. 62.
[26] Stiverson and Jacobsen, op. cit., p. 62.
[27] Stiverson and Jacobsen, op. cit., p. 63.
[28] Paca to Johnson, June 11, 1778, Gilmor Papers, vol. 3, fol. 75, MS 387.1, Maryland Historical Society.
[29] Stiverson and Jacobsen, op. cit., p. 91.
[30] Stiverson and Jacobsen, op. cit., p. 93.
[31] Stiverson and Jacobsen, op. cit., p. 92.
[32] Mark P. Leone, "Interpreting Ideology in Historical Archaeology: Using the Rules of Perspective in the William Paca Garden in Annapolis, Maryland," in *Ideology, Power, and Prehistory*, ed. Daniel Miller and Christopher Tiller (Cambridge University Press, 1984), p. 33.
[33] Joyce Appleby, *Capitalism and a New Social Order* (New York: New York University Press, 1984), p. 12.
[34] Appleby, op. cit., p. 18.
[35] James Deetz, *In Small Things Forgotten: An Archaeology of Early American Life* (New York: Anchor Books, 1977), pp. 62–64.
[36] Richard L. Bushman, *The Refinement of America* (New York: Vintage Books, 1993), p. xii.
[37] Bushman, op. cit., p. xii.
[38] Bushman, op. cit., p. xii.
[39] Bushman, op. cit., p. xiv.
[40] Bushman, op. cit., p. xiv.
[41] Bushman, op. cit., p. xv.
[42] Bushman, op. cit., pp. xvi–xvii.
[43] Rhys Isaac, *The Transformation of Virginia 1740–1790* (Chapel Hill: University of North Carolina Press, 1982), pp. 37–38.
[44] Isaac, op. cit., pp. 38–39.
[45] Isaac, op. cit., p. 39.
[46] Ronald Hoffman, *A Spirit of Dissension* (Baltimore: Johns Hopkins University Press, 1973), p. 169.
[47] Hoffman, op. cit., p. 269.
[48] Lewis Binford, "Archaeology as Anthropology," *American Antiquity*, Vol. 28, no. 2 (October 1962), pp. 217–226.

[49] Isaac, op. cit., pp. 73–74.
[50] Leone, op. cit., p. 26.
[51] Leone, op. cit., p. 27.
[52] I would like to thank head gardener Elizabeth Brewster and her staff at the William Paca Garden for allowing me to study the garden archives and to spend time on the garden grounds.

MALGORZATA ZURAKOWSKA

THE FOURTH DIMENSION OF ART

> To see a World in a Grain of Sand
> And a Heaven in a Wild Flower,
> Hold Infinity in the palm of your hand
> And Eternity in an hour.
>
> (William Blake, "Auguries of Innocence")

One day, a very religious painter was about to start painting God. Overwhelmed by the grandeur of the subject matter, and in order to show his full respect to the Lord, he decided to paint on his knees. He set up the easel to the appropriate height, laid down all of his brushes, palette and colors, and after a short prayer, was ready to start. Suddenly he heard a voice coming from above, telling him, "Don't paint me on your knees, you idiot, but rather paint me WELL."

This anecdote sets the stage for the topic of my inquiry: "art as numinous experience."[1] I will try to address the question of what art is for and what it does, through the prism of new paradigms in science, from quantum mechanics on the one hand to the relativity of time on the other. The former reintroduced the human mind to the mystery of uncertainty. The latter turned the static, spatial concept of the Universe upside down. The paradigm shift from three-dimensional to four-dimensional reality acknowledges time as the key element in the cosmic drama of life. In Arthur Eddington's words: "The quest of the absolute leads into a four-dimensional world."[2] Transposing his point, the quest for the fourth dimension leads to the absolute.

Every day, every minute, every second, we experience the miracle of life, both the marvel of our own life and the revelation of the existence of the Universe. We are immersed in cosmic energy and suspended in Being and Time. Our journey seems to have a direction, an arrow pointing towards some spiritual purpose, which we grasp in flashes of intuition, but whose sense we lose in everyday life. It is not often that we consciously ask ourselves, what is the very purpose of our existence within this cosmic creative clock? Confusion arises at the very beginning of our quest to unveil the mystery of time.

At the beginning of the third millennium, a unified vision of time is, more than ever, beyond our immediate grasp. Showered by contradictory scientific theories, expressed in the highly specialized and cryptic language of mathematics, physics, cosmology and molecular biology, we pick here and

A.-T. Tymieniecka (ed.), Analecta Husserliana LXXVIII, 219–226.

there through superficial, pseudo-scientific jargon, totally losing the perspective of the holistic, core sense of time as a vehicle for life and creation.

The human experience of time has always been one of the biggest sources of mystery. In archaic societies, the idea of time was primarily associated with cycles. The solstices, tides, seasons, and movements of the heavenly bodies made archaic humans believe that everything is subject to eternal renewal in the mega-cycles of cosmic re-creation. The idea of circular time was also common in Greek cosmology. It was Judeo-Christian tradition that first broke away from the old idea of eternal repetition, setting up the foundation for linear, irreversible time as a crucial feature of Western culture. That shift, from a circular to a linear paradigm of time, had enormous philosophical and sociological implications. The notion that linear time guarantees infinite progress, with its promise of improvement of life on Earth, seems to reign in Western culture and thought.

My thesis is that Western culture should open up to a more complex idea of time, and integrate it. Otherwise, the Western culture of Marshall McLuhan's "global village" that is fixated on a linear model of progress, is like a train that passes by more and more new exotic stations, but eventually goes nowhere.

In order to avoid getting completely lost in the different facets of time, let's have a brief look at two major opposite philosophical streams which originated in Greece. One regarded time as intrinsic and fundamental to the Universe, while the other made a distinction between Being and Becoming, two concepts which reappear in various aspects in modern scientific theories. Being is a fundamental world, eternally the same, while Becoming (the realm of time) is subject to opinion, coming to be and ceasing to be, but never fully real.

Mathematical time is indifferent to the fact that we are born to age, and eventually to die. The same process reversed from the deathbed to the mother's womb would be mathematically sound. And as completely counter-intuitive as it is, Isaac Newton's equation of motion describes a perfectly reversible world. Newton's time was considered absolute, like space. All events were regarded as having a distinct and definite position in space, and as occurring at a particular moment in time. Albert Einstein disproved the Newtonian idea of the absoluteness of time. Einstein's theory does not introduce a direction of time, thus theoretically allowing for a mirror-world to exist.

As much as we are intuitively aware of the arrow of time, so is thermodynamics, with its Second Law, which states that the passage of time is connected with increasing entropy. On the cosmological level, this means that the process initiated at the birth of the Universe during the Big Bang will stop when there is no more energy left for exchange. Cosmologists call this scenario "Heat Death of the Universe." Another equally possible scenario is that the Universe will expand until ubiquitous gravitational forces eventually halt the expansion and draw together all matter ineluctably into one huge final singularity mirroring the Big Bang: the Big Crunch. It is legitimate to speculate, as some cosmologists do, that what follows the Big Crunch is the Big Bang, and the whole process starts all over again.

* * *

"In any attempt to bridge the domains of experience belonging to the spiritual and physical sides of our nature, time occupies the key position."[3] With these words, astrophysicist Arthur Eddington perfectly states one of the main aspects of this paper, that is, the role of time in art.

Art represents union between matter and form. They are interrelated. One cannot think of art without matter – which seems to exist objectively – like without canvases and paints in the case of painting, or without sound in the case of music. On the other hand, color or sound do not constitute art by themselves. In order to lift them up to the level of art, one must bring them from the chaotic natural state to deliberate order. Hence art cannot exist in a timeless, absolute and uninhabited universe. It has to happen, it has to be called into existence. The artist is a crucial element in this process of "becoming" of Art, as a "choreographer" that masters time to amalgamate matter and form. Furthermore, the triangle of artist-art-work-of-art exists only in time, which one needs in order to "have an experience" (to use John Dewey's terminology). For an artist, the quest to achieve the power of a magician who transfigures what is common and trivial into an authentic work of art, is a long one. It takes disciplined persistence, self-consciousness, self-criticism, and a lot of hard work. This never ends, as the ultimate goal of an authentic artist is to keep creating ever-new organized realities.

If one is not careful, one can become like someone who examines a mirror in all of its material aspects – how the frame is constructed, chemical elements of crystal glass and silvering underneath, optical rules of reflection and refraction. This is what postmodernism's idea of art is. However, in this model one misses the extraordinary potential of a mirror: the opportunity to perceive infinity, which reveals itself when two mirrors face each other. The former allows nothing but material truth, while the latter offers a glimpse into other dimensions. By extrapolating this metaphor into my inquiry, only if we evoke new dimensions of art can we experience the epiphany of its numinous nature. In the context of my reflections on art I use the word "numinous" in its broadest spiritual sense, departing from Rudolf Otto's religious association.

In this process of unveiling new dimensions of Art, time is the crucial factor. One can get caught up in the race of efficiency and immediate gratification, allowing the passage of time to dominate. As a result, art, copying corporate culture, is reduced to objects produced in order to be possessed. The whole focus is on the material side of existence, cutting off what is eternal and spiritual, and as such, immeasurable. The time we spend to produce art only for the sake of its material qualities, I will call horizontal, and it is not within the scope of this paper.

We need what I call "vertical time" to connect with our centers. Time could be the most tragic experience of human life. The passage of time, and our awareness of its fleeting moments, enslave us in helplessness. Yet the very ability to consciously experience vertical time can also offer us freedom. The human quest towards eternity can be fulfilled by spiritual insights conveyed through symbols in art. In modern Western society, the term "spirituality" often has suspicious connotations, which are rooted in the highly rational tradition of Western philosophy, with verifiability being the only concern. The Newtonian paradigm reinforced this narrow perspective, which eventually excluded mystery and banned mysticism, as they did not fit into scientific methodology. The world was assumed to be measurable, defined, and predictable. Insufficient, perfect tools were the only limitations to the success of unveiling the remaining secrets of the Universe. Werner Heinsenberg, with his theory of uncertainty, once and for all spoiled that perfect illusion. Setting the foundation for quantum mechanics, he opened the door to scientific recognition of unpredictability. The world of solid material objects that we believe reality consists of, dematerializes under certain circumstances and becomes a web of possibilities. But what does this have to do with the numinous nature of art? What I am suggesting here is that the mystery of unpredictability and the prerequisite of consciousness, according to the quantum mechanics model, are crucial aspects of the creative process. Extrapolating science on art will not help us in understanding particulars, but can get us closer to an intuitive grasp of the Universals.

In the act of creation, full control of an artist over the means is a necessary condition but is not sufficient, as there is never only one perfect solution at any given moment in the process. I will turn for a moment to my own artistic experience. Years ago, I deliberately chose to express myself through mezzotint, considered to be the most laborious medium in plastic art. This intaglio technique, invented in 1640, is today almost unknown, and very rarely practiced. As much as it is appreciated for its visual quality, at the same time it is dreaded for its labor-intensive process. First, one has to spend hundreds of hours preparing the plate, to set the ground for the drawing. Then follow endless hours as the image is forced from a profound darkness by gradually bringing it to life, by adding more and more light. This is not a technique that offers fast gratification. This very slow process allows careful evaluation of every step. Still, no matter how this painstaking, slow process is mastered, every decision is some sort of "miracle of becoming." An artist is always facing myriad possibilities and confronting the ineffable. In Herman Hesse's words: "Within this fixed structure, or to abide by our image, within the complicated mechanism of this giant organ, a whole universe of possibilities and combinations is available to the individual player."[4] At every stage of the "becoming" of the work of art, the artist puts on hold one reality in order to evoke a new one. Furthermore, every decision opens a new path, which sets the conditions for further development. The process always involves an infinite web of possibilities. Every work of art is born as a result of a mysterious, impossible to explain continuum of choosing one reality over another potentially equally valid one, but lost forever. Both the agony and the ecstasy of creation come from uncertainty and unpredictability, and if pursued in the vertical time of human consciousness, a door is opened to transcendence.

In vertical time, using our subjective experience we can slow down, to connect with our inner selves and through our centers connect with the eternal universe. Thus in doing so, through art, we make a cultural shift from the existing dominant paradigm of fragmentation, to the underlying paradigm of the "whole."

In the emerging holistic model which modern science is leaning towards, the idea of time provides a link between science, theology, philosophy and art. This unified holistic vision leads to a new paradigm of approximations, which is replacing the previous paradigm of the absolute truth, born of the concept of absolute time.

Human existence, if devoid of transcendence, is primarily a horizontal struggle against the flow of time towards an unforgiving descent into decay and death. In such an existence there is no hope, and a deterministic perception of time takes away free will. Yet we have the human ability to transcend our mundane existence into the sphere of eternal values. To find the path of inner freedom, one has to open up to the ability to experience vertical time, and the universe as a web of possibilities.

* * *

On a starry night, when we look up to the glittering points covering the sky from horizon to horizon, we are not calculating the millions of light years separating us from the source of light, nor do we speculate on the nature of this light, corpuscular versus waves. What really overwhelms us is the mystery of its heavenly beauty. And this experience opens us up to eternity, which is beyond the comprehension of our rational mind and can be expressed only through symbols in the act of creation.

The different schools of twentieth-century science couldn't agree on many aspects, but there was a general agreement about one crucial point – that the human mind cannot intelligibly comprehend objective cosmic order. In other words, the evolution of twentieth-century knowledge pushed science from the position of the triumphant conqueror of the universe, to the much more modest position of wanderer.

Thus we are living in a schizophrenic reality. On the one hand, we live in a culture of a superficial fascination with progress, encoded in a linear perception of time, and we are culturally conditioned to naively believe that there is no limit to human knowledge and the materialization of the Universe. As a result, what really matters in art is either an externally existing final product, or, as in conceptual art, a lack thereof altogether. In the name of progress, the latter extreme divorces art from matter, and negates the temporal aspect of the amalgamation of matter and form. While the former, fixated on the same progress, focused on a maximally speeded-up process of art object production, totally ignores its spiritual dimension. But what are we speeding towards? Deprived of the ability to plunge into vertical time and creation, to breathe in and transcend all aspects of the human temporal condition, we are speeding to inescapable death. Material progress is becoming our only ticket to immortality, when the value of spiritual energy intrinsic in the vertical temporal creative process is not only unappreciated, but also wasted. Art becomes regarded mostly as a commodity, a toy for spiritually blind human beings. Or even worse, art is regarded as material objects to tranquilize anxiety caused by a terror of the flux of time and the pressure of an-ever increasing race to efficiency, as the only way to cheat death.

On the other hand, science, struggling with singularities and complex and often contradictory theories, disillusioned about linear conquering progress, is closer and closer to the philosophical ideas of the past, respectfully reconsidering the concepts of Being and Becoming.

The paradox of Western civilization in its present state seems to lie in roles that are reversed. Art that is supposed to be avant-garde for human consciousness, still dwells on nineteenth-century ideas of progress, and by doing so is conservative and academic, frozen by a fixation on change. The whole notion of postmodernism is based on undermining the past and limiting its focus on defining the present. It has become science's role to play devil's advocate by questioning its own boundaries. By going backwards, from the position of know-how to the position of doubt, science is really going forward as an avant-garde. Physics, the Queen of Hard Science, is rediscovering a holistic idea of the Universe, while trendy art is more and more preoccupied with detail. Where art dwells on self-examination with linguistic scrutiny, science is leaning towards ontological questions of uncertainty. While art still strives for absolute truth, science embraces mystery.

Western civilization and culture, with their conquering and controlling attitude, have divorced themselves from Nature. Despite ecological disasters, our minds are still geared to operate under the illusion of a Promethean better future. As a reflection of this cultural fixation, "newness" is considered the highest value in art, with complete disregard for the eternal and universal. However, new emerging paradigms of human creativity are calling for a reconciliation of the linear and cyclical time frames. Both temporal models, integrated with each other, offer a shift in cultural thinking towards ecology, empathy and spirituality. Eternal values re-experienced again and again in our limited human lives are the only chance to open a new cultural vision rebonded with the Universe. The Holistic concept, lived in vertical, specifically human time perception, offers hope to our suicidal culture misled by the mirage of limitless progress.

Max Planck once wrote: "Science . . . means unresting endeavor and continually progressing development toward an aim which the poetic intuition may apprehend, but which the intellect can never fully grasp."[5] And in Albert Einstein's words: "The most beautiful experience we can have is the mysterious. It is the fundamental emotion which stands at the cradle of true art and true science."[6] In the frontiers of science and in spiritual art, reality exists as a reflection in the mirror of consciousness. When science strives to reach its limits, the evident parallel with art begins; in both, we are confronting the ultimate mystery. Art as numinous experience reveals itself when we free ourselves from the straitjacket of past attitudes and open up to an unorthodox paradigm of vertical time, and humbly accept mystery and uncertainty as the immanent element of a fourth dimension of Art.

The remaining question is, if the fourth dimension of Art leads us to the absolute, where would we take the quest of Art's hyperspace? I leave this fascinating question for my future inquiry.

Royal Canadian Academy of Arts
Toronto, Canada

NOTES

[1] "Art as numinous experience" is my polemic paraphrase of the title "Art as Experience," by John Dewey. (The Berkley Publishing Group, 375 Hudson St., New York, NY 10019, 1980).

[2] Arthur Eddington, *The Nature of the Physical World* (Cambridge: Cambridge University Press, 1928).

[3] Ibid., p. 91.

[4] Herman Hesse, *Glass Beads Game* (Henry Holt and Company, Inc. 1969), p. 15.

[5] Max Planck, *The Philosophy of Physics* (New York: Norton, 1936), p. 83.

[6] Quoted from S. M. Ulam, *Adventures of a Mathematician* (New York: Charles Scribner's Sons, 1976), p. 289.

RANDALL RHODES

THE RUIN AESTHETIC: CONSTRUCTING THE DECONSTRUCTIVE LANDSCAPE

A DIDACTIC POEM

A ruin's time visible, tangible,
A memory stored within the cosmos.
A ruin's the disjunctive product of
The intrusion of nature upon art.
A ruin's a shattered masonry pile
Pointing to an infinite regression,
Obsessive regression into the soul.
A ruin's attended by dread and death,
Inducing righteous and perverse pleasures,
Irresistible decay. Diderot's
"La rêverie et la mélancholie."

Brooding o'er time-shook arches, stone defac'd,
Eighteenth century British dilettantes,
Patrons, Painters, and landscape architects
Debated o'er the (re)presentation
Of the actual and the fictional
Adorning the prospects in garden parks.
These broken texts of human history
Diasparactive tropes of incompleteness
Much more than antiquarianism,
More than English erudite researches
But, rhapsodies o'er splendrous artifacts
And sanctuary from their worthless age,
Poetic, wistful, backward gazing dreams.
View'd, colored through a Claudian mirror
Aestheticized via the picturesque,
Glamourized by the beautiful, sublime,
Jéjeune Edens in temporal arrest.
A romanticized presence of the past,
Much more than a vulgar heap of stones, for
What sinks from life, strikes as a site for life.

A.-T. Tymieniecka (ed.), Analecta Husserliana LXXVIII, 227–239.

Fig. 1. Ruins at Ephesus, Turkey. Photo: Author.

In Gibbon's, 1787,
Decline and Fall of the Roman Empire,
Pope Eugenius and his entourage
Ascended the Capitoline and view'd
The citadel of the earth, savaged, spoil'd.
How has it fallen! how changed! how defaced!
Here once was luxury, culture, drama,
Gladiatorial sport and commerce.
The marble theater's vacant, shapeless.
Fashion ill supplied the sacred place, and
Beauty scorn'd to dwell where use is exiled.
These marble architectural glories
Lie prostrate, naked, broken like a corpse.
The walls veiled in green, their edges effaced,
Framed by the organic grip of nature
Victory obliterated by vines.
Heritage is doomed, fortune has spar'd none.
Monumental flattery mix'd with dust.
Desolate, where Ruins now dreary dwell.

A ruin is a broken, faded text,
Slowly drifting toward anonymity.
A ruin's suspended between a past
Long lost and a future not yet possessed.
A ruin lies at a marginal site
At the crossroads of the pluralities
And oppositions of the "Heres" and "Nows."
Inherently unstable, covertly
Contradictory, the ruin's caught in
A poetic authorship, shifting webs
Ruled by the economy of the sign.
A ruin's a deconstructive construct,
A narrativized unraveled fabric
Without conclusion or resolution.
A ruin's an eidetic vision,
An inward and ideal discourse of the
World, and the void of the divine author.

We ask: Where now art thou? Here still am I!
The secret justice of loss, destruction

Righteous, perverse pleasures in surviving
The hoary, irresistible decay.
The subject's masochistic joy, rising
On viewing the righteous retribution
On antiquity's festivals of vice.
These pagan and vainglorious cultures
Left to progressive decay, negation
Ruined residues, bodies-in-pieces.
The subject's own interiority
Mirrored in this self-disintegration
Doomed to remain incomplete and floating,
Removed in time and space from his *eidos*
He ardently reaches forward his grasp
Seeking the absolute in this fragment
The primacy of the diasparact,
Beaudelaire's "l'infini dans le fini."

The subject senses incurable fault,
Irrevocable loss, nihilistic,
Self-defeating. I too will be so?
Is it permitted for man to look up
And pose the ontological questions?
Where is man to find his measure? In stars?
Or, in the ruins his being-toward-death?
Tis all in pieces, all coherence gone.
The philosophy of despair, Goodman,
Hölderlin, Freud's death instinct. As Donne wrote:
For "everything is a remembrancer;
For everything is a judge upon man.
O, lest the world should glorifie its selfe,
With an opinion of eternity.
We admit this observation as true:
There's a reproofe, a rebuke, an Ecce,
Mortality, a sensible decay,
Framing every living thing in the world."
For history, with all her volumes vast,
Insolvent, audacious, hath but one page.

A ruin's a-being-as-having-been,
Drawn into the present, while suffering

An erasure to absolute zero,
Decay, loss, the Absolute Flux of Time.
A ruin's a masterpiece, forever
Two-faced, as it looks back to its master,
As it looks ahead to its mastery.
A ruin is but time easing itself
Of endurance, progressive corruption.
Albeit, as Milton had once observed:
"Saturn's no slower than he used to be."
The subversion overcomes the object
While also dispossessing the subject.

Perhaps, the deceitful pomp of the world
Subordination of faith to reason,
Man's fall from higher and better estates
Have caused the attendant Divine rebuke.
Nations wandering in a labyrinth
Of errors, religious despotism,
The soul's progressively evil nature.
Volney cried: "O tombs! what virtues are yours!
You appal the tyrant's heart, and poison
With secret alarm his impious joys.
He cowardly flees your hoary aspect
And there erects his throne of insolence.
Tis man's discord, rapine, cupidity,
Ignorance, such are the evil genii.
For how perceptual are the errors
Which drive man's restless mind to exhaustion.
These are the celestial anathemas
Converting the splendor of a city
Into a solitude of waste, of death,
A solitude of mourning and ruins.
But as in the bosom of man have sprung
The evils which have afflicted his life,
There he also is to seek remedies,
To retrieve a true life-significance,
To reconstruct his own shattered ruins.

Endeavoring to locate the deeper
Self, the truth of these life situations,

He stretches his own sensibilities,
Activating vital forces within.
To reintegrate, relate, and retrieve
Perceptions that appropriate and grasp
The object's essence beneath the fragment.
Colored by the affective aesthetic
Response, the object *in concreto*
Becomes then, the object in cognition.
The interplay between Nature and man
Nostalgia for the natural object
Nostalgia for the object's origin
An accumulation of memories
An archaeological museum
Of fragments from the past, symbols, mythoi.
Sense perceptions of nature are no more
For they had only served to trigger the
Contemplative mode, the authentic self.
An epiphany of the creative
Psyche: "Etre with a capital E."
The catalyst excites, strikes the subject.
Then, a desire for satiation and
Submission overwhelms, the mind wanders,
Leading to an absorption, a final
Disorientation with the real world.

This eidetic, emotive, aesthetic,
And intentional revelation serves
As a human-self-life-world expansion.
The mind rises, swells to lofty transport.
The nervous fibers stretch beyond normal
Tone, suspending the motions of the soul.
Volney's poetic, profound reverie:
"Absorbed in meditation, a crowd of
New reflections paired in upon my mind.
For everything bewilders my judgment,
Trouble and uncertainty fill my heart."
As a spectator of the universe,
The aura of the beautiful, the great,
The sublime, and the divine stand supreme,

Planting in the soul an invincible
And eternal love for what we've been born.

Thus, the sublime is a discursive text
One animated by a turn of phrase
A grandeur and nobility of thought
With power to elevate and ravish.
Yet the physical basis becomes more
Forceful when the subject is conversant
With the terrible and obscurity,
Privations, vastness, and the infinite.
The eye is in constant irritation.
The mind reels from rapid alternations
'Tween forces of repulsion, attraction.
Stretching and shrinking, the rubberized soul
Passes beyond the bounds which limit it.
Claiming for itself the dignity and
Importance of the things it contemplates.

Beyond the romantic awakening,
The essential excellence celebrates
The emblematic, alternative trains
Of subjectivity, delivers an
Aesthetic via sunshine and verdure
Affecting the beautiful, picturesque.
Speculum and speculation converge
Like Narcissus gazing, meditating
At the transitive inverted landscape.
The imagination is colored by
Picturesque prospects like pictures installed
In the empty cabinet of the mind.
A reflection of the processes by
Which the world-picture is perceived, explored.
Rosa's *banditti* and ruins.
Reni's gentle tints and variations.
Numberless and untouched varieties,
Insensible transitions, flowing lines,
The terror and awe of deformity,
Within the shadow of antiquity.

Tinting the landscape *au choix*, as Knight wrote:
"How best to bid the verdant Landscape rise,
To please the fancy, and delight the eyes;
Its various parts in harmony to join
With art clandestine, and conceal'd design;
T'adorn, arrange; – to sep 'rate, and select
With secret skill, and counterfeit neglect;
I sing. – Do thou, O Prince, the song attend;
Instruct the poet, and assist the friend:
Teach him plain truth in numbers to express,
And shew its charms through fiction's flow'ry dress."

Landscape architects unite the powers
Of the painter of picturesque prospects
With those of the practical gardener.
Exploiting sublime sensual delights,
A new garden ought to be designed, not
In imitation of an old garden,
But of an old one in fictional ruin.
Partial incrustations, mosses, gnarl'd trees,
Rugged lanes, uncultivated thickets,
Weather stains, frame this image of decay.
Ruins, actual and artificial,
Appear in the prospect as mirages
Mellowed, softened by the onslaught of plants.
With creepers fastened, root'd 'tween the joints.
In George Crabbe's poem 'bout the *Borrough*,
Nature and time have worked congenial
Over the course: "till in unnotic'd Years
The stoney Tower as grey with age appears;
With coats of Vegetation, thinly spread,
Coat above coat, the Living on the Dead:
These then dissolve to dust, and make a way
For bolder Foliage, nurs'd by their decay:
The long-enduring Ferns in time will all
Die and depose their dust upon the Wall;
Where the wing'd Seed may rest, till many a Flower
Shew Flora's triumph o'er the falling Tower."

All dressed and cultivated, a ruin's
A representation (re)presented.

A conflict 'tween the *genius loci*
And the feigned, mythic literary past,
A frisson with the contemporary.
Translating old Homer into English,
Filtered through modern sensibilities
Deceptive associations of time.
Here, a fragmented cultural other
Labeled solely for modern consumption,
Freed from all fact patterns and coherence.

At a bend in a deep river valley,
Yorkshire, lay the ruins of Rievaulx,
A thirteenth century ruined abbey
In a state of progressive decay since
Its abandonment by the Cistercians.
A victim of rapine and robbery,
Hallowed, haunted, the dispossess'd gives grief.
A skeletal frame sold to Sir Duncombe,
Whose nephew saw its capabilities
As ornament for his landscape estate.
An ecclesiastical improvement,
An addition of a Christian note
Among the classical and natural.
Yet awakened from its exquisite death,
It suffered a cleansing, 'twas sanitized,
The signs of temporality removed,
The shattered, disjointed pieces, the moss,
And vegetation, deformed for tourists.

Visitors peered through the presbytery
Flanked by the two skeletal buttresses.
Gazed at the arches, played at time travel.
For this writing of the site sacrified
Beauty for the archaeological,
Ideas for a preference for form.
Lost was the transcendental signified,
Lost was Aram's romantic Abbey as
Quoted within Gent's *History of Hull*:
"Where Prayers were read, and pious Anthems sung,
Now Heaps of Rubbish the Apartment throng,
Up roofless Walls the clasping Ivy creeps,

Where many a Bird of Prey in Safety sleeps:
Or finds in dreary Caves a kind Retreat,
And broods on Rapine in her gloomy Seat, . . .
Here some vast Tow'r, or broken Arch behold,
Tenaceous Ivy with long Wreaths unfold:
Long Isles lie cover'd with spread Ruins o're
The Hollow Winds thro' empty Caverns roar:
The Beasts obscene to gloomy Vaults retire,
And Trees for Praises fill the Sacred Choir."

Constructed ruins were more so a hackneyed
Device, mannered, studied and forced onto
Situations where they may not belong,
To distinguish, to break, or adorn scenes,
To indulge the enthusiastic fit.
Fabricated follies intended as
Experimental, imaginative
Seemingly displayed as contrivances,
Serving as a foil to conventional
And functional architectural forms.
Preferred from the oblique angle of view,
The round versus square, rough versus polished.
Singular, extraordinary,
While animating neutral prospects and
Stimulating the sublimest studies.
To foster a living Arcadia,
Landscapes of Virgilian poesis,
Echoes of Hadrian's Tivoli, and
Shades of Pliny's Laurentian Villa,
A didactic discourse to be read by
Georgian *cognoscenti*,
Veterans of the Italian Grand Tour.

The unexpected, useless, whimsical
Mirrored propensities for retirement
And the obligations of public life.
Private follies were libraries, retreats.
Emblematic in program were public
Sites which forced certain associations
Upon the mind via the picturesque.

Rotundae from Croome Court to North Ireland
Triggered righteous and perverse pleasures.
At Croome: meditation, contemplation.
At County Derry, o'er the Atlantic,
In memorial of a young woman,
Masochistic joy, the inscription reads:
"Sweet it is, when on the high seas the winds
Are lashing the waters, to gaze from the
Land on another's struggles." Vainglory.
And at Painshill Park by Charles Hamilton,
Who had made two Grand Tours and sketched the sites,
There stands a faux ruined mausoleum
Housing funerary urns, artifacts,
A true sarcophagus, Roman relics,
Markers of lost historical splendor
And of spiritual dereliction.

The trains of the Gothick were more potent.
Constructed in ruin, Wimpole Tower
Evoked "the true riot of the Baron's War."
Of one hundred and forty feet in height,
Horton Tower was designed to function
As an observatory for scanning
O'er the park as well as the heavens above.
And King Alfred's Tower at Stourhead
Of one hundred and sixty feet in height,
While commanding, as well as providing,
A stimulating prospect, honored the
Lawgiver who secured political
Rights under the English constitution.
A Whig commentary on corruption.
A site of battle, Alfred and the Danes.
Temporal flux between the "Heres" and "Nows."

As professed by Thomas Whately in his
Observations on Modern Gardening,
The sight of ruins inspires reflections
On the change, decay, and desolation
Before us, around us, ahead of us,
And the memories of ole former times,

Coeval circumstances. For he wrote:
"Giving way to their effects . . . we follow
The track they have begun . . . it suffices
That the scenes of nature have a power
To affect our sensibilities for
Such is the constitution of the mind.
Passions are roused, their courses unrestrained.
Fancy is on the wing, flight unbounded.
We are led thought above thought till we rise
From the familiar to sublimest
Conceptions, rapt in the contemplation
Of whatever is great and beautiful
Attributed to nature, man, and God."

A ruin's a time-shook pile of rubble,
One large mass, an ideal of greatness
Sparking enquiries into former states,
On trophies of power, on time and life,
On the elasticity of the soul
And on the phenomenological
Ascent to a poetic consciousness
Above, beyond, the worldly rational.
For the mind, at once, must not hesitate
It must be detoured from delving into
The reality of the resemblance.
Out of this quiet and solitary
Contemplation of the deconstructed,
There arises transport to the sublime,
"la rêverie et la mélancholie."

As if loitering over his own grave
The subject confronts his being-toward-death,
Aware of his self-disintegration
As he tumbles into the unstable
Webs of signification with which his
Self-affirming world-picture is painted.
A ruin's time visible, tangible
A memory stored within the cosmos.
A ruin's the disjunctive product of
The intrusion of nature upon art.

A ruin's a shattered masonry pile
Pointing to an infinite regression.
A ruin's a sacred, noble object
Adrift 'tween the Absolute Flux of Time
And the dread silence of the subject's soul.

Frostburg State University
Frostburg, Maryland

REFERENCES

Crabbe, George, *The Borrough* (London: Murray, 1810).
Donne, John, *Donne's Sermons* (Oxford: Clarendon Press, 1959).
Gent, Thomas, *Gent's History of Hull* (Hull: Peck, 1869).
Gibbon, Edward, *The Decline and Fall of the Roman Empire* (Lausanne: 1787).
Gilpin, William, *A Dialogue upon the Gardens at Stowe* (London: 1748).
Harris, Victor, *All Coherence Gone* (Chicago: U of Chicago Press, 1949).
Hipple, Walter, *The Beautiful, the Sublime, and the Picturesque in 18th Century British Aesthetic Theory* (Carbondale: S. Illinois U. Press, 1957).
Hunt, John Dixon, "Picturesque Mirrors and the Ruins of the Past," *Art History* Vol. 4, No. 3 (Sept. 1981), 254–270.
Hussey, Christopher, *The Picturesque* (Hamden, CT: Archon, 1967).
Ingarden, Roman, "Aesthetic Experience and Aesthetic Object," *Philosophy and Phenomenological Research*, Vol. 21 (1960), 289–313.
Kaelin, Eugene, *Art and Existence: A Phenomenological Aesthetics* (Lewisburg: Bucknell U. Press, 1970).
Knight, Richard Payne, *The Landscape, a Didactic Poem* (London: Bulmer, 1795).
Macaulay, Rose, *Pleasure of Ruins*, ed. C. B. Smith (Norwich: Thames & Hudson, 1966).
Mallet, David, *The Works of David Mallet* (London: Millar & Vaillant, 1759).
Mason, William, *The English Garden: A Poem in Four Books* (York: Ward, 1783).
Milton, John, "Naturam Non Pati Senium" (1627), *The Works of John Milton* (NY: Columbia U. Press, 1931).
Monk, Samuel, *The Sublime* (Ann Arbor: U. of Michigan Press, 1960).
Mott, George, and Sally Sample-Aall, *Follies and Pleasure Pavilions* (NY: Abrams, 1989).
Price, Uvedale, *Essays on the Picturesque* (London: J. Mawman, 1810).
Ross, Stephanie, *What Gardens Mean* (Chicago: U. of Chicago Press, 1998).
Roth, Michael, *Irresistible Decay* (Los Angeles: Getty, 1997).
Taylor, Mark, *Erring: A Postmodern A/theology* (Chicago: U of Chicago Press, 1984).
Tymieniecka, Anna-Teresa, *Logos and Life* (Dordrecht: Kluwer Academic Publishers, 1988).
Volney, Constantin-François, *The Ruins: or Meditation on the Revolutions of Empires and the Law of Nature*, 1791 (NY: Truth Seeker, 1950).
Whately, Thomas, *Observations on Modern Gardening* (London: Payne, 1770).

SECTION IV

MARK MELI

JAPANESE AESTHETIC CONCEPTS AND PHENOMENOLOGICAL INQUIRY

The European discipline of aesthetics was first introduced to the Japanese reading public in 1872, in Philosopher Nishi Amane's short essay, "Bimyôgaku setsu" (Theory of Aesthetics).[1] There, Nishi presented something that can be read as a basically utilitarian view of the discipline,[2] with a discussion of the contrast between aesthetics and other branches of philosophy making up a good portion of the work's content. The first book-length translation of European aesthetic theory into Japanese was Nakae Chômin's 1883 translation of Frenchman Eugene Veron's *L'Esthétique*. As reflected in these two works, early aesthetics in Japan had a multifarious base, with works in various languages being read and presented to the public. From the 1890's, however, this began to change, and German aesthetics came to take an overwhelmingly central position. The most influential character in this change was probably Mori Ogai, well known today as one of Japan's greatest modern novelists, who spent some years in Germany studying the aesthetic theories popular at the time. Ogai was most taken with the thought of Eduard von Hartmann, and not only wrote various pieces discussing the latter's work, but also translated his *Philosophie des Shönen* with Ohmura Seigai in 1899. From that time, Hartmann's idealist thought held center stage in the Japanese aesthetic community for over a decade, until psychologistic, empathy-based aesthetics, primarily as elucidated by Theodor Lipps, became popular enough that we might say they began to take over that position. Abe Jirô's influential *Aesthetics* of 1917 was, according to the author, merely a restatement of Lipps' theory.[3] Moreover, in 1926 a translation of Lipps' *Aesthetik: Psychologie des Shönen und der Kunst* was published at a full 1608 pages. This fact becomes particularly impressive when we note that no complete work on aesthetics by Lipps has yet been translated into English. Lipps' influence is also widely seen in the discussions of traditional Japanese aesthetic concepts made by various authors working in the field of Japanese literature.

It was not until the beginning of the 1930's that phenomenology began to enter the mainstream in Japanese aesthetic thought, but from that time up to the present, phenomenology in one or another of its guises has been the favored philosophical methodology in Japan, and its importance in the realm

A.-T. Tymieniecka (ed.), Analecta Husserliana LXXVIII, 243–252.

of aesthetics has been no less impressive. Here I would like to single out for discussion one particularly interesting way in which phenomenological methodology has been employed in Japanese aesthetic research, namely, in the project of giving systematic aesthetic elucidation of traditional aesthetic concepts which, though well known in the art or literary world in Japan, had never come to be systematically or logically elucidated to any thorough extent by native means. I shall here look at two representative examples of this use of phenomenological method. These are Kuki Shûzô's *"Iki" no Kôzô (The Structure of "Iki")* from 1930 and Ohnishi Yoshinori's "Aware ni tsuite" ("Regarding Aware," pronounced "ah-wah-ray"), published in 1939.[4]

PHENOMENOLOGICAL METHOD IN KUKI

"Iki" is a term that was for all intents and purposes born in what is referred to as the "floating world" of early modern Japan's pleasure quarters in the Edo period (1603–1867). This is the world presented in the colorful woodblock prints or "ukiyo-e" from that period, so many of which depict the geisha and courtesans of those quarters as well as the kabuki actors and sumo wrestlers who formed other means of entertainment for the newly rich, but still disestablished, merchant class. Often rendered "chic" when translated into Western European languages, "iki" refers to the overall sense of style, including dress, speech, posture, and action, of the male customer as well as the female courtesan or geisha. According to Kuki, it consists in a kind of coquetry towards the opposite sex that is characterized both by a proud bravado and a "Buddhistic" resignation to one's fate in such a tenuous "floating world."

Kuki lets on in his prefatory remarks that he will be analyzing the concept of "iki" using a hermeneutic phenomenological approach, although he does so without explicitly speaking of phenomenology as a discipline or method. He proposes, first of all, to grasp the "cultural state of being" signified by "iki," "in its living form just as it is without damaging its concrete facticity,"[5] and attempting to recover the ethnic mode of being that the word expresses. Thus he aims to avoid, for example, any formal discussion of "iki" in the abstract sense in relation to "analogous phenomena in Occidental culture."[6] Kuki speaks of language as "nothing but the self-manifestation of the past and present mode of being of a people,"[7] and he sees in this word a concrete expression of the spiritual as well as the aesthetic life of the Japanese people. It is thus impossible that the term might be explained by analogy, unless merely through some coincidental similarity. The mode of being that is signified by this word has, and can have, no logical expression outside of its genetic ethnic context. Much has been made of Kuki's assertions by recent Western scholars who have read here ethnic essentialism or claims of Japanese uniqueness.[8]

Kuki also seems, in his approach to this word which is the concrete expression of the being of a people, to be implying something very similar to Husserl's notion of the direct intuition of essences, although he does not employ the Husserlian vocabulary. "Iki," he says, is not to be understood as a specific concept that is deductively grasped in its relation to some general concept. It must be intuited, rather, as "concrete, factual, and specific," "existentia." In this statement he also sounds uncannily reminiscent of Jean-Paul Sartre, his French tutor in Paris, to whom, it has been argued, Kuki himself introduced Husserlian phenomenology as well as the work of the young Heidegger:[9]

> In understanding *iki* we need the resolution to be like those heretics who solved the problem of *universalia* in the direction of nominalism. That is, we must not just handle *iki* as a specific concept and seek the "essential insight" directed at the abstract universal of the generic concept that comprehends it. Understanding of *iki* as experiential meaning must be a concrete, factual, and specific "being-comprehension." Before we enquire about the *essentia* of *iki*, we must first enquire about the *existentia* of *iki*. In a word, the study of *iki* must not be "formal." That it is to be "hermeneutic" is to be expected.[10]

Kuki also reminds us of Moritz Geiger in the next passage, where he seems, again implicitly, to be eschewing any kind of attempt to describe what "iki" is inductively by generalizing from the various objective expressions of the concept that can be found in Japanese culture. Instead of looking first at objective expression, we must instead explain "iki" as a conscious phenomenon.

> We must first comprehend *iki* as a mode of being which materializes as something which could be called a phenomenon of consciousness. Next, we must proceed to the understanding of *iki* as a mode of being which has objective expression. If we ignore the former or invert the order of our enquiry, our attempt to grasp *iki* will probably end up as an empty gesture. . . . In the first place, with objective expression as the subject of research, it will fail to grasp the ethnic specificity of *iki* even within the limits which concern objective expression – because it seeks the general characteristics of *iki* within those bounds. Again, the explanation of *iki* as a phenomenon of consciousness becomes abstract and formal. It cannot clarify concretely and interpretatively *iki's* mode of being, which is historically and ethnically stipulated, because this explanation presumes the comprehension of phenomena of consciousness to be the understanding of objective expression.[11]

Kuki here resembles Geiger when the latter eschewed inductive methods in the quest to find aesthetic essence. I shall treat this matter in more depth below. At this point I think we have satisfactorily seen the influence of phenomenological methods and ideals upon Kuki's own approach to the elucidation of "iki." Thus I shall next turn to Ohnishi Yoshinori's work.

PHENOMENOLOGICAL METHOD IN OHNISHI YOSHINORI

Next I shall look at Ohnishi's Yoshinori's explication of the term "aware," a much older term in the Japanese aesthetic vocabulary that is most often translated into English as "pathos." In its aesthetic usage, the term generally signifies a sad, fleeting beauty that moves its subject to the heart. The underlying motive driving Ohnishi's work on "aware" as well as that dealing with several other traditional Japanese aesthetic concepts is his hope to explain these as aesthetic – albeit specifically Japanese – categories, derivative from the general categories of the Beautiful, the Sublime, and the Humorous. Ohnishi seems to have been most influenced by Robert Vicher in his understanding of such categories. He states his goal in regard to "aware":

> We must here, however, ask whether or not this concept of aware can indeed be understood as a particular aesthetic category or aesthetic concept, and if it can be recognized as such, how we ought to consider its aesthetic essence. Supposing, moreover, that it belongs to the basic category of das Shöne, and can be thought of as a regional category deriving from it, we must ask in just what manner it derives.[12]

Even though Kuki never explicitly endorsed phenomenology by name, he did, as was seen above, frequently employ terms which quite obviously translate the vocabulary of the discipline (i.e. "phenomenon" (*genshô*), "hermeneutic" (*kaishakyuteki*) and the essence/existence pair). Ohnishi also never actually endorses phenomenology by name in his work, and neither does he regularly employ such phenomenologically sounding terms. These facts seem to have led several researchers to overlook the phenomenological influences upon Ohnishi's work. The references made to Moritz Geiger's aesthetics, however, and the similarities between many of Ohnishi's methodological statements and ideas put forth by Moritz Geiger and his teacher Husserl, show an unmistakable influence. This is seen particularly in three points: his anti-psychologism, anti-inductionism, and, at least in a nominal sense, his use of the notion of an "aesthetic essence."[13]

It is important to note that in 1937, two years before Ohnishi published his work on "aware," he published a book, now relatively obscure in comparison with his original work on Japanese aesthetics, entitled *The Aesthetics of the Phenomenological School (genshôgakuha no bigaku)*, which included detailed discussions of modern aesthetic currents in Germany before phenomenology, as well as thorough explanation and criticism of phenomenological aesthetics. The key figure in this discussion is Moritz Geiger, who is seen as the first and foremost phenomenological aesthetician. We find here proof of Ohnishi's thorough understanding of phenomenology, proof which works to make my present assertion of its influence upon his own thought more tenable by ruling out the possibility that the points where Ohnishi agrees with Geiger are mere accidents.

It is well known that Husserl himself rejected psychologistic explanations of cognition, replacing such notions with his model of intentionality. Geiger, in his own work, also fought against such interpretations of aesthetic experience (such as seen, for example, in the thought of Lipps). Such experience was, for him, not reducible to a discussion of the object's psychological effect upon the viewer: aesthetics was for him rather the science of objective aesthetic value. Ohnishi is seen to be in agreement in his call for the need to overcome psychological interpretations of "aware."

> [W]hether understood in its meaning of praise and affection on the one hand, or sorrow and pity on the other, it is because aware always expresses a kind of emotion or impression that research into this concept has advanced beyond the field of linguistics and has entered into psychological investigations of this emotional content. And even in cases when it has advanced further into aesthetic consideration, the problem has had a tendency to stagnate in a slanted psychologistic aesthetic standpoint, in which we can find a disposition that prevents a full unveiling of the concept's aesthetic depth and breadth. Of course generally there might arise many occasions wherein the psychological and the aesthetic points of view overlap, and it might also be possible to use psychology as one's sole method in doing aesthetics, but in an attempt to uncover directly the semantic content of this concept, such psychological aesthetics have been easily tugged into a subjectivist direction, and, from my viewpoint, the detrimental effects which such subjectivism poses to the complete elucidation of aware as an aesthetic category are by no means few.[14]

As seen in the quote previous to the last, Ohnishi is seeking the "aesthetic essence" of aware, and we see here that he is convinced that this essence, especially when seen as constituting an aesthetic category, cannot be found simply in the effect of such an experience upon the aesthetic subject.

Geiger has written, in a discussion focused upon what it would take to uncover the essence of the tragic, that a simple act of induction based upon a survey of various tragedies from different time periods and locales would not suffice. The reason is simple: without a notion of what the essence of the tragic is, one would not even be able to conduct such a survey, for one would have no criteria concerning which phenomena to include and which to exclude.[15] We have noted above that we can see such thinking reflected in Kuki's notion that we cannot answer the universal question of what "iki" is through generalizing from its objective expression, and Ohnishi likewise states that an empirical survey of the use of the term "aware" in traditional literature will not further his project:

> Since *aware* as an aesthetic concept developed mainly in our national literature, specially in the *monogatari* literature of the Heian court era with the *Tale of Genji* as the central example, it goes without saying that had we, from the very beginning, done a survey of the various occurrences of this concept in the literary works, and tested what the ancients had meant with this word, we could expect that such literary research would have been an appropriate procedure. However, to the extent that this word "*aware*" in fact expresses almost every kind of emotional meaning, and can be used in relation to almost every kind of emotional content . . . had we searched by such an inductive method only, outside of a rather useless classification of the concept's complicated and disorderly semantic contents, we probably would have been able to attain almost no results whatsoever.[16]

Ohnishi's reasons are somewhat different from Geiger's and Kuki's. His main objection to such a method seems to be the sheer number of examples of the word and the variety of meanings in which it is used. In the end, however, he is pointing at the same problem: all such a survey would yield is a classification, not an insight into the aesthetic essence of "aware" as such.

I shall stop here with my discussion of the manner in which the work of these two scholars was influenced by phenomenological methodology. I would next like to look at the more general question of what it means to treat an aesthetic term or, better yet, concept, as the object of a phenomenological inquiry.

INTUITING THE AESTHETIC ESSENCE OF A CONCEPT

In considering the manner in which these scholars employed phenomenology, an interesting problem emerges. Both of these works attempt to explicate a traditional Japanese aesthetic concept. This fact gives a different cast to the aesthetics we find here from that seen, for instance, in Geiger, who was primarily concerned with locating the value of art objects through the direct intuition of aesthetic essence.[17] In what would such direct intuition consist in cases when one's object is a concept or a term like "iki" or "aware"?

On the surface, there are similarities to the manner in which Kuki and Ohnishi approach their problems. Both freely quote from traditional texts in which their respective words appear, and yet each also seems to be aiming at something over and above what can be seen in those texts – that is to say, both go beyond positivistic philology in an attempt to arrive at "aesthetic essence." There are thus times, when reading both of these works, that one is led to wonder just what is being explained, times when there is no concrete object in sight and it seems as if we are merely being given access to the author's imagination. Kuki's use of textual evidence is unsystematic; Ohnishi drags in references at the end as evidence for his theory. So, we might ask, aside from the borrowing of certain ideas and terminology, where is the phenomenological method here, and what are its objects?

Ohnishi approaches "aware" as a term and concept that has long existed in various forms within the Japanese literary tradition, and on account particularly of this variety he seems to lack any single solid foundation from which to interpret the concept. For the most part, however, he eschews dealing with the use of the term in classical literature and focuses upon the explanation of "aware given" by the eighteenth-century National Learning scholar Motoori Norinaga (1730–1801). Norinaga's notion is that literature and poetry are born when an author is compelled to creative expression after coming to know "the 'aware' of things" in his or her contact with the outside world. Ohnishi's project then becomes the explanation of a properly *aesthetic* "aware" based upon Norinaga's writings, which he claims moved past simple linguistic interpretation on to psychological interpretation, but stopped after only hinting at a truly aesthetic one. I would claim that in this process, Ohnishi parts with the phenomenology that he references here and there in his work, and is basically impelled by contemporary notions of what aesthetics and aesthetic experience are (some of these notions do in fact show continuity with Geiger's own thought) and also by his fundamental project of interpreting "aware" as an aesthetic category. The question of how this very notion of an aesthetic category would stand up to phenomenological investigation is an interesting one in itself. Would it not, one might ask, have to be bracketed in a phenomenological reduction as a theory that seeks to somehow objectify the natural world? I shall leave this question open for speculation.

Turning to Kuki, I would like to propose that his guide in the investigation of "iki" as both "conscious phenomenon" and "artistic expression" is no text or artwork handed down from the past. It is rather his own experience in the geisha district of Gion, where he lived "iki" in his notorious lifestyle, wherein he sought traditional Japanese pleasures. Thus it was within his own bodily experience of "iki" that Kuki could, using Geiger's terms, intuit aesthetic value directly. Such an analysis was possible for him not only because he had privileged access to the world of "iki," but because the term itself describes a specific, limited aesthetic world: that of the courtesan/geisha and her customer, a world which was born in the Edo period and lives on to this very day. This stands in opposition to the status of "aware," a word with a much longer history that is based in no such regional artistic locale.

Should we assume that Kuki's real object of inquiry into "iki" is his own aesthetic experience in the Gion tea houses, then his project comes to look more similar to that of Geiger. Just as how for Geiger the universal essence of the tragic is present in every presentation of a tragedy, so should the essence of "iki" be present in every experience of the geisha world. According to Geiger, the ability to intuit the tragic in a single piece is dependent upon a thorough knowledge of the history of tragedy, and yet that essence must be intuited directly rather than arrived at inductively through a mere accumulation of historical facts. We have of course stumbled upon a difficult claim in Geiger's work, and actually in all phenomenology, regarding such intuition of universal essence, a claim which we have no time to discuss generally here. At any rate, looking at Kuki's standpoint, we see in it all the marks that "the object must have been brought to the proper condition for us to be able to intuit in it a universal essence, and the inquiring subject must have been brought to the proper condition before he is able to exercise the proper intuition."[18] This is seen in Kuki's numerous references to Edo period Japanese literature that describe the pleasure quarters. He is obviously familiar with the manner in which the word has been used from Edo period times up until his own day. Furthermore, his privileged position, while never explained or justified in any way, is visible in the manner in which he begins by giving the "Intensional Structure of Iki" as a conscious phenomenon. There is no textual basis anywhere for Kuki's assertion that "iki" is composed of the three moments of "coquetry" (*bitai*), "brave composure" (*ikiji*), and "resignation" (*akirame*), nor is there such a basis for his elucidation of the "extentional structure of iki" wherein he relates the term to various other aesthetic terms and finally comes up with a cubic diagram expressing the connections between seven of these other terms and "iki." I would like to propose that these structures are based upon something that might be read as Kuki's own personal intuition of iki's essence: his own privileged aesthetic experience.

Of course we come across a problem here. We have no way of verifying Kuki's claims unless we too are proficient in the world wherein "iki" is to be found. What we have in the end is Kuki explaining his own aesthetic experience to us, who are unqualified to evaluate it independently. We might wonder how this fact stands when considered in light of Kuki's prefatory statements, cited earlier, that "iki" as language reveals the concrete existence of the Japanese race. It seems as if not only has the inter-subjective phenomenon (which he seems here to be implying) been converted into something more subjective, but the focus upon language as the bearer of meaning has also, we might judge, strayed from the mark. To what extent

Kuki is actually presenting the meaning of "iki" as language is difficult to judge.

Ohnishi also falls short on a similar point. For him, "aware" is not grasped as language, but as concept, essence, and finally aesthetic category. In grasping aware in such a way, Ohnishi becomes painfully divorced from the literary tradition before Norinaga gave a theoretical elucidation of the term, the tradition in which the word had its genesis, the tradition which did the most to shape the meaning and future use of the term. On this account, Ohnishi ends up with a theory that fails in important ways to address the historical and traditional term. Ohnishi's free interchange of "word" (*kotoba*) and "concept" (*gainen*) in speaking of *aware* above may be significant. Ohnishi looks at the concept as something over and above "what the ancients had meant with this word." It is this abstract concept which he then treats, ignoring the more concrete phenomenon, the word itself wherein this conceptual understanding is rooted. His concept is based upon Norinaga's use of the term, to be sure, but that turns out to be no more than a starting point for Ohnishi's own aesthetic speculations.

The purpose of this essay has been first to demonstrate the influence of phenomenological aesthetic theory upon the work of Kuki and Ohnishi, and second to raise some questions concerning what it would be to phenomenologically approach the problem of explaining or defining traditional Japanese aesthetic concepts. I hope that in regard to the second and more important task I have at least succeeded in accurately pointing to some of the problems involved in such a project. The next logical step, that of attempting to solve such problems and develop a better approach to the elucidation of such concepts, must wait for another time and place.

International Research Center for Japanese Studies, Kyoto

NOTES

[1] Translated in Michele Marra, *Modern Japanese Aesthetics: A Reader* (Honolulu: Univ. of Hawai'i Press, 1999).

[2] See, for example, Hamashita Masahiro's essay on Nishi, in Michele Marra, ed., *Japanese Hermeneutics: Current Debates in Aesthetics and Interpretation* (Honolulu, Univ. of Hawai'i Press, forthcoming).

[3] See Makoto Ooda, "Yûgen and Erhabene," in *Culture and Identity: Japanese Intellectuals in the Interwar Years*, ed. Thomas Rimer (Princeton: Princeton Univ. Press, 1990).

[4] The former book has been translated into English by John Clark as *Reflections on Japanese Taste: The Structure of Iki* (Sydney: Power Publications, 1997). A large section of Ohnishi's revised thoughts on "aware" has been translated by Michele Marra in *Modern Japanese Aesthetics.*

[5] Kuki, op. cit., p. 33. I borrow Clark's translation here and elsewhere.
[6] Kuki, op. cit., p. 34.
[7] Kuki, op. cit., p. 28.
[8] See, for example, Meera Viswanathan, "An Investigation into Essence: Kuki Shûzô's "*'Iki' no Kôzô*," *Transactions of the Asiatic Society of Japan*, 4th Series, vol. 4 (1989), pp. 1–22, as well as Leslie Pincus, *Authenticating Culture in Imperial Japan: Kuki Shûzô and the Rise of National Aesthetics* (Berkeley: Univ. of California Press, 1996).
[9] See Stephen Light, *Shûzô Kuki and Jean-Paul Sartre: Influence and Counter-influence in the Early History of Existential Phenomenology* (Illinois: Southern Illinois Univ. Press, 1987).
[10] Kuki, op. cit., p. 34.
[11] Kuki, op. cit., pp. 34–35.
[12] Ohnishi Yoshinori, *Yûgen to Aware* (Tokyo: Iwanami, 1939), p. 106. All translations of Ohnishi are my own.
[13] My discussion of Geiger's aesthetics is based primarily upon the following text: *The Significance of Art*, trans. Klaus Berger (Washington D.C.: Univ. Press of America, 1986).
[14] Ohnishi, op. cit., pp. 109–110.
[15] Geiger, op. cit., pp. 7–10.
[16] Ohnishi, op. cit., pp. 219–220.
[17] Cf. Geiger, op. cit., pp. 7–10.
[18] Geiger, op. cit., p. 9.

MAX STATKIEWICZ

THE WISDOM OF THE MIRROR IN COCTEAU'S *ORPHÉE*

> Les mythes sont faits pour que
> l'imagination les anime.
>
> Albert Camus, *Le Mythe de Sisyphe*[1]

> Ist er ein Hiesiger? Nein, aus beiden
> Reichen erwuchs seine weite Natur.
>
> Rainer Maria Rilke, *Sonnets to Orpheus*[2]

An old dispute rages between *philosophia* (the love of wisdom) and *poetikê* (creative art), an old quarrel (*palaia diaphora*), says Socrates in Plato's *Republic*.[3] The love of wisdom is, for Socrates, the longing for the reality of Ideas, and creative art is the love of shadows and simulacra. He epitomizes the latter by the reflections in the mirror (*katoptron*), which thus stand for what is unwise and foolish in the world. A mirror suggests – better than any other device, better than painting, which is also analyzed in the tenth book of the *Republic* – the superficiality and the mechanical reproduction of images. Unlike philosophy (Platonic dialectic), art is unable to break the surface of "specular" appearances (*phainomena*) and to lead to the true reality of the things in themselves, the "invisible Ideas." The old quarrel cannot be easily settled since, according to Socrates in the *Republic*, the artists deny the existence of the real things behind the phenomena – deny the depth of the mirror. Hence, the necessity of the radical political decision: the banishment of art from the ideal city.

The *diaphora* of the "old quarrel" can be literally translated as "difference" or, better, by "*différance*" or "*différend*," in Derrida's and Lyotard's terminology respectively. The two latter terms indicate the difficulty of separating the domain of philosophy from that of art and the difficulty of judging equitably between them. Lyotard's *différend*, in particular, expresses the danger of an unjust settlement of the dispute. A *différend* is a conflict that cannot be equitably resolved for lack of a rule of judgement applicable to the adverse parties, that is, to radically different genres of discourse.[4] If, nevertheless, a judgement in such a *différend* is attempted, it results in a "wrong" (*tort*), an irreparable injustice, as opposed to a simple damage, which can be compensated. The history of Western thought often witnessed a wrong that rationalist philosophy inflicted on art. Aristotle's catharsis, Kant's frame of categories, Hegel's *Aufhebung* or

A.-T. Tymieniecka (ed.), Analecta Husserliana LXXVIII, 253–268.

Fig. 1. Orpheus and Euridice, Greek, classical period. The Louvre, Paris. Author's photo.

sublation, are examples of such a wrongful solution to the old dispute between philosophy and art. I would like to present here an artist's (poet's, painter's, filmmaker's) response to the imperial assault of rationalist philosophy. Indeed, Jean Cocteau's film *Orphée* (*Orpheus*) can be seen as such a response, which is performed through the magic of the mirror.

"Les miroirs feraient bien de réfléchir davantage," announces the voice on the radio at the beginning of the film. And indeed, the mirrors "reflect" more and more, in every sense of the word. They show the unfamiliar figures of those who look at them, thus aiming at the realization of the old philosophical maxim *gnōthi sauton*, or "know thyself." But they also open an unsuspected vista on "the other side of the mirror," that is, on the other side of life. Mirrors "reflect" *on* life and death, and since death arriving through the mirror happens to be a beautiful woman and the object of the poet's desire, they also reflect on love and death. *Orphée* is a love story but not the one we are familiar with. Orpheus is in love with "his" death, in love with the transcendent beauty, in love with eternity. I shall argue that to some extent Cocteau's Orpheus realizes the Platonic dream of the philosopher "striving for death."[5] One might think that it is paradoxical, at least from Plato's point of view, to conceive the philosopher's dream, realized by a poet, through sensuous love. Plato's own *Symposium*, however, allows for such "anagogic" employment of Eros. Admittedly, a mirror used in the process of transcending ordinary life cannot be a mere reflection, like the "mimetic" *katoptron* or mirror of the *Republic*; it has to "reflect more."

However, the poetic mirror of Cocteau seems to distort rather than transcend the ideal of philosophers. "Jupiter rend sages ceux qu'il veut perdre," Jupiter makes wise/prudent/reasonable those he wishes to destroy, pronounces the voice on the car radio in the middle of the film. The scene takes place in Orpheus's garage where the poet spends most of his time listening to radio messages. The Rolls Royce parked in the garage belongs to "the Princess," and Orpheus believes the messages come from her world. In fact, they are composed and aired by Cégeste, a dead poet, now in the service of the Princess. What is the meaning of the message on *sagesse*, which seems to reverse the common wisdom of the Greek tragedy? Does it come from another world, from the Princess, or is it just a projection of Orpheus' own feelings, the feelings of an anxious artist? Perhaps it is just a jest of Cégeste, that is to say, an amusement of Cocteau playing with the old Platonic tradition of opposing the philosopher and the poet.

For a poet, the temptation is strong to identify with the figure of Orpheus, the poet *par excellence*. If s/he is not explicit, the interpreter will often suggest an Orphic ideal behind the text. This view of the poet is common among the Romantics. For Novalis, e.g., the myth of Orpheus is the myth of the *Ur-Dichter*, a Messiah capable of revealing the truth of faith and of liberating man. Novalis sees himself as such an Orphic *poeta vates* of the Christendom.[6] Gerard de Nerval shares with Novalis the exalted "*Ich*" of the Orphic poet, whom he also sees as a singer of Christian faith.[7] This identification is not limited to Romanticism, however. Virgil – possibly the creator of the best known version of the myth of Orpheus and Eurydice – is perhaps also the initiator of "the poet as Orpheus" tradition.[8] In modern literature, Edward Waldo Emerson, Rainer Maria Rilke, Pierre Jean Jouve, Pierre Emmanuel, et al., are considered "Orphic poets."[9]

One is naturally tempted to see Cocteau, the author of *Le Testament d'Orphée*, in which he himself plays the leading role, "as Orpheus." Many critics maintain such identification, often supporting the analysis of Cocteau's work with his biography and the explicit statements in his letters and interviews.[10] Identification, however – a dialectical maneuver – is difficult to perform in proximity of the famous "Zone": a no man's land of the film *Orphée*, which tends to extend its principles or rather its anarchy, to the whole of Cocteau's work. Orpheus can be seen as a mirror image of Cocteau, but only if the mirror "réfléchit davantage"[11] – reflects more *of* and *on* its object, that is to say, if it allows one to see the poet in proximity to Death, in the paradoxical space of the "Zone," at once limited and full of possibilities. In other words, the myth of Orpheus will figure in the destiny of the poet who strives to transcend the strict dichotomies of Western philosophical tradition, such as reflection and its object, truth and illusion, image and simulacrum, the visible and the invisible, the serious and the frivolous, reason and imagination, life and death.[12] More than the theater play *Orphée* of 1926, Cocteau's film of the same title from 1950 and his *Testament d'Orphée* (1960) constitute a defense of the poet against the philosophers and philosophically minded critics. The cinematic transformation of the myth of Orpheus into the story of the poet fascinated with the realm of the invisible and with death contests Socrates's summary dismissal of Orpheus as a deceiver and anti-philosopher in Plato's *Symposium*.

One of Plato's paradoxes is that human beings can attain immortality ("as far as it is humanly possible") through death. Indeed, philosophy – the royal way to the Islands of the Blessed – is defined by Socrates in the dialogue *Phaedo* as a "practice of death" (*meletē thanatou*).[13] It consists in renouncing the life of the senses for the sake of true love, the love for the Incorporeal, the Invisible (*aeidēs*) often associated with Hades (*Haidēs*, poetically: *Aïdēs*). A poet is portrayed as lacking the courage to confront Hades and the Invisible. Another name for this love-death philosophy is "catharsis" or purification from sensuality;[14] and the scapegoat of the sacrificial ritual is often designated as the poet.[15] Plato's *Symposium* explicitly opposes the great lovers – Alcestis, Achilles, and the lover of wisdom Socrates – to Orpheus, the lyre player who did not have enough love and courage to face real death, but entered the invisible realm of the dead by deceiving (*diamēkhanasthai*) the gods.[16]

Deceit (*mēkhanē*) belongs, nevertheless, to the very nature of art, which consists in concealing its "artifice." This is not only the opinion of Socrates, but that of his great adversary, one of the first theoreticians of poetry, Gorgias, and of the great part of the Western tradition.[17] When, in the *Symposium*, Orpheus is said by Phaedrus to fall short of the real lovers, those who sacrificed their life for their love, this is a sort of recognition of his calling. Orpheus is condemned for being true to his profession of deceit: a lyre player (*kitharōidos*) and a poet. The very essence of his art – producing illusion instead of truth – poses the problem. The condemnation of Orpheus is not a personal affair; Orpheus is guilty of being an archetype of the poet in the Platonic sense of the word. His poetic virtue is a sin against true love and philosophy.

Plato's version of the myth of Orpheus might be considered an interpretation of the well-known Attic sculpture, a bas-relief from the end of the fifth century BC that can be seen in the Museo Nazionale of Naples. The sculpture shows a man (Orpheus) turning to see a woman, raising her veil, looking into her eyes. The woman puts her hand on the man's shoulder. At that moment another man (Hermes) is retaining the woman. If the identity of the two men seems certain – the conventional traveling hood of Hermes, the lyre of Orpheus – that of the woman is doubtful. Is it Persephone or Eurydice, or the phantom (*phasma*) of Eurydice, as Phaedrus in Plato's *Symposium* suggests. What need is there to escort a phantom, however? The presence and gesture of Hermes make sense only if he accompanies the queen of the Netherworld, or if he retains the "real" Eurydice. The latter possibility is supported, of course, by the most common version of the myth (that of Virgil's *Georgics* and of all its successors), while the former is suggested by Cocteau's film *Orphée*.

Virgil's *Georgics* is our oldest textual authority as to the myth of Orpheus in its canonical version.[18] Its fourth book ends with the story (told by Proteus, a sea-god) of the poet entering the Underworld in search of Eurydice, who died bitten by a venomous snake. Orpheus charmed the powers of Hades and obtained the release of his wife. The only condition imposed by the gods was that he not look at her during the upward journey. Impatient Orpheus, "overcome by passion, looked back" (*victus . . . animi respexit*) and lost Eurydice forever. After that, he remained devoted to the memory of his wife and disdained other women. Offended by his indifference, the Bacchantes killed and dismembered him during their rite. Orpheus's severed head, carried by the river Hebrus, continued to evoke Eurydice.[19]

Ovid's *Metamorphoses* contributed to the "canonization" of Virgil's story, even if Ovid's intention was initially to satirize the tragic treatment of the myth in the *Georgics*.[20] Ovid's only change in the plot line concerns the homosexual loves of Orpheus after the second loss of Eurydice; they are presented, somewhat paradoxically, as proof of Orpheus' faithfulness to his heterosexual commitment.[21] The eventual reunion of the two spouses in the world of shadows becomes a sort of happy ending. Ovid's treatment clearly moderates the Virgilian tragic climax when he ends his account of the myth with these words: "Here now side by side they walk; now Orpheus follows her as she precedes, now goes before her, now may in safety look back (*respicit*) upon his Eurydice."[22]

The gods' taboo and Orpheus' fatal glance remained the center and the "turning point" of most classical and modern versions of the myth, including Cocteau's play. Cocteau's film *Orphée*, on the other hand, seems to break with the dominant Virgilian "hermeneutics" of the Attic sculpture. It reproduces, or rather replays, the sculpture with all its mystery, literally and figuratively. A statue of a woman figure in Orpheus' garden, comparable to the one on the bas-relief, belongs to the setting of the film. At one moment, Jean Marais, playing Orpheus, and François Perrier, playing Heurtebise (Hermes), are placed in such a position that the bas-relief is fairly well enacted. Later on, before their journey into the "Zone," Heurtebise will ask Orpheus a question that might have been suggested to Cocteau by the sculpture: "Orphée, regardez-moi dans les yeux. Désirez-vous rejoindre Eurydice ou la mort?"; and again after Orpheus's speechlessness, "Je vous pose une question précise, ne l'oubliez pas. Est-ce la mort que vous désirez rejoindre ou Eurydice?" Orpheus' answer ("Les deux . . .")[23] still leaves the enigma intact and only emphasizes the difference between both the Platonic and the Virgilian "definite interpretations," on the one hand, and Cocteau's problematizing of the sculpture and of the myth, on the other.

In fact, Cocteau's film is not centered on the famous "turning point."[24] Literally, Orpheus does not need to turn; he catches sight of Eurydice in the mirror: the rearview mirror of the car. Thus, as a symbol of the wisdom of the *gnōthi sauton* ("know thyself"), the mirror here becomes definitely a symbol of transgression. More importantly, the involuntary character of the backward or mirror glance is put in doubt. Orpheus seems to wish the loss of Eurydice since this is the only way to rejoin the Princess. Actually, the breaking of the taboo does not change anything in Orpheus' attitude towards life and death. Orpheus loses Eurydice much earlier, or rather it is Eurydice who loses Orpheus to his death. Indeed, the real turning point (*peripeteia*) coincides – in the tragic manner – with a recognition (*anagnōrisis*): Orpheus' recognition of the Princess, the object of his passion, as his death. This recognition is preceded by a period of fascination in which Orpheus progressively loses the sense of boundaries, which the Ancients identified with wisdom (*sophia, sagesse*).[25] The initial characterization of Orpheus as a *poète sage* might have earned him the right to reenter the ideal Platonic city. A national poet recognized by authorities, friendly with the police, he would not present a danger to the established political regime. Small wonder that Orpheus is not accepted among the opponents of that regime, or any regime for that matter: the young avant-garde artists. The opening scene in the "Café des Poètes" is a caricature of the famous Parisian avant-garde gathering "du genre 'café de Flore.'"[26] At the same time it examines the established poet's distress. Orpheus suffers from the rejection of his fellow-poets. Asked about the cause of this isolation, an old friend pronounces the judgement: "Orphée . . . votre plus grave défaut est de savoir jusqu'où on peut aller trop loin."[27] This judgement is based on the absolute law for an avant-garde artist, a rule strictly opposed to the classical *mēden agan*,[28] and stated a moment earlier by the same character: "Aucun excès n'est ridicule."[29] Thus, Orpheus' sin is tantamount to a classical virtue – respect for limits. One of the possible meanings of "*sagesse*," cited in the beginning is precisely this *savoir* or knowledge of measure and limits, or in other (Platonic) words, the lack of daring (*tolma*) to risk everything.[30]

This portrait of a "classical" poet would not be complete without a rational element that Cocteau attributes to his Orpheus. "Vous cherchez trop à comprendre ce qui se passe, cher Monsieur. C'est un grave défaut,"[31] says the Princess, and the readers of the script would recognize this assertion as the epigraph to *Orphée*.[32] The respect for boundaries is also a respect for definitions (*horoi, logoi*). Thus, the avant-garde artists' and the Princess's critique of Orpheus seems to reject a "poète sage," a poet-philosopher in the limited, "Socratic" (of the so-called "Socratic dialogues," obsessed with rational definitions) or Cartesian sense. On the other hand, this critique coincides with the explicit taunt of Plato's Phaedrus in the *Symposium*.

Orpheus's *sagesse*, or *sophia*, hardly agrees with that exalted wisdom of the Socrates from the *Symposium*, the *Phaedo* or the *Republic*, the wisdom of the lover and philosopher, an adventurer ready to cross the boundaries, ready to die for his love, ready to proceed to the heart of Being and even "beyond Being" (*the epekeina tēs ousias* of the *Republic*).[33]

What awakens Cocteau's Orpheus from his "slumber of glory" (Je somnolais sur mes lauriers. Il est capital que je me réveille . . .),[34] and from wisdom in the "limited" sense of the word, is his encounter with the Princess. When addressing Orpheus for the first time, she commands: "Eh bien, ne restez pas là comme une borne!"[35] – an apparently common expression for "Do not stand there like a lamp-post or dummy." However, the French word *borne* literally means "limit," and it is worth noting that the Princess pronounces "*borne*" when one would expect her to ask Orpheus his name. Later on, when Orpheus returns home after the night spent in the Princess's chalet, and in proximity of the liminal "Zone," he is accused of "dépasser les bornes."[36] Again, a common expression of annoyance with excessive behavior also functions here as a pointer towards the transgression of limits. The nightly experience, the acquaintance with Death, and the estranging effect of the mirrors/passages renders Orpheus resistant to the limitations of his rational bourgeois worldview. "Ma vie commençait à se faisander, à puer la réussite et la mort,"[37] he confesses to Eurydice, in order to justify the change in his heretofore "*sage*" behavior and his interest in the messages coming from the car radio, that is to say, his interest in the new, the extraordinary, the "unknown" (Je traque l'inconnu).[38]

Thus, when seen in the light of the apparently meaningless radio message concerning "*sagesse*," the film *Orphée* refers less to the Virgilian interpretation of the myth of Orpheus and Eurydice than to Plato's critique of the poet in the *Symposium*. Cocteau shows the poet going through a crisis "in the middle of his life," from which he is saved by meeting the Princess. The death of his wife, Eurydice, is only a consequence of this crisis (she has an "accident" on the way to see her friend, whom she wishes to consult on the subject of Orpheus' unfaithfulness) and a means for its resolution (Eurydice's death becomes an occasion for Orpheus to reunite with *his* death, that is to say, with the Princess). Eurydice herself points out the radical difference between the traditional Virgilian version of the myth – still retained in the 1926 play – and the film:

Eurydice (soulève son visage): Il me hait. . . .
Heurtebise: S'il vous haïssait, il n'aurait pas été vous arracher à la mort. On le citera en *exemple*. . . .

Eurydice: Ce n'est pas pour moi qu'il est venu. . . .
Heurtebise: Eurydice! . . .[39]

Heurtebise's reproaching exclamation is not sincere. He is well aware of the Princess and Orpheus' relationship. His own feelings toward Eurydice should make him particularly sensitive to inter-worldly love affairs. Heurtebise knows full well the untruth of the "exemplary" version of the myth of Orpheus in the context of Cocteau's cinematic remaking of this myth.

Formally, however, Cocteau follows the traditional myth after all. When the Princess carries Eurydice away into the Zone, she leaves behind her gloves and thus enables Orpheus to cross through the mirror and to descend into the Underworld. The difference with the tradition resides in Heurtebise's role in organizing the "descent." He is the leader, Orpheus remaining a docile follower. The trial in the Zone eventually reveals the "truth" behind the myth: the reciprocal love between Orpheus and his Death. Even though the verdict still belongs to the tradition – Eurydice may come back to life on condition that she avoid Orpheus' gaze – the ensuing comedy measures the distance between the Virgilian tradition and Cocteau's film. Here, the playful character of Cocteau's treatment of the myth reaches its summit.[40] The tragic myth apparently becomes a *comédie de boulevard*. A "fatal" sight in the rearview mirror of the Princess's car puts an end to the unbearable (from more than one point of view) situation. The "second descent" – after Eurydice's disappearance – seems to be once again Cocteau's response to Plato's accusation: Orpheus is now ready to die "for real," although, again, rather for the Princess than for his wife. When this actually happens,[41] he is rewarded with immortality.

Presented in this way, Cocteau's transformation of the myth of Orpheus answers Plato's indictment by reversing to some extent his painting of, and his scorn for, the poet in the *Symposium*. Orpheus is not presented as a deceiver. Only the Princess and her assistants, Heurtebise and Cégeste, practice deceit. When Orpheus finally takes the initiative, he dies the "real death" and thus proves his courage and readiness to sacrifice his own life. On this superficial level of the literal allegations in the *Symposium*, Cocteau's defense seems skillfully conducted.

The Platonic denunciation of the poet's "trickery" (*mēkhanē*) is another matter, since Cocteau himself – especially Cocteau the filmmaker – finds himself under indictment. Indeed, Cocteau does use all sorts of sleight of hand, although in a more discrete fashion than in the theater play, which at times becomes a true prestidigitator's performance. The principal instrument of illusion common to both the play and the film is the one associated in Plato with the poetic craft of producing simulacra, namely the mirror (*katoptron*).[42] However, the genius of Cocteau and the motion picture technology allow for the complexity of a mirror effect unknown to Plato. In the film, the mirror displays both "reflection" and "passage," and it is a source of self-knowledge and at the same time a means of transcending the limits of the self (in other words, transcending narcissism, often interpreted as a mirror effect). The old wisdom of "*gnōthi sauton*" ("know thyself"), associated with Socrates,[43] when taken in its original Greek sense – "know your limits and especially your mortality" – is thus "refined" by Heurtebise:

Je vous livre le secret des secrets. . . . Les miroirs sont les portes par lesquelles la mort vient et va. Du reste, regardez-vous toute votre vie dans la glace et vous verrez la mort travailler come les abeilles dans une ruche de verre.[44]

Nonetheless, the common – although not exactly Platonic – wisdom of the mirror is immediately contradicted:

Orpheus: Heurtebise! Il n'y a plus rien à faire.
Heurtebise: La [Eurydice?] rejoindre.
Orpheus: Aucun homme ne le peut . . . sauf s'il se tue.
Heurtebise: Un poète est plus qu'un homme.[45]

And a poet's mirror is more than a common mirror, one might add in Cocteau's name. True, the hubris of the poet is expressed here not by Orpheus, but by Heurtebise. However, Cocteau is as much "Heurtebise" as he is "Orpheus," and he would not refuse the epithet "divine" (*theios*), which Plato seriously attributed only to the philosopher.[46] The poet is more than an average man or woman because s/he is able to transform death into an object of his/her desire, to see it as a mysterious woman or man, as a guide beyond the limits of the mortal world. The "trickery" of the open mirror allows the poet to communicate this desire to the spectator. The legerdemains in the film – the permeable mirrors, the gloves, the fades, the vertical upward movements of rising bodies – are not only means of deception, but of revelation as well; they are shown to be more than just an artful *mēkhanē* of the Platonic poet-illusionist. In the words of Chris Marker, a contemporary critic of *Orphée*, Cocteau is rather a "reverse illusionist" who "uncovers his gimmicks, opens his trunks, reveals that his white magic is really black magic, his pigeons alive, his ladies actually cut to bits, and that he truly treads on Death."[47] Cocteau's supposed *mēkhanē* reveals, indeed, the infinite potential of the human imagination in its play with boundaries. For the poet, to defy death does not mean to shun it, as Orpheus supposedly did (at least in Plato's view), or to choose it as a mortification of the senses (as Plato's philosopher was supposed to do), but – like Cocteau's Orpheus and Cocteau himself, and like the spectators of his film – to open oneself, through the mirror, to the seduction of a space (metaphysical, aesthetic, political) without limits.

When Orpheus' friend pronounces the diagnosis of the poet's ailment in the opening scene of the film, in the "Café des Poètes," he proposes a remedy as well. "Étonnez-nous, Orphée" ("Amaze us, Orpheus"), he says.[48] In other words, he encourages Orpheus to become a philosopher, if we are to believe Socrates, Aristotle, or Brecht and take the ability to become amazed and to amaze (*thaumazein, verfremden*) as the beginning and principle (*arkhē*) of philosophy and philosophical theater and art.[49] The protagonist of the film tries to follow this advice, and has some success. However, the poet who really succeeds in performing the Brechtian and Socratic task of *Verfremdung* and of *thaumazein* is Cocteau himself. He does it through his confrontation with popular myth and unpopular philosophy. He reveals "endless" (*sans fin*) possibilities of the Mirror, Death and the Imagination. In his treatment, the myth of Orpheus loses its "common sense" of a love story with a melodramatic or a happy ending, and acquires an apparently paradoxical sense of philosophical drama.

Cocteau's own "earthly" ending is, of course, a scornful critique of the "Glück-like"[50] versions of the myth of Orpheus. The real ending takes place in the realm of the limits – the Zone – where the Princess, the poet's death, and Heurtebise, the true lovers and heirs of Alcestis, Achilles, and Socrates, are walking to their fate.

The film ends with an image of the Zone:

> Vaste décor de la zone détruite. On voit la Princesse et Heurtebise qui rapetissent, encadrés par les motocyclistes. Leurs ombres défilent sur le fond. Ils disparaissent pendant qu'il apparaît le mot:
>
> *FIN*[51]

The word "end" (*fin*), however, does not mark the de-finite limit (*finis*) of Cocteau's version of the myth. Ten years later, in *Le Testament d'Orphée*, he replays this apparently final scene and modifies both the earthly and unearthly endings of the film *Orphée*. The earthly immortality of the couple did not work: "[La] survie [d'Orphée] était un mirage. Sa tête divine est morte et Eurydice remise aux enfers."[52]

One might think that Cocteau has returned *finally* to the orthodoxy of the myth. This would imply, however, that the categories of the earthly system of mythology operate in the Zone. In fact, these categories do not pertain to the setting of *Le Testament d'Orphée*. Cocteau himself is now in a Zone-like world of cinematic studios; he is condemned by the Princess herself (or is it by Minerva?)[53] to the *peine de vivre*. The "life penalty" is, indeed, the fate of "Orpheus," not the comfortable life of an established poet, of course, but life in the nocturnal shadow of the Zone and Death, a life of struggle with the "*mensonges du jour*."[54] When, at the end of *Orphée*, the Princess sacrifices herself in order to return Orpheus to his former comfortable life, she unwillingly fulfills the augury announced in her name by Cégeste: "Jupiter rend sages ceux qu'il veut perdre." Paradoxically, when separated from Death, Orpheus' life "stinks of success and death." He is but a wise poet (un poète sage) and a philosopher in the common sense of the word. Heurtebise understands better than the Princess her mistake when, in the penultimate line of the *Orphée*, he says: "Il fallait les remettre dans leur eau sale."[55] This ironic comment could have been made by Socrates himself, the Socrates of the *Phaedo* that is, a philosopher for whom the purification of the soul meant precisely the preparation for the de-finite liberation from the senses, in other words, the "life with Death" (*meletē thanatou*). "L'eau sale" represents here, as it did for Socrates, the routine, the habit, the familiarity of everyday life untouched by poetical and philosophical reflection. It represents the common mirrors, which reflect at the same time too much and not enough ("réfléchissent trop et pas assez"), mirrors that "renversent prétentieusement images et se croient profonds."[56]

Le Testament d'Orphée definitely amends the "happy ending" of *Orphée*. The news of Orpheus' death coincides with Cocteau's rebirth as a poet in the zone "entre deux eaux," or, to use the language of the current judges of the Zone, the Princess and Heurtebise, "entre deux règnes."[57] Eventually, Cocteau himself, struck by the spear of Minerva, the goddess of wisdom and philosophy, becomes a true "Orpheus," a poet and philosopher. The blow is, indeed, only another way of executing the Princess's condemnation to the "*peine de vivre*" (life penalty), which, for the poet and true philosopher, must be at the same time the "*peine de mort*" (death penalty), or at least, as in *Le Testament d'Orphée* itself, a semblance of death. For the Orphic poet must be a true "passenger," a "passer through" the magic mirror or, in Rilke's words, a dweller in the *Doppelbereich* of life and death, that is, of the dual realm of the Zone, and an explorer of the *Zwischenraum* of the imagination – the intermediate space between philosophy and poetry or art, where myths are destroyed and recreated.[58]

Comparative Literature
Stony Brook

NOTES

1 "Myths are open to the work of imagination." Albert Camus, *Le Mythe de Sisyphe*, Les Essais XII (Paris: Gallimard, 1942), p. 163.

2 "Does he belong here? No, out of both realms his wide nature grew." Rainer Maria Rilke, *Sonnets to Orpheus*, trans. M. D. Herter Norton (New York: W. W. Norton & Company, Inc., 1942), p. 27.

3 Plato, *Republic* X, p. 607b-c.

4 Jean-François Lyotard, *Le Différend* (Paris: Les Éditions de Minuit, 1983), p. 9; *The Differend: Phrases in Dispute*, trans. Georges Van Den Abbeele (Minneapolis: University of Minnesota Press, 1988), p. XI.

5 See note 12 below.

6 Novalis, *Schriften* I (Dresden: Wolfgang Jess Verlag, 1929), p. 263 and *Fragmente* (Stuttgart: W. Kohlhammer Verlag, 1960), p. 32; cf. Walther Rehm, *Orpheus, Der Dichter und Die Toten* (Darmstadt: Wissenschaftliche Buchgesellschaft, 1972), pp. 16 and 95f.; Walter A. Strauss, *Descent and Return: The Orphic Theme in Modern Literature* (Cambridge, Mass: Harvard University Press, 1971), pp. 45ff.; Gwendolyn Bays, *The Orphic Vision: Seer Poets from Novalis to Rimbaud* (Lincoln: University of Nebraska Press, 1964), p. 13.

7
Et j'ai deux fois vainqueur traversé l'Acheron
Modulant tour à tour sur la lyre d'Orphée
Les soupirs de la sainte et les cris de la fée.

Gerard de Nerval, *Oeuvres*, vol. 1 (Paris: Garnier, 1958), p. 695.

8 M. Owen Lee, *Virgil as Orpheus: A Study of the Georgics* (Albany: State University of New York Press, 1996).

9 See, e.g., Eva Kushner, *Le Mythe d'Orphée dans la littérature française contemporaine* (Paris: A. G. Nizet, 1961); Henri Le Maitre, "Les Mythes antiques en France," in *Mosaic* 2/3 (Spring 1969), pp. 25–41; Walter A. Strauss, *Descent and Return* (note 1 above); R. A. Yoder, *Emerson and the Orphic Poet in America* (Berkeley: University of California Press, 1978).

10 See, e.g., Milorad, "Le Mythe orphique dans l'oeuvre de Cocteau," in *Jean Cocteau 1: Cocteau et les mythes*, ed. Jean-Jacques Kihm and Michel Décaudin (Paris: La Revue des Lettres Modernes, 1972, No. 298–303), p. 124. Arthur B. Evans considers Orpheus the "mouthpiece" of Cocteau and he often uses the hyphenated expression "Orphée-Cocteau" in his *Jean Cocteau and His Films of Orphic Identity* (London: Associated University Presses, 1977), pp. 109ff. and 124f., *et passim.*

11 "Les miroirs feraient bien de réfléchir davantage," announces the voice on the radio during the initial visit of Orpheus in the Princess's chalet – Jean Cocteau, *Orphée: The Play and the Film* (Oxford: Blackwell's French Texts, 1976), p. 78. "It would be better if mirrors reflected more – Jean Cocteau," *Three Screenplays: L'Éternel Retour, Orphée, La Belle et la Bête*, trans. Carol Martin-Sperry (New York: Grossman Publishers, 1972), p. 115.

12 The *dichotomies* perfected in the French Cartesian tradition.

13 Plato, *Phaedo*, p. 81a.

14 Plato, *Phaedo*, pp. 65c–66a.

15 E.g., Plato, *Republic*, p. 398a.

16 Plato, *Symposium*, p. 179d.

17 For Gorgias, see Fragment 11 ("Encomium of Helena") and Fragment 23, in Herman Diels, *Die Fragmente der Vorsocratiker* vol. II, 6th edition by Walther Kranz (1952, reprint, Zürich, Hildesheim: Weidmann, 1985), pp. 288–294 and 305–306, respectively.

[18] Conon's *Narrationes*, a Greek text from the second half of the first century BC or the beginning of AD, which also has the backward glance, is generally not considered to be Virgil's source; see Raymond J. Clark, *Catabasis: Virgil and the Wisdom-Tradition* (Amsterdam: B. R. Grüner, 1979), pp. 122ff.
[19] Virgil, *Georgics* IV, 453–527; Latin text with an English translation by H. Rushton Fairclough (1916, Cambridge, MA: Harvard University Press, 1998), pp. 229–232.
[20] See Charles Segal, *Orpheus: The Myth of the Poet* (Baltimore: The Johns Hopkins University Press, 1989), pp. 54 and 81ff.
[21] Ovid, *Metamorphoses* II, book X, 3–85 and book XI, 1–92; Latin text with an English translation by Frank Justus Miller (1916, Cambridge, MA: Harvard University Press, 1994), pp. 65–71 and 121–126, respectively.
[22] Ovid, *Metamorphoses* XI, 61–66, p. 125; the light tone of Ovid's poem and its ending announce such "lived happily ever after," mostly operatic, versions of the myth as those of Renucci, Monteverdi, and Gluck.
[23] *Orphée*, p. 102.

Heurtebise: "Look into my eyes. Whom do you want, Eurydice or death? Mark that I am asking you a precise question. Is it death you would find or Eurydice?"
Orpheus: "Both . . ." (*Three Screenplays*, p. 151).

[24] There is one in the theater play, both literal and metaphoric. See *Orphée*, scene IX, pp. 33ff.
[25] Significantly, the message on *sagesse* reaches Orpheus in his garage just before the peripety and the recognition of the Princess.
[26] This is Cocteau's own reference in the script to the famous Parisian Café on Boulevard Saint Germain, frequented by avant-garde artists and philosophers, Sartre among others – *Orphée*, p. 69; *Three Screenplays*, p. 103.
[27] *Orphée*, 71. "Orpheus . . . your most serious fault is knowing just how far one can go too far" – Jean Cocteau, *Three Screenplays*, p. 106, translation modified.
[28] "Nothing too much."
[29] *Orphée*, 71. "The excessive is never ridiculous." *Three Screenplays*, p. 106.
[30] Plato, *Symposium*, p. 179d.
[31] "You mustn't try too hard to understand what's happening, my dear Sir; it's a very serious fault" (*Three Screenplays*, p. 116); the epigraph to the film script (*Orphée*, p. 59) is curiously left out of Carol Martin-Sperry's translation.
[32] The statement is repeated verbatim in *Le Testament d'Orphée*.
[33] Plato, *Republic*, p. 509b.
[34] *Orphée*, p. 90. "I've been resting on my laurels . . . it's essential that I wake up." *Three Screenplays*, p. 135.
[35] *Orphée*, p. 74. "Well, don't just stand there like a dummy!" *Three Screenplays*, p. 109.
[36] *Orphée*, p. 84. "Your behavior is really the limit." *Three Screenplays*, p. 123.
[37] *Orphée*, p. 88. "My life was beginning to fester, to stink of success and death." *Three Screenplays*, p. 131; this phrase hardly needed improvement since Cocteau's theater play of 1926; see *Orphée*, p. 12.
[38] *Orphée*, p. 88. "I am on the trail of the unknown." *Three Screenplays*, p. 131.
[39] *Orphée*, p. 115.

Eurydice: He hates me.
Heurtebise: Were that so he would not have snared you from death. He'll be held up as a wonderful

example. . . .
Eurydice: He didn't do it for me. . . .
Heurtebise: Eurydice! . . . (*Three Screenplays*, p. 172).

40 This scene fairly reflects the dominant atmosphere of the theater play *Orphée*.
41 When Orpheus is attacked by a crowd of young poets and the "Bachantes" from the women's club, he refuses Heurtebise's offer to save him.
42 See Plato, *Republic*, p. 596d and *Theaetetus*, p. 206d.
43 It is discussed, e.g., in Plato, *Charmides*, p. 164e ff., *Protagoras*, p. 343b, *Philebus*, p. 48c.
44 *Orphée*, p. 101; "I will tell you the greatest secret. . . . Mirrors are the doors through which death comes and goes. Besides, look at yourself in the mirror throughout your life and you will see death at work like bees in a glass hive." – *Three Screenplays*, p. 150.
45 *Orphée*, pp. 101–102.

Orpheus: Heurtebise! There's nothing we can do!
Heurtebise: We can join her.
Orpheus: But no man can . . . unless he kills himself.
Heurtebise: A poet is more than a man. – *Three Screenplays*, pp. 150–51.

46 Plato, *The Sophist, ab initio*.
47 Chris Marker, "The Most Advanced Frontiers of Demiurgy," in *Esprit*, November 1950, reprinted in René Gilson, *Cocteau: An Investigation into His Films and Philosophy*, trans. Ciba Vaughan (New York: Crown Publishers, Inc., 1969), p. 158.
48 "Étonne-moi," was the advice the young Cocteau received from Diaghilev, with whom he was collaborating as a poster designer and scenarist for the *Ballets Russes*. He understood it better later – after the famous disaster of the premiere of *Le Sacre du Printemps* – as "the insult to habit, without which art stagnates and remains a game." See Margaret Crosland, *Jean Cocteau: A Biography* (New York: Alfred A. Knopf, 1956), pp. 34–36.
49 See, e.g., for Socrates, Plato, *Theaetetus*, p. 155d; for Aristotle, *Metaphysics*, p. 982b; for Brecht, *A Short Organum for the Theater*, paragraphs 42ff.
50 This spelling appears in the French version of the screenplay, *Orphée*, pp. 77–78.
51 *Orphée*, p. 124. "Vast scenery of the destroyed area in the Zone. The Princess and Heurtebise, flanked by the motorcyclists, walk away into the distance. Their shadows move across the background. As they fade out, the words THE END appear." *Three Screenplays*, p. 187.
52 Jean Cocteau, *Le Testament d'Orphée*, in *Romans, poésie, œuvres diverses* (Paris: Librairie Générale Française, 1995), p. 1352. "Orpheus' survival was a mirage. His divine head is dead and Eurydice sent back to Hades."
53 The Princess appears when Cégeste, the poet's guide, announces Pallas-Athena, that is, Minerva. See Cocteau, *Romans, poésie, oeuvres diverses*, p. 1342.
54 "The lies of the day." See Cocteau's Preface to *Le Testament d'Orphée*, in *Romans, poésie, oeuvres diverses*, p. 1321.
55 *Orphée*, p. 124. "We had to return them to their muddy (literally, 'dirty') waters." *Three Screenplays*, p. 187.
56 "Pretentiously inverse their images and believe themselves profound"; one thinks, of course, of the famous *camera obscura* as a metaphor for ideology.
57 "In between," that is, "between two realms." *Romans, poésie, oeuvres diverses*, p. 1351.

[58] Cocteau's transformation of the myth of Orpheus is a response not only to Plato's dismissal of the poet from the philosophical life, but to all later repetitions of this gesture, including the one Sartre directed explicitly toward Cocteau: a "très brillant causeur," who had "peu d'idées." Jean-Paul Sartre, "L'Autoportrait à 70 ans," in *Le Nouvel Observateur*, no. 55 (30 June–7 July 1975): p. 76.

JEANIE S. DEAN

SHE LOOKS IN THE MIRROR: THE IDEOLOGIC SHIFT OF THE FEMININE GAZE IN THE FILM *THE FRENCH LIEUTENANT'S WOMAN*

During the 1960s and 1970s, a time of volatile political activism of liberation movements of racial and sexual equality, Hollywood was ending a fifty year golden period. In this period female film roles were limited to marginal stereotypic sex objects, as noted by critics Laura Mulvey, Molly Haskell and others. In the Hollywood classical era – prior to 1981 – a plot formation about a woman striving for personal freedom rather than for a husband was ridiculous, if not untenable. *The French Lieutenant's Woman (TFLW)* is one of the earliest commercial films to represent the shifting consciousness of feminine aspiration for sexual liberation and equality. As a major Hollywood/British commercial film, *TFLW* breaks with the prevailing ideology of the stereotypic female characteristic of the time. Through its unique use of mirrors in nine significant scenes, the 1981 release of *TFLW* into mainstream film subverts the classic ideology by disturbing the illusion of the film narrative and inverting the gaze of the female role.

The mirror, an instrument of reflection and illusion, is a mythic symbol of narcissism and feminine narcissism in particular. Film critics and feminist film theorists took up the mirror as a potent metaphor for dual representation to reflect ideology and the origination of meaning in film. The mirror is a cogent representation of the film image as signifier, according to Christian Metz's theory of semiotics. With its capacity for projection the cinema, like a mirror, projects cascading signifiers in a "chain of mirror-like effects," states Metz in *Imaginary Signifier*.[1]

Psychoanalytic theory and semiotics merged in film theory during the 1970s and 1980s. For Jean-Louis Baudry, the film apparatus reproduces something like dreams, and displays the subconscious for the spectator. He asserts that "dream . . . is a state in which mental perceptions are taken for perceptions of reality."[2] The film viewer, like the dreamer, experiences in the darkened theater via the dominant visual image a regression to an anterior phase of being. Cinema thus exposes the unconscious self to the self. Through the play of the imaginary narrative, an experience of dream-like regression is reproduced by the film apparatus which displays the unconscious desire of the unconscious.

A.-T. Tymieniecka (ed.), Analecta Husserliana LXXVIII, 269–294.

Semantics and psychoanalysis blend in the theories of Jacques Lacan. Accordingly the self is a subjective entity generated by a cognitive function called the imaginary in what Lacan calls the mirror-phase of child development. This phase begins when the child at about eighteen months old sees the mother, identifies with her visual image and gains a sense of the unified body. The concept of self is therefore derived in the child's mind by mirroring the mother before physical control is attained. The imaginary function is defined as a generative structure giving cause to the formation of things, including identity.[3] Freudian psychology also references a later mirror-like stage in development when the male child recognizes sexual difference, usually through the mother, and experiences this knowledge of difference as a traumatic realization, signifying a threat of castration. The importance of this mirror-phase will be considered in this analysis of the use of mirrors in the 1980 film *TFLW*.

In the 1960s and 1970s, the *Cahiers du Cinéma* group applied psychology to define ideology and thereby influenced film theory. Ideology is recognized as an unconscious expression of the communal politic. Ideologies are formations that may be associated with periods of history, and lead to the social institutions of the time. In its ambiguous quality of representing illusion and the subconscious or unconscious ideology, the mirror eventually became a metaphor for feminist discussions about the unseen yet ever present assumptions about gender in society, history, science and art. The mirror supported deconstructionist analysis of patriarchal structures in the ideology of both critical academic disciplines and political or social institutions.

From semiotics and psychoanalysis, feminist theory emerged to challenge the phallocentrism of both film and film theory. Laura Mulvey's seminal essay, "Visual Pleasure and the Narrative Cinema" (1974) is considered to be the shot that launched feminist film theory. Her work analyzes the structure of three gazes involved in film spectatorship, that of the camera, the actor and the spectator. Her analysis is grounded in Freudian constructs.

This paper explores the mirror as dual signifier in the film *The French Lieutenant's Woman (TFLW)*, focusing on its nine mirror scenes and dual narratives. Building on Laura Mulvey's analysis of the film-gaze and the *Cahiers du Cinéma* theories on ideology, *TFLW* is a prototypic film indicative of a changed ideology for feminine roles in film through redirection of the film gaze. This specular reversal of gaze is achieved when the actress looks in the mirror. Ultimately, a mirror-within-the-mirror effect emerges between Feminist Film Theory, changing ideology and *TFLW*'s mirrors. The mirror metaphor animates ideology through both the art form and its critical analysts. The refractive duality of the mirror as a deconstructionist tool and as the object of film signification adds a semiotic complexity that cannot be fully resolved within the scope of this paper. However, in both the film form and its critical analysis the mirror as signifier is consistent with Lacan's mirror-phase of development. A brief discussion of the cinematic construction and the narrative text are given for an introductory perspective.

TFLW LITERARY TEXT

The novel *The French Lieutenant's Woman* (*TFLW*) by John Fowles projects itself into the center of any semiotic discussion. The text is constructed as a self-conscious challenge to narrative conventions, in which the author intrudes on the fictional story. All fiction is premised on a suspended belief in the imaginary reality of the story. Narrative conventions are destroyed in Fowles' *TFLW* with both its interactive narrator, and the narrative device of three potential endings.

TFLW is a romance set in the country village of Lyme, England (the same locale as Thomas Hardy's nineteenth century romantic novels) in the year 1867, and in 1967, the year of the author's intrusive writing. The characters' quest for liberation from Victorian social custom is entwined with the novel's quest for freedom from literary convention and modern notions of scientific and historic accuracy. The story, by mocking its own artifacts, seeks a means to define personal identity.

Questions of meaning for signification are at the heart of semiotics. Fowles's trend-setting work is a novelistic inquiry into many of the same questions of semiotics and film theory. Both the novel and the film version of *TFLW* work as interesting mirrors for several questions in film theory including: authorship as signifier, narrative convention as film apparatus, textual analysis as signified meaning, reader identification as psychoanalytic theory and lastly personal freedom as gender bias in the enactment of meaning.

Peter Schillingsburg in his 1997 semiotic study *Resisting Text*, uses *TFLW* to exemplify issues of textual complexity, with regard to both history and authorship. He asks, referring to *TFLW*, "Is it Fowles's novel, Harold Pinter's screenplay, director Karel Reisz's film or Meryl Streep's movie?" Shillingsburg says, "Perhaps we go too far when we say it's a Meryl Streep movie . . ."[4]

Shillingsburg's innocent comment reveals a subtle layer of lingering presumption of male superiority in textual analysis. In terms of Author Theory, why not call it a Meryl Streep movie? A female actress like a director or male actor with a long career can generate a signifying signature in a body of work known as her films. Obviously the assumption of origination in Author Theory is appropriately challenged, yet the significant participants in a film are still used to identify a film by class or type. Noted film critic Molly Haskell, in *Holding My Own in No Man's Land*, reviews the careers of a dozen actresses, including the body of work by Meryl Streep.

With *TFLW*, additional layers of complexity occur in a 1990s novel or film reading of this 1867 story set in 1967. For instance, are the historic nuances and novelty of the film/novel narrative lost on post-modern readers? Are its simplistic notions of liberation pertinent to readers in a time when women's liberation has presumably been accomplished? Are its questions about interpretation prosaic or outdated to readers in an era of quantum physics and deconstructionist philosophy? Does the work possibly reflect a view of changing perspectives, identifying a significant time in history when questions of textual meaning, identity and gender were first asked?

TFLW FILM TEXT CONSTRUCTION

The motion picture's unique textual construction of placing the actors in the text is original and historic for its time, although the device may seem adolescent or "camp" to contemporary film viewers. Other important films have used a similar self-referencing device, such as Godard's *Contempt* about filmmakers making a movie. A variety of movies of the 1980s and 1990s explore this concept in a virtual reality motif, in which characters enter a video game reality or encounter the creators/authors of the games.

A note on the cinema production: *TFLW* was nominated in 1981 for Academy Awards for Best Actress, Film, Screenplay, Art Direction, and Costume, and received Golden Globe Awards for Best Film, Screenplay and Actress. This work may have launched the careers of actors Meryl Streep and Jeremy Irons, as well.

Notable for its exceptional aesthetics in all cinematic categories, the film *TFLW* endures. A few examples are offered: The cinematography for the Victorian text achieves a distinctive nineteenth-century look, with sepia tones and candle or gas-fixture lighting. The innovative musical score and soundtrack create a unique interplay of visual and auditory messages to queue scene changes, subvert narrative with self-referencing signals, or enhance emotional tones. As additions to film sensation, the soundtrack authenticates and emphasizes different noises from the two centuries. The Victorian scenes contrast sounds of the "quiet" with minimal background noise such as ticking clocks, rustling leaves or turning wagon wheels, against the modern noises of helicopters, cars, and telephones.

The screenplay, while radically departing from the book version, adapts cinematic conventions to render the novel's dual text as a movie-within-the-movie and in so doing maintains an admirable fidelity to the spirit of the original text. The film/sound editing is a masterful creation and blending of the two narratives into a tightly woven configuration. One outstanding edit sequence portrays the two modern actors rehearsing a scene for the Victorian movie in which the woman trips and falls. The edit adroitly cuts from the actress in mid-fall in the modern rehearsal space, to the nineteenth-century woods, setting of the actors in their Victorian roles, and continues with the second story line. The cinematic features of *TFLW* in and of themselves could be the subject of a fruitful critical analysis of film aesthetics.

CHANGING IDEOLOGY, MIRRORING AND A DUAL FILM TEXT

In its time, *TFLW* was a film of distinction that represents a type of ideological breakthrough discussed by French *Cahiers du Cinéma* critics Jean-Luc Comolli and Jean Narboni in "Cinema, Ideology and Criticism." They conclude that films as by-products of the economic and social system, reproduce reality mirroring the operative ideological systems, and thus indicate the status of prevailing social institutions in a given time. They state: "The film is ideology presenting itself to itself, talking to itself, learning about itself . . . we can see that the filmmaker's first task is to show-up the cinema's so-called depiction of reality."[5] The narrative text of *TFLW* embraces this ideological en-garde with two-fisted enthusiasm.

Harold Pinter's screenplay adaptation transforms the intrusive narrator of Fowles's novel into the lead actors of the Victorian movie text. A second story of an illicit love affair between the married lead actors (Mike and Anna) frames the Victorian story line. This Victorian text introduces Charles Smithson an aristocrat scholar paleontologist, the male lead, played by Jeremy Irons, as he makes an important decision. Charles summons his coach to go and propose marriage to Ernestina Freeman, the shallow and petulant daughter of a wealthy industrialist. With birds chirping in the garden of her father's estate, the couple clumsily yet fervently embrace to formally seal their engagement.

The scene changes to what appears to be the Victorian character, Charles, asleep in bed. The soundtrack nonchalantly informs the viewer of a jarring juxtaposition of narrative continuity by the familiar ring of a twentieth-century telephone. The viewer must then reconcile the impossibility of a telephone in this nineteenth-century text. The mirror of film has projected the viewer into the present; the viewer sees the actor in another role, as Mike, a film actor in the movie. Mike wakes and groggily answers the telephone while the camera pans to indicate a woman asleep next to him. The phone call is for Anna (played by Meryl Streep), because she is late for her make-up call. Anna wakes and says, "Now they know about us," meaning their illicit love affair. Yet from this moment, the viewer knows this film is now telling another story. The viewer is now engaged in a story that reframes assumptions about film narrative. Of course Anna dashes off for her call.

The love story of the co-stars mirrors the romance of the Victorian characters, mirroring issues of changing feminine roles, while simultaneously breaking the rules of story-telling. The mirror is also a metaphor of the fiction of historical ideology and fact, projected through time. In the film, the mirror raises questions of reader/spectator interaction with the fiction. The question of narrative form is self-consciously asked. Does the movie end happily, or is it a tragedy?

According to Comolli and Narboni, critics for *Cahiers du Cinéma*, commercial films are emblems of ideology. As products of prevailing institutions, they enact, reveal and/or change ideology. These critics apply a continuum of five categories to discuss how a film relates to social and political ideology. The first category reinforces the status quo of existing social and political ideology. Category two films openly attack ideology on political and cinematic grounds. Category three films are mainstream and not directly political, yet go against ideology through their use of form and are of primary interest to critics at *Cahiers du Cinéma*. Category four films appear to attack political ideology but actually reinforce it. Category five films appear to belong to the prevailing ideology yet unconsciously undermine the cinematic framework from within. *TFLW* fits the third category, as a film that hails an ideological shift with its use of film technique.

As such *TFLW* is not an overtly political or feminist film, although its story is about liberation and self-definition centralized on the feminine. The French Lieutenant's woman, Sarah Woodruff (Meryl Streep), is not a stereotypic Victorian woman, or a vacuous Hollywood romantic character. Her desire is not for the man of her dreams, but for personal expression and autonomy. She is artistic and educated, and aspires to break out of the limited sphere of activity allowed a single woman in colonial England. Her desire is to be loved without being shackled.

The mirror projects illusion or a facsimile of reality. The film's opening shot and first mirror scene discloses its artifice to underscore this point. In a medium close-up a hooded figure of unknown gender wearing an elaborate Victorian cape, is being groomed. A small compact mirror is held up. The camera and the viewer look over the hooded figure's shoulder into the upheld mirror. From this perspective, two of the three film gazes discussed by Mulvey are merged in a single viewpoint in the mirror. For a disconcerting moment the unknown actor looks directly into the camera lens – a film taboo, but through the camera focused on the mirror. She looks in the mirror, and her gaze meets the viewer's gaze. This first scene asserts the film illusion and the break with convention. In the film's second shot the camera pans backward to display a film set. The intriguing figure is now revealed to be a woman.

The director calls "action," the film begins, the credits roll and the cloaked figure strides alone out from the film set onto a long breakwater while waves crash and the wind violently blows her billowing cape. The drama begins, and engages the viewer in the filmic spectacle. As Baudry describes, the viewer has entered a dream-like regression into the subconscious. This elaborately staged and filmed scene is a conspicuous yet subtle application of film art, to play the illusion to its full potential – the camera and spectator gaze – are unified in the same line of perspective. The third gaze, that of the actor, is now reversed for a moment as the film explicitly states itself to be a film of a film production.

The opening scene of the actress prepping, the director's call "action," and her walk on the breakwater, while the credits roll, describes the first text. The spectator has seen the mirror, seen the actor's gaze pointed at the camera, and yet is not sure what meaning is in the image. The hooded figure reaches the end of the breakwater and looks out to sea. She stands undaunted at the edge, facing the danger of her precarious situation. The transition from the first text then to the second Victorian story is virtually subliminal.

When the credits complete, the scene shifts from the breakwater to prosaic shots of an English Victorian village (Lyme) and the home of the young paleontologist, Charles Smithson, as he decides to propose marriage to Ernestina Freeman. Soon the viewer has forgotten about the enigmatic opening scene of the actress walking on the breakwater and is absorbed into the eighteenth-century romantic narrative, until later disturbed by the wake-up call and the continuation of the first sub-text with Mike and Anna, the film actors. Of course Anna, the actress who controls this drama, has rushed from the hotel room to the film set for her scenes as the estranged Sarah Woodruff.

The film meeting of the Victorian characters, Charles and Sarah, is delayed for dramatic effect. On a windy day, Charles and Ernestina stroll alongside the breakwater discussing their marriage plans, when Charles sees the lone woman, Sarah, standing at the edge of the breakwater. Charles asks Ernestina about the strange woman, and she says, "Oh that is Tragedy, the French Lieutenant's whore, er . . . woman." The wind is blowing hard, Charles feels compelled to walk out on the breakwater to bring the unknown woman to safety. On this precipice of land and sea, a metaphor for the separation of gender, class, and dual texts, they meet.

Sarah looks at Charles as he runs toward her, dodging the crashing waves to call her back to safety. Her gaze is vulnerable, challenging and arresting. They stand staring at each other. The photographic still of this scene is the signifying moment of the film, used as the film's poster image. The viewer may have forgotten the opening scene of the actress preparing for the shoot of this scene, but now, is returned to the breakwater and this moment where the characters meet and the viewer like Charles encounters a jarring juxtaposition of unreconcilable images.

Much of the nineteenth-century story occurs in the village outskirts of the surrounding woods called the undercliff, where Charles explores the exposed rock formations and notices the mysterious Sarah walking alone. They engage in conversation as Charles tries to understand her predicament and why she stares out to sea. She wishes to expose herself and confide in someone. The locale of the woods and the undercliff have erotic undertones. It is where the Victorian romance is engaged and where Sarah violates convention by walking alone. The dangerous undercliff at the shore of the sea is a representation of the portal of feminine consciousness and sexuality that the film explores. The latent fears and menacing consequences of empowered female sexuality represented in Freud's nineteenth-century psychoanalysis come to the surface at the undercliff in *TFLW*. Sarah then pursues Charles by sending him a note asking him to meet her.

MULVEY'S ANALYSIS OF THE FILM GAZE

Questions about the film gaze dominated film theory during the 1970s, in the search for the cause of meaning by identification of the unconscious, using Freudian and Laconian psychoanalytic methods. Mulvey uses these methods to postulate that films prior to 1974 are grounded on the male gaze, and in that gaze, the active female spectator is a virtual non-entity. According to Mulvey, the function of the male gaze, in both the narrative and the spectator, has a common origination. The male gaze of the actor has the power to enact the action, and by watching the empowered actor the viewer's gaze neutralizes the repressed castration fears, represented in the female. The male gaze upon the female object, driving the film narrative, creates a fetish object of the female character, and disavows the castration fear, embodied in her differentiated gender status.[6]

Mulvey turned psychoanalysis on its head in her radical analysis of the film apparatus and gaze. Using the most phallocentric of its analytic tools, Freud's psychoanalytic method, she exposed film's ideological representations of male dominance. The female role as fetish object is either a perfected form in the screen goddess, or punished and purified by the film drama, particularly in horror and morality genres. The male spectator reenacts the Freudian projection of castration fear and suppresses it through fetishism, by his participation and identification in the cinema narrative. In *TFLW*, the lead character Sarah Woodruff looks out to the sea of possibility envisioned in Mulvey's 1974 concept of a new cinema. Sarah is an artist and she gazes at her world. She actively recreates it with drawing, while her counterpart, Anna the actress, recreates the film illusion.

MIRRORING IN AVANT-GARDE FILM

After Mulvey, feminist theory turned to the matter of the male gaze. Ann Kaplan addresses the privileged status of the male gaze in Hollywood films in "Is the Gaze Male?" She identifies the early 1980s New Wave film movement in Europe and America. Kaplan analyzes New Wave Films that explored the changing female gaze and roles. She cites Von Trottas's film, *The German Sisters*, about women actively engaged in the struggle to define their lives.[7] Kaplan asks the question: when women are dominant, are they then characterized in the masculine position? She concludes that many late 1980s Hollywood films do not present a new vision, but rather represent the dominant submissive ideology with a minor role reversal. *TFLW* however, as a major studio film, presents a woman who maintains her sensitivity, while striving for control of her life and sexuality.

Not only does the *TFLW* film hold a mirror to film's illusion when the perspective shifts to modern times with views of the actors behind the scenes, but it also effectively reverses the direction of the gaze cast in film. How do the mirror's scenes in *TFLW* reverse the direction of the gaze in film and simultaneously reflect a change in the ideology of film portrayals of women? For early Feminist Film Theorists, the mirror is a potent icon that discloses the ideology of gender. Interesting intersections occur between the hermeneutics of the theorist and the filmmaker's art in the analysis of this film. The concomitant use of the mirror as metaphor in this ground-shifting film parallels the discourse of film theory itself.

As stated, *TFLW* reversed the direction of the gaze in film in its opening scene when the actress projects a dead-eye look into the camera lens through a ladies' make-up mirror. This look disturbs the viewers' sensibilities about reality, illusion, film and women. Anna the actress poses with fabric before a full-length mirror, saying, "I shall like Sarah in this." Here, the woman directs the action. Anna the actress actively creates her illusion of Sarah, the character of the second text. Simultaneously, Sarah's look consumes Charles. His attachment is the cause of his social ruin in Victorian society. In the modern story, the actor Mike is consumed by Anna. In the final scene, Mike watches Anna drive away, calling the name "Sarah," Anna's characterization. Not even the actor can separate himself from the illusion of the apparatus.

The mirror-phase of development is highly important to psychoanalysis and film theory. Kaplan describes two avant-garde feminist films about this issue. *Sigmund Freud's Dora* (1979) by Tyndall, McCall, Pajaczkowska and Weinstock, addresses psychoanalytic theory and cultural discourse in the arts. Sally Potter's film *Thriller* studies how art makes the woman a victim by murdering her in order to create the spectacle of the drama, in the tragic opera *La Boehme*. *Thriller*, like *TFLW*'s Sarah/Anna pair, creates a diachronic lead character of Mimi I, a speaking subject outside the narrative, and Mimi II, the heroine of the story, with significant mirror scenes.[8]

As Kaplan observes in her analysis of *Thriller*, the mirror in this film signifies the entry of both the feminine role and the filmmaker into the mirror-phase of Laconian self-awareness. Both avant-garde films (*Thriller* and *Dora*) exemplify new trends in film texts in the early feminist era and use the mirror literally and symbolically to project, expose, and re-direct feminine iconography.

Eventually the feminist perspective explored in early New Wave films is brought to a wider audience in the commercially successful *TFLW* in 1981. Like these avant-garde films, *TFLW* uses the mirror as the instrument of transformation. Feminism embraced the reflexive qualities of the mirror as a weapon to assail the gorgons of patriarchy with the disclosure of their monstrous presumptions in art and society. Concomitantly, as an emerging socio-political trend, feminism itself was in its mirror-phase of expression and dissemination

MIRROR METAPHORS IN FEMINIST FILM THEORY

In retrospect, the mirror-phase aptly describes the formative period of Feminist Film Theory itself during the 1980s. Examples of feminist explorations employing the mirror now abound. French cultural philosopher Luce Irigary used the mirror metaphor in her *Speculum of the Other Woman* (1974) for its premise that western ideology effaced woman as the specular image of man. The work takes its title from the curved mirror of self-examination used in gynecology. Like Mulvey, Irigary affirmed in 1974 that there is no place for the feminine other than one of repression and censorship. Irigary was expelled from academic circles because of her contentious and sexualized writing.[9]

A survey of feminist film theory titles underscores the mirror as feminist referent. For instance, there is *Re-Vision Essays in Feminist Film Criticism* by Doane, Mellencamp and Williams; Kaplan's *Women in Film. Both Sides of the Camera*; *The Future of an Illusion: Film, Feminism and Psychoanalysis* by Penley; Rosenburg's *Women's Reflections: The Feminist Film Movement*; *The Acoustic Mirror: The Female Voice in Psychoanalysis and Cinema* by Silverman; *The Power of the Image* by Kuhn and Gilbert and "The Mirror and the Vamp: Reflections on Feminist Film Criticism" by Knight.[10] The mirror is now used to refer to a branch of feminist film theory and is the source of heated discussion. Toril Moi, in "Am I that Name? A Reply to Deborah Knight," challenges Knight's designation of two camps in feminist film theory, the humanist "mirror critics" and the anti-humanist "vamp group."[11]

ANALYSIS OF THE FILM GAZE

The question of the woman's gaze is crucial in the synthesis of feminist film theory. As Mulvey states, "It is the place of the look that defines cinema, . . . and the possibility of varying it and exposing it." According to film theorist Linda Williams, the woman in later films is allowed to look, but most typically in the horror film. Her gaze then reinforces Mulvey's theory of castration in reverse. For instance, the figurative castration fear is enacted by the killer upon the woman in *Dressed to Kill*, with the implication that her desiring look caused him to kill.[12]

According to Williams, the female gaze serves to reflect a monster to her. When Christine looks in the mirror in *Phantom of the Opera*, "she sees a monster that offers a distorted reflection of her own image. The monster is thus a particularly insidious form of the many mirrors patriarchal structures disclose of the view of woman."[13] The monster to the male spectator is the ugly consequence of the difference recognized in Freud's theory of castration.

Whereas most horror films such as Hitchcock's *Psycho* and others perpetuate the identification of woman's sexuality with a terrifying monster, vampire films portray feminine sexuality in a different light. Williams suggests that in the instance of vampire films, "the woman's look at the monster offers a potentially subversive recognition of the power and potency of non-phallic sexuality."[14]

The question of the female look is re-viewed in a historical context with Miriam Hansen's analysis of female spectators and Rudolph Valentino. The "Valentino effect" on female audiences is a well-known historic phenomenon of consequence. Because of the challenge that the "Valentino effect" presents to Mulvey's theory of a non-existent female viewer and gaze in film, Hansen's research offers an important dimension.

Hansen notes that although promoting an active female viewer, Valentino films reinforce the existing gender rules for gender gaze. The woman Valentino looks at is his chosen partner, yet the woman who gazes at Valentino is a vamp, destined for condemnation in the film story. She further notes that the code of male prerogative is upheld in the narrative convention of the "first look." She concludes, "Valentino's appeal depends on . . . the manner in which he combines masculine control of the look with the feminine quality of 'to be looked-at-ness.' . . ."[15] Hansen also observes the parallel trends of the 1920s Valentino craze with the changing social role of urban women entering the industrial work force.

DIRECTION OF THE GAZE IN *TFLW*

The role of the gaze in film is significant in the viewing of *TFLW*. In its diegesis, the pre-eminence of the first male look is usurped. It is the female lead Sarah and the actress Anna who have the first look and motivate the narrative story. Sarah states her control of the look during the film climax after the lovers' sexual tryst when she states, "I wished it to be so. I have long imagined a day like this . . . from the moment I saw you."[16] She claims her sexual desire and control of the gaze, the monster fearfully represented in the mirror of horror and vampire films. Sarah initiates the action when she passes a note to Charles, walks alone in the woods, and later leaves Charles.

Sarah's look is the central construct of the film's action. The look is played with full force in Streep's performance. The mysterious hooded figure in her billowing cape as the nineteenth-century romantic heroine reveals the twentieth-century fascination for the camera's vivification of the imaginary. In their first encounter, Sarah looks at Charles from behind her caped persona with disturbing sensuality and vulnerability. That look is the film poster icon. Her look is the sizzle that sells the cinema and it unveils a chimera of fantasies of a romantic bygone era within a modern-day illusion. Her look entices the viewer and Charles, the viewer's signifier, into the illusion. It disturbs Charles. The viewer is simultaneously disturbed and entranced by the film's deliberate breaking of the Victorian narrative with scenes of the modern-day romance.

Sarah's look, rather than repudiating Freud's castration fear, reinforces it. The mirror, with its monstrous connotations, is upheld. For when Charles succumbs to Sarah's look, he becomes a social outcast, cut off from the proper world he had known. It is a world shown in vignettes of jolly good old boys drinking at the club, rational inquiry into Cartesian causation, and imperious bourgeois superiority to servants, women and other classes. However, as an outcast from this superficial and privileged social circle, Charles does gain the ability for self-reflection which Sarah represents, as he wanders through the Victorian underworld looking for her.

If the look of the woman represents the odious other who threatens castration, her look also threatens the destruction of gender and racial supremacy. On another level, the film reveals the collapsing ideology of white male dominance in the hundred years between 1867 and 1967. This smug Victorian world of Charles Smithson the colonial aristocrat is glaringly offensive to the egalitarian senses of modern viewers in the aftermath of the socialist, racial and gender movements at the end of the twentieth century.

MIRRORS SCENES IN *TFLW*

The mirror as a symbol of feminine vanity and weakness is also a weapon of artifice to create herself. A close look at the nine mirror scenes of *TFLW*, shows seven are controlled by the female lead. The first is the film's opening scene with Anna preparing for the shoot. The actress's face is exposed in the small hand-held mirror on the film set, but her body is cloaked and the gender is uncertain. What is seen is the face in the mirror, the expression of self and identity. This actor violates a filmic convention and makes eye contact with the camera in a fleeting focused moment of acknowledgment through the mirror. The mirror is seen over the actor's shoulder, from the same vantage point as the character gazing. Here the film image claims the gaze of the viewer and the woman/actress looking, as a shared perspective. The gaze is not a lascivious viewing of the female sex object fetish, but instead a woman's gazing, in an act of her own creation. In her eye contact with the camera, the actress casts her eye upon the viewer and the mirror of film is reversed in a new projection.

The second mirror scene discloses the rigid Victorian standards of behavior that preceded twentieth-century industrialization and women's emancipation. Sarah is an educated woman and a destitute outcast. She interviews for a position, as a companion, for the widow Mrs. Poultney, a hypocritical arbiter of nineteenth-century morality. Mrs. Poultney's mirror is the field of Sarah's angst. The view of Sarah held by Victorian standards does not match her self-view.

In this scene the audience learns about Sarah's situation. She stands facing the seated grand dame. The camera shows Sarah's face, while Mrs. Poultney is witnessed as a hazy reflection in the full-length mirror behind Sarah's shoulder. This uncanny disturbance of speaker and camera position is visually significant. Mrs. Poultney states, "I'm informed you have an attachment to a certain foreign man and you stare out to sea waiting for him. Such staring is provocative. . . ."[17]

Sarah is hired with the restriction that she may not walk in the woods or on the undercliff. The distorted mirrored display of Mrs. Poultney indicates that her knowledge about Sarah is skewed and false, in fact a fabrication perpetuated by Sarah. Sarah later discloses to Charles the fiction of her relations with the French Lieutenant. The status of shamed outcast gives Sarah freedom from the normal social conventions allowed for women. She says, "I have freedom . . . no blame can touch me . . . I have set myself beyond the pale . . . I am nothing, hardly human."[18]

Charles discusses Sarah's bewildering nature with the local Dr. Grogan. Citing Freudian psychoanalysis, Grogan explains the three kinds of melancholia. The first is natural; one is born with it. The second is temporary, caused by trauma, and the third, as in Sarah's case, is unexplainable. Sarah's angst is unfathomable to the colonial gentlemen's assumptions of privileged class and male superiority. Sara tells Charles, "You cannot imagine my suffering. . . . Why am I not born Miss Freeman?"[19]

Charles encourages her to leave and go to London. The scene changes to the third mirror scene within the modern text to explain Sarah's dilemma. The two actors are studying the script together in Anna's room. Anna reads aloud from a book – Fowles' novel: "In 1867 there were 80,000 prostitutes, and one in six homes was a brothel, in London."[20] A disturbingly loud motorcycle drives by and Mike gets up to close the window adjacent to Anna's dresser and mirror.

Anna's dresser mirror is covered with notes to herself. This is not the emblem of vanity, but of self-direction. The camera pans back to Anna. She is not shown reflected in the mirror. The mirror is the projection of history and the literal facts of history examined in the film's projection. The mirror is silent yet evident. The scene returns to 1867, and Sarah tells Charles, "If I go to London I would become what they call me."[21] The text holds a mirror to history, contradicting clichés about Victorian morality, while showing the modern Anna's freedom.

The fourth mirror scene depicts Anna posing in a full-length mirror consulting with the costume designer. She is planning her characterization of the role of Sarah. The large mirror is minimized in the brief shot and barely seen from a side angle. Her image in it is skewed and almost imperceptible. Anna is confident, and pleased with her decision. The short scene reinforces the subliminal presence of the mirror as an instrument of reversal.

The fifth and most significant mirror scene at the film climax discloses Sarah drawing a self-portrait. She draws incessantly in moments of solitude. Sarah looks into an ornately framed picture-size mirror, and she is illuminated by a thin wavering candle. The camera frames Sarah's face in the mirror, as if a portrait. She studies herself not in vanity, but with penetrating examination, and draws vigorously.

While gazing, Sarah clutches herself in a mournful embrace of anxious distress. The camera lingers for nearly forty seconds on her fervent image framed in the mirror, and never before has film beheld a female character so genuinely engaged with herself. The film shot is almost a still and it is a shot of long duration with minimal action, creating a sense of intense emotion and personal reflection.

The film text does not display the actress; instead the film depicts the mirror filling the frame with the reflection of the actress's gaze within it. Although the haunting seductive gaze of the hooded woman on the breakwater is the film marketing icon, this is the signifying gaze in the film. The disconcerting moment of the film's first mirror scene is now completed in Sarah's sustained gaze of self-reflection. The scene is performed and filmed with an arresting intimacy, showing a lament for the universal oppression of women as a class. Sarah's plight is every woman's struggle. The accompanying artist drawings repeat the vulnerable picture seen framed in the mirror, framed in the film.

While Sarah reflects and draws, the maid urgently knocks on the door, and calls her to come and see Mrs. Poultney at once. The grim desperation of Sarah's nineteenth-century position is confirmed when Mrs. Poultney discharges her for "being seen" walking alone in the woods. Sarah sleeps that night in a barn. These fourth and fifth mirror scenes purposefully describe women engaged in creation, introspection and self-awareness. This is not the Hollywood stereotype. These women are not primping in preparation for a man's gaze. They have intent, and are interested in themselves and their creations. These are images of women in their own right, mirroring the 1970s feminist's affirmation of women's rights.

Out of chivalry or lust, Charles finds Sarah and gives her money to leave the village of Lyme. She rents a room in Exeter and sits by the fire. The sixth mirror scene at the inn foreshadows a change in direction for Sarah. She looks in the mirror and models a new shawl for herself, enjoying her peace and comfort. Charles finds her at the inn and they quickly consummate their repressed Victorian passions. As Charles is a moral man, he promises to marry Sarah and leaves to break his engagement with Ernestina, telling Sarah he will return later.

Ernestina is first pleased, though petulant, by Charles' unexpected visit and his request to meet in the imposing library. A large wall-length mirror above the sitting couple frames their conversation. However, neither character is reflected or revealed in this mirror which dominates the set. Ernestina's crestfallen mood turns to angry revenge when Charles tells her he cannot marry her. In this seventh scene, the mirror is the unseen real woman that Sarah strives to be. Although vain, pampered and protected by social class, Ernestina does not hold the mirror's power of self-realization and determination. Likewise, Charles cannot see himself, and must now enter his own journey of self-reflection.

When he returns to the inn, Sarah has left. She disappears into obscurity for three years. He searches for her without success and finds himself in solitude staring out to sea, as Sarah stared when they first met. Sarah allows Charles to find her by sending his solicitor a letter. When she asks for his forgiveness and explains her need for freedom, Charles sputters angrily, "Your freedom!" He is outraged and strikes at her, but a kiss seals their reunion and happily concludes the Victorian love story. As the couple romantically ride away together in a boat, the breakwater of the land and sea where they met has now been bridged.

The male role has one mirror scene. Mike the actor wakes up one restless night. He stands and gazes out the window into the darkness of uncertainty. His face is reflected in the window glass. His reflection is shaded and imperfect however, without the clarity of a mirror. He returns to bed and looks at Anna sleeping. He, the actor, is gazing at the source of Mulvey's film fetish, the unobtainable real actress, and he knows their affair is temporary and fleeting. Charles uses several lenses, a magnifying glass, a microscope and a telescope in his scientific examinations, but he does not have the power of reflection or projection of the mirror. He cannot separate from the assumed premise of gender superiority.

Mirrors represent reversals. In the story, the film gaze is shifted. A woman is not looked at but instead looks at herself and appraises her situation. The actress looks at the viewer and creates her identity. In *TFLW*, Sarah stands at the undercliff, appearing to wait for a man who she knows will not return, and plays herself, a compromised woman, in order to gain a small amount of personal freedom. There is also a reversal of roles in the text. It is Charles who is left estranged, waiting and searching for Sarah, when she leaves, changes her name and eludes Charles for three years while she seeks a place for herself.

The final mirror scene involves Anna the actress in her dressing room. The film shoot is complete. It's the cast party and she prepares to leave to return home to her spouse. The mirror's projection through time is complete, from Victorian oppression to sexual liberation. The woman faces herself in the mirror, and the actress removes her costume and leaves. The camera focuses on the actress looking at herself in the mirror. Here we see a side profile of both the actress and the reflection in the mirror. The film text claims and discloses its dual image and its tool of projection.

Mike looks for Anna in her dressing room. He fondles her discarded wig, the effigy of her fetish role. Hearing a car engine, he runs and looks out the window. As he watches Anna's car leave, he calls the name of her illusion, Sarah. The film ends with his longing for a woman who, if she ever was, no longer is. The window was his mirror of reflection in the sleepless night with Anna. Now he gazes through the glass to find she is gone.

The male gaze is confounded, ignorant and frozen in Mike, the actor. Mike succinctly represents the shock of the male spectator toward this shift in gaze and ideology, with his calling for the fetish object of the cinematic illusion. Mulvey's prediction about male viewer response to an empowered female in film is proven in *TFLW*'s final scene. Mulvey states, "Simultaneously, the look of the audience is denied an intrinsic force: as soon as fetishistic representation of the female image threatens to break the spell of illusion, and the erotic image on the screen appears directly (without mediation) to the spectator, the fact of fetishisation . . . freezes the look, fixates the spectator and prevents him from achieving any distance from the image in from of him."[22]

The gaze of the Victorian character Charles, however, is not vilified in the texts, for his story ends happily. When Sarah returns, she has discovered herself, an artist and governess, happily situated among people who encourage her talent. The reunited couple exit together in a quaint rowboat. *TFLW*'s dual endings are self-conscious eventualities, playing the odds of fictional outcomes. The enigmatic drama of the film's opening is completed with its contrary endings as a disconcerting reminder of the film's veritable artifice.

CHANGING IDEOLOGY

As an exploration of changing roles, the *TFLW* effectively mirrors changing social conditions for the status of women, and the concomitant change in the male spectator. The 1981 scene of Sarah studying herself in the mirror to draw a self-portrait is a pivotal moment in the twentieth-century film art, shifting the axis of the film gaze, as described by Laura Mulvey in 1974.

Sarah's self-portrait mirror scene is also reflexive of a 500-hundred-year old illusion in visual art. In the 1980s, a team of computer analysts at MIT led by Lillian Schwartz examined the mystery of Mona Lisa's illusive smile. In the digital analysis, the facial features of Mona Lisa were compared to self-portraits of the artist Leonardo da Vinci. Researchers concluded the faces were identical.[23] In da Vinci's immortal painting, Mona Lisa's illusive smile is but a mirror reflection, a self-study of the artist, in a feminine guise. Sarah's self-reflection represents the woman's purposeful mastery of her gaze in film and her empowered action to create her own image, and this turn of the eye is duplicated in the actress Anna's modern liberation.

TFLW invokes a historic use of mirror in the film text signifying the liberation of woman's gaze and the advent of empowered female characterizations. The mirror is also an icon of the first wave in Feminist Film Theory. In turning the mirror on herself, the woman in film was released from the status of fetish object and ornament.

As woman's gaze provokes images of the monster of the empowered feminine, and as the mirror is an instrument of feminist liberation, Medusa has fully turned the mirror against the ideology of patriarchal privilege. When she looks in the mirror in the 1981 film, *TFLW*, Sarah/Anna portrays self-referencing feminine empowerment. This shift in the direction of the feminine film gaze reflects an ideological shift of the role of women in film and society at the end of the twentieth century.

NOTES

1 Metz, "Identification Mirror" in *Film Theory* 805.
2 Baudry, "The Apparatus" in *Film Theory* 769.
3 Dayan in *Film Theory* 120–121.
4 Shillingsburg 113.
5 Comolli in *Film Theory* 755.
6 Mulvey in *Film Theory* 838–841.
7 Kaplan 105.
8 Kaplan 155.
9 Briganti, "Irigary" 2.
10 See Works Cited.
11 Moi 1.
12 Williams 95.
13 Williams 88.
14 Williams 94.
15 Hansen in *Film Theory* 588.

[16] Pinter 72.
[17] Pinter 16–18.
[18] Pinter 33.
[19] Pinter 49.
[20] Pinter 44.
[21] Pinter 50.
[22] Williams 844.
[23] Schwartz and *Los Angeles Times*.

WORKS CITED

Baudry, Jean-Louis. "Ideological Effects of the Basic Cinematography Apparatus," in *Film Theory and Criticism*, 5th ed. Eds. Leo Braudy and Marshall Cohen. New York: Oxford University Press, 1999.

Briganit, Chiara; Robert Con Davis. "Luce Irigary," *Guide to Literary Theory and Criticism*, 1997.

Comolli, Jean-Luc; Jean Narboni. "The Apparatus: Metapsychological Approaches to the Impression of Reality in Cinema," in *Film Theory and Criticism*, 5th ed. Eds. Leo Braudy and Marshall Cohen. New York: Oxford University Press, 1999.

Dayan, Daniel. "The Tutor Code in Classical Cinema," in *Film Theory and Criticism*, 5th ed. Eds. Leo Braudy and Marshall Cohen. New York: Oxford University Press, 1999.

Doane, Mary Ann; Patricia Mellencamp; Linda Williams. *Re-Vision Essays in Feminist Film Criticism*. Frederick, MD: The American Film Institute, 1984.

Hansen, Miriam. "Valentino and Female Spectatorship," *Film Theory and Criticism*, 5th ed. Eds. Leo Braudy and Marshall Cohen. New York: Oxford University Press, 1999.

Haskell, Molly. *From Reverence to Rape: The Treatment of Woman in the Movies*. New York: Holt Rinehart and Winston, 1974.

Haskell, Molly. *Holding My Own in No Man's Land: Women and Men and Film and Feminists*. Oxford: Oxford University Press, 1997.

Kaplan, E. Ann. *Women and Film. Both Sides of the Camera*. New York: Metheun, 1983.

Kellman, Stephen. "The French Lieutenant's Woman," review in *Salem Press*, 1997.

Los Angeles Times. " Reason for the Smile: It's a Self-Portrait. Computer Says 'Mona Lisa' Is 'Da Vinci. Dec 18, 1986; p. 2.

Metz, Christian. Excerpts from "The Imaginary Signifier," in *Film Theory and Criticism*, 5th ed. Eds. Leo Braudy and Marshall Cohen. New York: Oxford University Press, 1999.

Moi, Toril. "Am I that Name? A Reply to Deborah Knight," *New Literary History*, 26.1: 1995, 57–62.

Mulvey, Laura. "Visual Pleasure and Narrative Cinema," in *Film Theory and Criticism*, 5th ed. Eds. Leo Braudy and Marshall Cohen. New York: Oxford University Press, 1999.

Pinter, Harold. *The Screenplay of the French Lieutenant's Woman*.

Schwartz, Lillian, Laurens R. Schwartz. *The Computer Artist's Handbook*. W. W. Norton, 1992.

Shillingsburg, Peter. *Resisting Texts. Authority and Submission in Constructions of Meaning*. Ann Arbor: The University of Michigan Press, 1997.

Williams, Linda. "When the Woman Looks," in *Re-vision Essays in Feminist Film Criticism*. Editors Doane, Mary Ann; Patricia Mellencamp and Linda Williams. Frederick, MD: The American Film Institute, 1984.

The French Lieutenant's Woman. Dir. Karel Reisz. Perf. Meryl Streep and Jeremy Irons. United Artists 1981.

SPECULATIONS ON VISION

Van Gogh cut off his ear
because he couldn't stand to see
unlike Salvador Dali
who ritually
plucked his eye
in a movie

Oh the masterpiece
by Leonardo Da Vinci
Mona Lisa's mystery
the secret posed to history
her illusive smile is merely
a mirror's memory

Look, Pablo Picasso's
abstractions and blue cubes
are the real close-up view
of two
lovers face to face
eyeball to eyeball

See?

This poem anticipates by ten years the 1986 controversy reported in the *New York Times* and *Los Angeles Times* about art historian and Bell Labs computer graphics engineer Lillian F. Schwartz's claim in *Art & Antiques (1987)* that digital analysis proves the *Mona Lisa* to be none other than Da Vinci's self-portrait or a "mirror's memory".

Lillian F. Schwartz and Laurens R. Schwartz. *The Computer Artist's Handbook* (W. W. Norton, 1992).

"Reason for the Smile: It's a Self-Portrait Computer Says 'Mona Lisa' Is 'Da Vinci'"; *The Los Angeles Times*: Dec 18, 1986; pg. 2.

"Artist's Theory Gets the Brushoff"; Ann Herold, *The Los Angeles Times*: Dec 19, 1986; pg. 2.
"We consider it to be a lot of nonsense and not worth the time to discuss it," said a spokeswoman for the European Paintings Department of the Metropolitan Museum of Art in New York. "Every artist paints images that look, to a certain extent, like him. That's a truism in art", said James Beck, chairman of the Art History Department at Columbia University and author of a book on Italian Renaissance painting.

THE GIRDLE OF VENUS

Dedicated to the gallant women of art and the academy at the World Phenomenology Institute with their scarves valiantly wrapped.

The Girdle of Venus
A sash a shawl a scarf a shield
Born of the ocean indomitable as the sea
She carries no sword of lightening bolt speed
She wears no crown nor rides a steed
Her only badge of power is a simple shield
The girdle wrapped about her undulant hips veils
the divine womb primordial void where
life is born to guard the source of creation
What power rests hidden in her modest array?

A sash a shawl a scarf a shield
Grandma's bright babushka worn everyday
under overcast clouds and hot summer's blaze
Gypsy girl's scarf tossed with dashing flare
tightly coiled binds her free flowing hair
unwinds to bandage a wound with gentle care
knots together baggage and goods for the fair
To save a life, a cloth is even ready there
to smother a burning blaze suddenly flared

A sash a shawl a scarf a shield
Basket full of ripe fruit, her braided gear
carries home, the ground spring water clear
Handkerchief in sister's purse is ever handy here
to wipe baby's fat cheek or a sadly falling tear
Coyly tossed, her hanky beckons love near
Worn by a champion knight on his spear
her kerchief is sweet token of his lady dear
In a man's pocket, sign of the goddess near

A sash a shawl a scarf a shield
Golden kite catching fancies in the wind
Bright canvas portraying the artist's vision
Thin wispy ream of delicate onion skin
leaf laden with power of the word written
The printer's platen of firm fabric casting
ink's impression, words for mass production
The simple candle wick and electric filament
sends light into night's dark fear bending

A sash a shawl a scarf a shield
Winding sheet of father's coffin
gentle glory for the soul's procession
stitched and sewn with fond attention
weaving kinship with tradition
Heirloom quilt on mother's bedstead
where family's intimacies are said
from childhood's loud romping play
to the quiet mournful last dying day

A sash a shawl a scarf a shield
The peasant's cloak and nomad's turban
unfurls into a warm blanket for cuddlin
shading curtain, and towel for her skin
or fashions a bed on the earth to rest in
The humble rag–tattered cleaning cloth sweeps
the table scraps of litter and human debris
trails the unending parade of human need
Its thankless service never quite complete

A sash a shawl a scarf a shield
Taut sail propelling a mighty ship
launching sailors on their questing trip
over the waters from which Venus sprang
Atop the ship's mast her flags are waving
revered symbol, honor of the nation
On the deck a sailor signals trim pennants
sounds the mute call speaking with silence
Her transforming arts endure reminding

A sash a shawl a scarf a shield
Simple mat at friend's door a welcome sight
Omate rug tapestry, heroic lore recites
Plaited patterns of her world in woman's hand
paisley, striped, mosaic, vista and tartan plaid
prints of sky, sea, earth, color and heaven
Woman's fingers working the Goddess's hands
weaving primordial primitive feminine
power to create transform and nourish
The Girdle of Venus
A sash a shawl a scarf a shield

THE CATHEDRAL OF SOUND OR ODE TO THE ONE ELEVATOR IN CURTIN HALL

Dedicated to Herbert Blau

Going up
We with our etude
pump pump
to the ground
What are the words
to tell you the sound
this poem writing me
            all the time
some ignite the paper
a few resound

Going up
The boys made me cry
oh how the boys made me cry
springs of Winters'
wry erudition on the slide
We enter enveloped in
hushed cacophony of
breaths forced intimacy
broken into
by a guarded thought

Going up
the flights to the ninth floor
damn why does it take
so long
We enter the cathedral of sound
where the choice of words
doesn't mean much
and the height of her hemline
doesn't distinguish the one
who looks up her footnote

Going up
Heaven's bent
on the knee
laughing on the lap
of the 18th century
within the knock knock
of please please
She stoops speaks
now turned on words weeds
husk of incantations repeated

Going up
Whew
It's crowded in here
Nice weather today
broken silence spoken
there's a poem in this elevator
We come to this place
come and go
to the cathedral of sound
from the hallowed halls below

Going down?
Would this place be the same
without you
Would this edifice stand
without me
as surely as it does
where we query and ponder
the question more important
than
you know . . .

Going up
Thoughts burnish
with sounds recital
monumentalize moments
of immortal triviality
from quiet repetitions of
rhythm's conviction,
desire's recital
the poem writing me
all the time

Going up
Come again?
as mother would say
now unheard
Can we fit one more in?
leaning on the sound
We hold up the edifice again
scattered together
the hours gone when we meet
ascend from here
at the top at the peak

PAUL MAJKUT

THE PHENOMENOLOGICAL *FLÂNEUR* AND ROBERT IRWIN'S "PHENOMENOLOGICAL GARDEN" AT THE GETTY CENTER

The concept of "interactivity," under the name "conditional art," plays a central role in the phenomenological theory underpinning Robert Irwin's Central Garden at the Getty Center in Los Angeles, but the interactions that make up the experience of visiting that space need not be understood as evidence of phenomenological intersubjectivity. Descriptions of the phenomena of interactive relationships, whether in museums, gardens, theme parks, or cyberspace, often result in an attempted "ontology of flesh" that is detached from lived nature.[1] This is pointedly true when privately owned and controlled "public space" frames the cultural and social consciousness of interactive experience. Such is the case of the Central Garden, where an "artist who produced no art objects" has constructed a complex object of art in which the intersubjective, lived experience of animated flesh is purposely replaced by a manipulative interactivity based on – if a parallel may be permitted – an "ontology of meat," that is, inanimate flesh.[2] That is to say, the interaction between visitor and elements in the Central Garden is a purposeful manipulation in which the visitor becomes a moving, mechanical part among parts in a complex and compound object of art that is constituted as a subjectless totality. When, however, an element unaccounted for in Irwin's mechanistic understanding of phenomenology is introduced, the illusory shimmer of the garden machine dims and the delusion of intersubjectivity is clearly observed as lifeless interactivity. That element is the unobserved critic as phenomenological *flâneur*.

Nowhere in Irwin's reflections on phenomenology does he take into consideration the unobserved observer of his efforts and their effects on visitors to the Getty Center. But it is this idling spectator, strolling through the complex mechanism of the Central Garden in order to criticize its bourgeois pretensions, who exposes the noematic, mercantile ghost at the other polarity of the experience of the garden-machine as the self-tribute to J. Paul Getty, the *auteur* of the whole, of which the garden is a decorative distraction. Ironically, naïve "viewers engaged in perceiving themselves," posited by Irwin as an ideal audience, have no selves to perceive if the artist has successfully achieved his goal of creating "objectless art," for, by attempting to

A.-T. Tymieniecka (ed.), Analecta Husserliana LXXVIII, 295–304.

exclude the art object, Irwin would unwittingly also exclude its phenomenologically necessary correlate, the art subject. The attempted exclusion of the art object and the implied exclusion of the art subject should not, however, be confounded with the *epoche* delineated by Husserl. The *epoche* does not suggest leaving the object "unproduced" or "uncreated," but simply brackets or sets aside the question of its transcendence, neither affirming nor denying its facticity. The *epoche* is an epistemological change of attitude, not ontological empowerment. Its purpose is solely to set aside the question of the existence of transcendent objects and psychophysical subjects. Irwin's "objectless art," on the contrary, rests upon the unphenomenological presumption that the artist has the power *to leave objects uncreated*, an unrealizable attempt to appropriate power over the process of intersubjective consciousness and the structure of intentionality.

The Central Garden is located in what is uncritically thought of as "public space." To the degree that a space is "public," that space invites discussion in terms of that very public itself. Adequate description draws our attention to the publicness of the space under discussion and, having now entered a universe of descriptive discourse, these terms acquire the gravity of those used for the site and the *auteur* of the site. If the notion of an artistic *auteur* suggests an author behind a narrative and the garden walk of the phenomenological *flâneur* itself suggests a narrative in which landscape is the naming of places within a site, an obvious tension is created between the two for control of the narrative, that is, for omniscient authorship or authority.

Still another term, one better reflecting the anonymity of the disparate, as-yet ungrouped population of civil society that congregates in specific locations, is "crowd," that is, anonymous civil society coming together in specific places at specific times for specific reasons, for example, "the sports event crowd," "the theater crowd," "the crowd at the Central Garden," and so on.

The use of the term "crowd" in this context comes with literary credentials that begin with Edgar Allan Poe. Nowhere is the anonymous individual in the crowd better described than in his narrative birthplace, "The Man of the Crowd,"[3] in which the ineffable sadness of La Bruyère's remark, "Ce grand malheur, de ne pouvoir être seul," resonates as an acute diagnosis of late twentieth-century life. Public spaces such as the Getty Center, depersonalized shrines to desanctified anonymity that cause spectators to squint,[4] serve as loci for this intensely anonymous experience. Dehumanized by Irwin's "objectless art" (insofar as the attempt is successful), this experience stems from the quest for "subjectless art." Phenomenological description, resting on intentionality, provides that every object has a subject, every subject has an object. To deprive art of its objectivity is to simultaneously deprive it of its subjectivity, that is, to attempt to "produce" a subjectless art. But the relationship between *l'objet d'art* and *le sujet d'art* is just that, a relationship; phenomenologically considered, the attempt made to deobjectify that relationship must be coupled with its desubjectification, a phenomenological impossibility.

While the efficacy of Irwin's artistic project in the Central Garden to tamper with the apodictic and inviolable noema-noesis structure of consciousness is dubious (at least within a phenomenological framework), the attempt itself is not merely a capricious act of artistic inspiration by which epistemology is set on its head. In its full sense, the Central Garden sets up an experience for visitors similar to the dehumanized, depersonalized mercantile exchange found in a theme park. Mickey Mouse or Robert Irwin may welcome the visitor, but both are literally and figuratively puppet spokesmen for the true *auteurs* of their respective spaces, Walt Disney and J. Paul Getty. But authority over the public ownership of these privately controlled spaces may be denied to these *auteurs*, for, as Stefan Morawski observes in an essay of combative insight, "*Flâneurie* is still possible when one does not surrender to the Disneyland" (187).

The generalized-standardized "man of the crowd" to whom *auteurs* of public space (philanthropists, architects, city planners, landscape artists, et al.) address themselves is himself an *a posteriori* abstraction of statistical behavior, not eidetic intersubjectivity. He is the consumer monitored and modified regularly by the ongoing quantification of continuous visitor use. The interactivity found in theme-park and "spectator sport" magnitudes is one of "crowd control," "traffic flow," and "efficiency engineering." The relationship between the "public" and the private *auteurs* of public space may, in turn, be viewed in a phenomenology of crowds by a "phenomenological *flâneur*," a disinterested witness not accounted for within Irwin's understanding of phenomenology.[5]

The garden path in the Central Garden, then, describes an interactivity of the corporate *auteur*, the anonymous and naïve man in the crowd, the phenomenological *flâneur*, and the public space of the Getty Center. The last of these elements calls for special attention because of the contrast Irwin makes between "public art" and "art for the public," the former proposed as objectless, the latter always *l'objet d'art*.

A broad consideration of "public" must include consideration of ownership and control unless philosophical use of these concepts is to be precluded by a legalism that conflates "open to the public" and "controlled by the public." The public space of the Getty Center is, in fact, *semi-public*. The struggle for control of the Center's public space was, at one level, a controversy between Richard Meier, the architect of the Getty Center, and Irwin. At another level, more significant because it is the source of control, it was a struggle between factions on the Board of Trustees of the Getty Foundation. But, in either case, what became a *cause célèbre* in the world of art never included the *voces populi* for which the space was said to have been built. Without public control, no space is authentically "public."

What becomes immediately and eidetically visible to the phenomenological *flâneur* is that the social and cultural consciousness of the public man in the crowd visiting the Getty Center is formed by and rarely rises above indistinct, inadequate and confused perception. The late modernist architecture that encloses the Central Garden on three sides is overwhelming in size and geometric complexity, towering height, and the triumphalism of monumental statement. It is, above all, this monumental statement that is of interest. It is this statement that is the first line in the discourse of the space, the narrative of the landscape of the Getty Center in which the Central Garden is thematically crucial. In this narrative, the man of the crowd is not only overwhelmed by indistinct visual complexity and detail, but also by the social and cultural statements that engulf the whole interactive landscape-architecture mechanism by which his subjectivity is transformed and reified into a part of the interactive whole and by which his subjectivity is despiritualized.

The Central Garden is perhaps admirable because Irwin identifies epistemological experience as a basis for aesthetic judgment, but the garden itself provides no ontological ground for mechanical interactivity, no intersubjective experience. Consumer stimulus-response interaction with the artifacts of corporate culture satisfy only convinced apologists for late twentieth-century corporate reification of experience justified by a metaphysics of inanimate commodities, not intersubjectivity.

Of phenomenal perception, Irwin holds that there "is no place that you can confront this possibility more so than in a garden. A garden is by its basic nature phenomenological and it defies most of those ideas of confinement" (Interview), and a walk-as-narrative through the confined space of the Central Garden is the method of choice for the phenomenological *flâneur*-journalist. The narrative of this garden walk is not unlike the narrator's account of following an anonymous man in Edgar Allan Poe's "The Man of the Crowd." The phenomenological *flâneur* narrating the *flâneurie* of this garden walk notes in agreement with Poe's storyteller that

> At first my observations took an abstract and generalizing turn. I looked at the passengers in masses, and thought of them in their aggregate relations. Soon, however, I descended to details, and regarded with minute interest the innumerable varieties of figure, dress, air, gait, visage, and expression of countenance (1489).

And, of the individual visitor who makes up the public crowd at the Getty Center, another observation that Poe makes is helpful: "He refuses to be alone. He is the man of the crowd. It will be in vain to follow; for I shall learn no more of him, nor of his deeds" (1494).

In *Being and Circumstance: Notes towards a Conditional Art*, Irwin explains that he creates no art objects, yet in another place he announces that "what our perception presents us with (at every moment) is an infinitely complex dynamic whole envelope of the world and our being in it" (10). Merleau-Ponty, summarizing the relationship humanity has with the teeming world of nature known through flesh, notes that "existence is spatial, that is, that through an inner necessity it opens on to an 'outside'" (293). In *A Phenomenology of Landscape*, Christopher Tilley, relying on Merleau-Ponty, develops this understanding, noting that the "kinetic activities of human beings orientate apprehension of the landscape and create it as human. Space is existential, and existence is spatial in that it opens onto an 'outside', a series of reference points" (13). These referent points are the names given to places within a space, in this case, locations within the Central Garden. In this sense, "landscape is the naming of places" and the *flâneurie* of the unobserved phenomenological observer and phenomenological description is a form of narration.

Irwin maintains that the Central Garden is a "sequence of experiences" that is an "exuberant response to site," and that his Central Garden attempts to mirror the complex geometry of the architecture, using the organic form and texture of plant materials as an inanimate construction kit and palette derived from nature. Irwin maintains that the complexity found in his garden is "not loaded with metaphors and politics," but that gardens are narrative walks. But if landscape is naming, and the *flâneur*'s Adamic power to name is not to be denied by institutional *auteurs*, then the description of the Central Garden becomes a struggle to give appropriate names. From the perspective of the *flâneur*, it is the very complexity (or *les objets d'art complexes*) that is the narrative theme of the Central Garden, not the monadic self found at the other pole of phenomenologically intersubjective experience.

Irwin believes that he stands in a pivotal position in the history of Western art, arguing that "the history of modern art is essentially phenomenological history, that is, moving away from intellectual structures. . . ." His own concept of conditional art allows "all viewer decisions" to be "made on this set of circumstances: viewers engaged in perceiving themselves" (Interview).

Assuredly, phenomenology provides insight into noetic experience, but this understanding is partial without noematic considerations. Such incompleteness leads to neglect and, at times, critical indifference to the noematic objects of conscious, and in this case animate, nature. Self-absorbed "perception of perceiving" easily results in the uprootedness of the viewer from what is viewed in the natural world.[6] The "radical change in our lives" that Irwin believes occurs when a viewer perceives himself perceiving, is all but impossible within a consciousness absorbed in itself as a social or cultural construct uprooted by perception misdirected from nature. It is neither ecologically nor ethically reasonable to suggest that "the determination of value lies directly in the deliberations of the experiencing individual" (Being 17). Deprived of a secure epistemological ground in intersubjective or social experience, detached from the objects of the natural world, the spectator-visitor is thrown back on perceiving himself perceiving, but this deprivation is an artistic imposition. The phenomenological art of a garden separates itself from nature only at the cost of uprootedness from the environment, reflecting, as is the case of the Central Garden and the architecture surrounding it, the dominant corporate aesthetics of our time, an aesthetics based on the control, manipulation, and exploitation of human and natural environments that have become alien, strange, and mysterious. If, as Irwin maintains, "art is knowing in action" (Being 24) and that action is the subjugation of nature for artistic purposes, then the Central Garden is not simply, as Irwin claims, "site-adjusted, site-dominant, and site-specific," but site-dominating. "The subject of art," contrary to Irwin's formula, is not only "the human potential for an aesthetic awareness," but also awareness of the source of that awareness, nature (Being 25). Further, the relocation of art from object to "knowing in action" can be accepted as an opening to ethical consideration of the relationships of *auteur*, visitor, site and *flâneur*.

It is perhaps in this vein that Irwin insists that his art goes beyond the frame of traditional painting, seeing a garden as "an observation on nature, and the phenomenological property of nature that goes beyond a painting, which does not have the same quixotic, the same wonder that gardens do"[7] (Interview). While Irwin avoids the distinction between observation and manipulation of nature, understanding the garden itself as "an observation on nature," a more serious difficulty is the attribution of a "phenomenological property of nature" to a garden. It doesn't require phenomenology to go beyond the frame of a painting, but when "phenomenological nature" itself is forced into the antiquated frame of privately controlled "public space," the distinction that a visual artist makes between nature and frame is lost. Rather than removing the frame of a traditional medium from his art, more easily accomplished than suspending the framing-consciousness that goes with this medium, Irwin imposes that very frame on nature in the form of a geometrically composed garden which becomes a composition that may be understood as a framed narrative.

J. Paul Getty, whose wealth made the Getty Center a reality, believed that "few human activities provide an individual with a greater sense of personal gratification than the assembling of a collection of art objects that appeal to him and that he feels have a true and lasting beauty." While there is no reason to doubt the philanthropist's gratification, the man's aesthetic sensibility also allows him the observation that "fine art is the finest investment" (Walsh 8). The Getty Center may be seen as expressing the manners and aesthetics of triumphant corporatism that not surprisingly finds aesthetic value in the same place from which its wealth and power arise: the subjugation and exploitation of nature.

Tilley enumerates a number of qualities descriptive of capitalist space: desanctified, controlled, partitioned, economic, useful, disciplinary. Landscape in this worldview is a background for action, as opposed to the sedimented, ritual, mythological, cosmological space of non-Western landscape and pre-modern practice that is intersubjectively interactive (20–1). And, along the same lines, Martin Filler notes of the Getty Center's "late-modern architecture" that

> Conceived amid the certitudes of 1980's prosperity and conspicuous consumption, which it echoes in its extravagant architecture, the Getty arrives on the world cultural stage with much more than a touch of relic about it; yet it lacks the intimations of immortality its sponsors so clearly intended (Filler 34).

The relic, of course, is Getty himself. "Only a high-ranking or wealthy person," Tilley explains, "is likely to be daring enough to invent a path or plant a relationship not established before" (30).

Michel de Certeau discusses visiting a garden in terms of "rhetoric" and "tropes" of walking. Of this pedestrian rhetoric, Tilley says,

> If writing solidifies or objectifies speech into a material medium, a text, which can be read and interpreted, an analogy can be drawn between a pedestrian speech act and its inscription or writing on the ground in the form of the path or track. Both are sedimented traces of activity, and both provide ways to be followed. . . . Just as the writing of a text is dependent on previous texts (it has

the characteristic of intertextuality), the creation and maintenance of a path is dependent on a previous networking of movements in particular. . . . In *The Practice of Everyday Life*, de Certeau discusses walking as a mode of existence: Spatial and textual stories are embedded in one another. Narrative structure and emplotment can easily be claimed to have ontological significance for humans, to be as universal as language itself. Stories organize walks, making a journey as the feet perform it, organizing places by means of displacement that are described (29–30).

To the contrary, Irwin holds that "seeing is forgetting the name of the thing one sees" (Interview). While there is a neoplatonic charm to Irwin's thought, Tilley tells us that "a landscape is a series of named locales," suggesting that "perceptual space is egocentric" (34). Because the individual perception Irwin describes is visual, the result is an artist who claims to have ceased creating art objects but has failed to make a similar reduction of the existential *cogito*. But if the Getty Central Garden is "existential space in a constant process of production and reproduction through the movements and activities of a group. . . . It is created through life-activity" (16–17), and the activity of this group, composed of the anonymous men and women of the crowd that is the public, is phenomenologically intersubjective, and thus not the production of artistic ego.

NOTES

1 Anna-Teresa Tymieniecka observes that "humanity seems to have stripped itself of all flesh – to the point that not even a skeleton remains. Today the flesh of humanity is so disjointed that it cannot maintain any posture or direction. This is our disarray" (29).

2 This paper relies on a discussion of body found in Ideas II ("Body and Spirit as Comprehensive Unity: Spirtualized Objects," 248–259) in which Husserl notes the distinction between the thingly, inanimate, physical corpse-body (Körper) and the animated flesh of the lived-body (Leib).

3 Poe also wrote insightfully of gardens and landscapes in "The Landscape."

4 Martin Filler describes the Getty Center as a "shrine to visual art that makes you squint" (29).

5 The transcendental ego of Husserl's pure phenomenology might be identified with the phenomenological *flâneur*, but for reasons unrelated to the present paper I am not prepared to take that step. Without going into detail, the phenomenological *flâneur* is a person, the transcendental ego as presented by Husserl is not.

6 Tymieniecka: "Rootedness involves on the one hand endowing a ground of one's own with meanings that are one's own and on the other sharing these meanings with others" (30); "The nomads of old at least had a rhythm and cadence to their lives, a measure. Often it was Nature that provided that rhythm. But for us the firmament is not firm" (31).

7 Irwin "Interview": "The whole of art in public places. . . . There is a history of modern art . . . that has gone through all of its homework, the reason for the ambition of art in public places is really, at this moment in time, to test the principles of modern thought and modern art. We know what the philosophic ground is, we know what the conceptual ground is, we have a new visual

language, we have a new visual syntax, and now it's become, how do you practice from that point of view?"

WORKS CITED

De Certeau, M. *The Practice of Everyday Life.* Berkeley: University of California Press, 1984.

Filler, Martin. "The Big Rock Candy Mountain": Review of *Building the Getty* by Richard Meier; *The J. Paul Getty and Its Collection: A Museum for the New Century*, by John Walsh and Deborah Gribbon; *Making Architecture: The Getty Center*, by Harold M. Williams, Ada Louise Huxtable, Stephen D. Rountree, and Richard Meier. In: *The New York Review of Books.* 18 Dec. 1997: 29–34.

Husserl, Edmund. *Ideas Pertaining to a Pure Phenomenology and to a Phenomenological Philosophy.* Trans. R. Rojcewicz and A. Schuwer. Kluwer Academic Publishers, 1993.

Irwin, Robert. *Being and Circumstance: Notes towards a Conditional Art.* San Francisco: The Lapis Press, 1985.

——. Interview with Robert Irwin: San Diego Artist. With Dirk Sutro. *These Days.* KPBS. San Diego. 8 Oct. 98.

—— *Robert Irwin: The Beauty of Questions.* Dir. Leonard Feinstein. Center for Media and Independent Learning, 1997.

Merleau-Ponty, Maurice. *The Phenomenology of Perception.* Trans. Colin Smith. London: Routledge & Kegan Paul, 1962.

Morawski, Stefan. "The Hopeless Game of Flanerie." *The Flâneur.* Ed. Keith Tester. New York: Routledge, 1994. 181–197.

Poe, Edgar Allan. "*The Man of the Crowd.*" Ed. by Nina Baym et al. New York: W. W. Norton & Company, 1994. 1489–1494,

Tester, Keith, ed. *The Flâneur.* New York: Routledge, 1994.

Tilley, Christopher. *A Phenomenology of Landscape: Places, Paths, and Monuments.* Oxford: Berg, 1994.

Tymieniecka, Anna-Teresa. "Measure and the Ontopoietic Self-Individualization of Life." *Phenomenological Inquiry* 19 (1995): 26–51.

Walsh, John and Deborah Gribbon. *The J. Paul Getty and Its Collection: A Museum for the New Century.* Los Angeles: J. Paul Getty Museum, 1997.

SECTION V

HOWARD PEARCE

THE DREAM OF ASCENT AND THE NOISE OF EARTH: PARADOXICAL INCLINATIONS IN EURIPIDES'S *BACCHAE*, SHAKESPEARE'S *THE TEMPEST*, AND STEVENS'S "OF MODERN POETRY"

> Be not afeared: the isle is full of noises,
> Sounds and sweet airs that give delight and hurt not.
> Sometimes a thousand *twangling instruments*
> Will hum about mine ears; and sometime voices
> That, if I then had waked after long sleep,
> Will make me sleep again; and then, in dreaming,
> The clouds methought would open and show riches
> Ready to drop upon me, that, when I waked,
> I cried to dream again.
>
> (William Shakespeare, *The Tempest* 3.2.132–40)

> The actor is
> A metaphysician in the dark, twanging
> An *instrument*, *twanging* a wiry string that gives
> Sounds passing through sudden rightnesses. . . .
>
> (Wallace Stevens, "Of Modern Poetry" 18–21)

"To the deep beat of the drums sing and dance to Dionysus" (*Bacchae*, Kirk 156); ". . . the timbrels thunder-knelling . . ." (Way 156); ". . . the sound of the booming drum . . . ." (Internet)[1]

In Wallace Stevens's "The Noble Rider and the Sound of Words," the Platonic dream of ascent toward the Heavens elicits a demurral in the modern spirit. We "recognize" but do not "participate in" the figurative ascent in the *Phaedrus*, in which the noble soul ascends by means of the "noble . . . winged horses . . . of the gods . . . soars upward, and is the ruler of the universe."[2] For us, Stevens says (the poet's imagination working perhaps like the light of the sun on lcarus's wings) "The figure becomes antiquated and rustic"; our transit describes an arc downward, to earth: "We droop in our flight and at last settle on the solid ground." Although the impulse to descend to earth is recurrent and overtly thematic in Stevens's poetry, it might be observed as recurring suggestively in the plays of Shakespeare and Euripides to be discussed here.

The descent to earth, as imaginative *eidos*, is perhaps as essential, in the Husserlian sense of "essence," as the idealization of other worlds – "heaven,"

A.-T. Tymieniecka (ed.), Analecta Husserliana LXXVIII, 307–324.

the pastoral retreat, the ideal Republic, the Golden Age – typically but not necessarily imagined as departures from this world to another (higher) sphere. Such ideas of other places than the present one may be thought of as generally evoking possible other worlds.[3] Phenomenological essences for Edmund Husserl, as Emmanuel Levinas observes, are not "empirical concept[s], wherein the essential and the accidental are mixed," or "formal like the objects of logic," but "a whole series of conditions which the study of the essences of contents reveals. This study [the third of Husserl's *Logical Investigations*] . . . concerns the a priori conditions for objects; above and beyond the objects, the intuition of essences concerns their conditions of existence."[4] Pertinent to a consideration of what I refer to as "paradoxical instincts" is Levinas's remark that the pursuit of essences involves "the notion of the ideality of all cultural objects – a symphony, for example."

The "cultural objects" I here construe as instances manifesting these instincts can be observed ranging across Western culture from classical Greek to English Renaissance to American modernism. Recognizing that historical distinctions must be, and have been, made, I nevertheless propose that contrary impulses occur and recur as essential to thought about temporality and transcendence and can be seen to be thematic in particular instances – Euripides's *Bacchae*, Shakespeare's *The Tempest*, and the poetry of Stevens. Observing their persistence must entail a precise and attentive look – going *zu den sachen selbst* – at the ways in which the motifs appear thoughtfully in these "cultural objects." To notice the "descent" toward earth is to highlight what is perhaps cast in shadow by the major conflicts and themes in the earlier works, but on the other hand is thematically primary in the works of a modern poet, affirmed as a reluctance to pay tribute to Plato's "dear, gorgeous nonsense" (643). Stevens has been seen both as developing out of American pragmatic attitudes and as demonstrating an affinity for the early twentieth-century phenomenology of Husserl and the ensuing hermeneutics of Martin Heidegger.[5]

* * *

The devolution, Stevens's "drooping in our flight," is realized in all three of the instances – *Bacchae*, *The Tempest*, and Stevens's poems – as an acceptance of the conditions of earth, both on the one hand the sensual and intellectual possibilities of life and poetry, the possibility of transcending the subject-object divorcement, of touching another human being, and on the other, the necessity of time-bound existence, death, and the limits of knowing. Heidegger has referred to this essential condition of being as entailing us in the "withdrawal," which is the "event" that "could be what is most present in all our present, and so infinitely exceed the actuality of everything actual."[6] The possibility of "touching" *in extremis* is the affirmation of Stevens's "Lebensweisheitspielerei," which occurs in a world from which the "proud and the strong / Have departed" (*Collected Poetry and Prose* 429–30), a world of "stellar pallor." The capability of touching in this "poverty / Of autumnal space becomes / A look, a few words spoken," and it is both "complete" and essential to being on earth, a person's touching another "With what he is and as he is."

I have selected as epigraphs passages that vividly express the condition of earth's aesthetics as opposed to divine harmonies. Historically, or mythically, the descent might be seen for Stevens as, in "The Sense of the Sleight-of-Hand Man," an insistent plummeting to the earth from the exalted images of "floods of white" that "Came bursting from the clouds" (205) and "the wind" that "Threw its contorted strength around the sky": "Could you have said the blue jay suddenly / Would swoop to earth?" The blue jay is appropriately the bird to swoop to earth, since it is notable among birds for its raucous sociability. In "Of Modern Poetry," the world has, again, undergone a change, lapsing from the stage upon which "the scene was set" and in which the actor "repeated what / Was in the script" (218). The "modern poet" is a "metaphysician in the dark, twanging a wiry string that gives/Sounds passing through sudden rightnesses." I would set alongside Stevens's "twanging . . . instrument . . . that gives / Sounds passing through sudden rightnesses" the "noises" of Shakespeare's Caliban, the "sweet airs that give delight and hurt not," which, Caliban says, are "Sometimes a thousand twangling instruments" that "hum about mine ears."[7] The echo of Caliban's "twangling instruments" in Stevens's "twanging . . . instrument" implies Stevens's choice of Caliban as analogy to the modern "metaphysician in the dark"; but I am proposing to argue only analogy rather than historical influence. If the alternating ideas of movements toward heaven and earth are accessible in light of Husserl's *eidos*, they need no explanation in terms of historical probability or necessity. If Stevens is thought to have chosen Caliban, however, as an idea of this depreciated concept of human being, his poem would become a commentary on, a reading or misreading of, an essential element in Shakespeare's play. It would point up the essential earthiness of both Caliban and the modern poet; it would seem to suggest too that Caliban's experience of the island's "noises" confuses or fuses what is natural and what is created (by Ariel, by Prospero) and plays toward the modern poet's creation, which eludes placement as being either subjective or objective, internal or external. In *The Tempest*, then, Caliban's sense of identity on the island expresses that harmony of earth that is "rightness" for Stevens's poet or actor, both the aesthetic sensitivity and the noisiness of earth.

The impulse to soar, to rise toward idealized otherness, is implicit in Prospero's first having abdicated his Dukedom by turning to his books. "Neglecting worldly ends," he was "all dedicated / To closeness, and the bettering of [his] mind," and his "library / Was dukedom large enough" (1.2.89–90, 109–10). Whereas Prospero moved away from the world of ducal responsibility into a world of Ideas, which becomes indeed another "world" of the text, Gonzalo's idealization of a world is expressed in terms of the concretization of a world as Golden Age in Ovid's *Metamorphoses*. Gonzalo's "commonwealth" would have no "magistrate," no "riches, poverty, / And use of service," no "occupation; all men idle, all; / And women too, but innocent and pure" (2.1.145–48). His dream is of a state that would "excel the golden age" (164). This ideal primordial world is of course antithetical to the ordered, controlled state of Plato's Republic. It repeats the opposition of exalted, designed, architectural worlds and the world that rests comfortably and fruitfully in earthly harmony. Prospero's imposition of control and discipline on such a world in the island is finally ceded, and the island is abandoned to its "natural" state, which is bountiful in Caliban's description of how he will provide "crabs . . . pignuts . . . clust'ring filberts . . . and . . . young scamels" (2.2.163–68) for his newfound confederates.

The idealization of another world as Christian Heaven, the idyllic pastoral, or the Platonic world of Ideas can be reflected in common earthly realizations that rise upward toward the ideal, as the architecture of Plato's Republic is to be realized in the social design of the city state. With the ascendancy of Platonic Ideas in the Italian Renaissance, city states were visually imagined and shaped on an eternally true geometric pattern.[8] These essential exaltations of the idea of human nature and a supreme world are contributed to the education of Miranda under Prospero's tutelage, and she recognizes, as if manifesting Plato's idea of anamnesis, that dream of the ideal in an ordinary mix of human beings. When she sees for the first time the other company to the island, after having intuitively recognized Ferdinand as divine, she declares in these famous words, "How beauteous mankind is! O brave new world / That has such people in't!" (5.1.183–84).

Her antithesis, Caliban, is, however, no mere insensate brute. He is, like Stevens's metaphysician in the dark, paradoxically brutish and aesthetic. The "thousand twangling instruments" that "hum" for Caliban, "give delight and hurt not" (3.2.133–35). Although the "isle is full of noises," they are constituted as "sounds and sweet airs." The noises of earth cause him, ironically, to dream, implying that they are the natural phenomena of the island combined with the magical powers of Prospero (the present god of the island supplanting the witch Sycorax, Caliban's mother) to cause sleep, and that sleep can realize even in Caliban the impulse to dream that "the clouds . . . would open and show riches." Whether of the natural world or of Prospero's doing, the oxymoronic sweet noises manifest a benevolent and generative influence. This counter-impulse in Shakespeare's play, the sensitivity to and appreciation of earth's aesthetic, both when it "hums" and when it "twangs," is not refutation or disappreciation of the ennobling instinct, but joins with it as paradoxical essence in *The Tempest*; and it echoes in Stevens's poetry. In both, an inherent humanism recognizes and experiences sympathetic harmony with the proper environment for human beings, instantiated as the music of earth. I think it not inappropriate to suggest that accepting this state of human being is central to the insistence by Martin Heidegger that earthly "paths" are the felicitous ways of *Da-sein*, the "being there" and "there-being" of the human.

This linking of discordant or opposing impulses appears as oxymoronic in *The Bacchae*. As the winds in *The Tempest* have, in pitying Prospero and Miranda, done them a "loving wrong" (1.2.150–51), the "sacred orgy" of the Bacchantes can be read as either destructive or restorative. Pentheus's craving for a voyeur's watch on the women's rites assumes that what he sees will be sexual, and titillating, and scandalous. He has heard tales of their ecstasies, their serving Aphrodite rather than the new god. Teiresias warns him from the beginning, of course, that the rites do not corrupt the women, that chastity is a matter of character (314–17). Furthermore, the messenger's report of what takes place on the mountain Cithaeron confirms that Pentheus's imagination is untrustworthy. Pentheus thinks of their activities as "villainous" (Kirk 232; "pestilent," [Way 232]) and "pernicious rites" (Kirk 260; "pestilent rites" [Way 260]), "rituals" in which "nothing wholesome remains" (Kirk 262; "revelry" from which "no good . . . comes" [Way 262]).

Not only do the women perform miracles in the natural world, causing milk, wine, water, and honey to rise from the earth, the messenger's description of the scene and events on the mountain having a surreal effect, but also their activities, until interrupted by intruders, are idyllic and harmonious with nature. They are asleep among boughs and leaves. Pentheus is told that their activities are not as he thinks; they are not abandoned, drunken, "hunting desire" (Way 688; "seeking to gratify their lusts alone amid the woods, by wine and soft flute-music maddened" [Internet]), but resting "on the ground / Flung careless . . . [but] modestly" (Way 686; "carelessly, but decently" [Kirk 686]). The view of the women seems to echo in the stipulation that Gonzalo in *The Tempest* makes about women in the Golden Age: "All men idle, all; / And women too, but innocent and pure." Women's idleness, unlike men's, seems to require some reassurance of its being virtuous. Even when they hear the herdsmen's cows lowing they are a beautiful sight, springing up in a "strange, fair array of ordered ranks" (Way 693; "a miracle of discipline to behold" [Kirk 693], "a wondrous sight of grace and modesty" [Internet]). The messenger assures Pentheus that if he could have seen them he would honor their god Dionysus (Kirk 712–13; Way 712–13). To see them is to see their rightness in that natural harmony. Although their subsequent deeds are terrifying, they are, in this blessed state, in harmony as a group and at one with nature.

The descent to earth and the corollary transformation of character in that reality is not, of course, akin to the music of the spheres, the rise to heaven, or the ennobled escape from the environs of earth, its noise, its danger, its darkness. The music of the Bacchantes has, like that of Caliban's island and Stevens's poet-actor-musician, the raucous affinity with the earth, the sound of noises rising up from it. In the opening stasimon, the Chorus sings, not of the elevated image of the individual ego characterized by Pentheus in his arrogance, self-aggrandizement, self-assurance and certainty of things, but of the virtue of love, participation, and community – the dance of Dionysus and his followers accompanied by the "deep beat of the drums" (Kirk 156); ". . . the timbrels thunder-knelling . . ." (Way 156); ". . . the sound of the booming drum . . ." (Internet).[9] The drum (timbrel) has been "won / By the madding Satyr-band" (Way 130; the "ecstatic Satyrs / obtained it" from Rhea [Kirk 130–31]; "frantic satyrs won it from the mother-goddess for their own, and added it to their dances in festivals" [Internet]), having been appropriated, then, by the most unrefined of creatures. The music is, nevertheless, like Caliban's, sweet: the sound of the drums is "blended . . . with sweet-crying breath / of Phrygian flutes" (Kirk 127–28): "They blent with the cry ringing sweet and high / From the flutes of the Phrygian land" (Way 126–27).

On the other hand, the idea of an exalted world is attached to the architecture of the city Pentheus rules, his sense of rightness and privilege being essential to the protective and benevolent social design that is mimetic of cosmic logos. Plato's concept of the duplication of ideal reality in the state, the Republic, is implicit in Pentheus's attitude of superiority and control, that idea of architectural instantiation of cosmic form revived in the Italian

Renaissance. The Dionysian principle literally demolishes and burns the palace, and the release of the sacred into this world is not an event of justice and order but a renewal of earth's power, the reaffirmation of earthly truth. The Chorus asks, "What is wisdom?" and concludes that "him whose life day by day / is happy do I count blessed" (Kirk 897, 910–11; "Happy is he whose bliss from day to day doth grow" [Way 911]). To accept being of the earth is to accept "what will suffice" (Stevens, "Of Modern Poetry," 218–19) and that resounds as a joy in mortal music.

Essentially, then, the affirmation of earth fosters its presence as sweet noise, a counter- or alter-beauty, and the twang, the drum noise of Satyrs, the raucous harmony, is heard as the music of the Earth. This music is not mere cacophony, meaningless sound, however, but speaks a truth as essential as the structures idealized as above and beyond the earth, ascendant. The primal truth thus spoken might be, like the essence that is discoverable in a phenomenological approach to things, a gaining of a purity and an exactitude of knowledge. In Stevens's poetry, this apprehension of an essence entails the recovery of a kind of innocence, the return to a world that retains its *virtú*. If we "must become an ignorant man again / And see the sun again with an ignorant eye," we are by the death of Phoebus (and "the death of one god is the death of all") sent "back to the first idea" (Stevens, "Notes Toward a Supreme Fiction," 329–30). This motion is not a presumed escape to simple, childlike innocence, a return to a past idealized condition, but is a disciplined achievement of a pure perception that is a new understanding. The precise awakening of knowledge in "Not Ideas About the Thing But the Thing Itself" begins with "a scrawny cry from outside / [that] Seemed like a sound in his mind" (451–52). The "scrawny cry" of the bird, appropriate to the music of earth, is both a "chorister whose c preceded the choir" and "a part of the colossal sun." The discovery of this *bild* in *bildung* is "like / A new knowledge of reality," a precise understanding acquired not in romanticizing the past but in paying attention to the present.[10] Levinas identifies this accomplishment as constitution of the "concrete" life: "To analyze the intentionalities which constitute any given object is to do phenomenology" (36), and we thereby learn to "consider the phenomena in their concrete freshness, in their irreducible originality. The whole of being, when considered as it emerges from our concrete life, is not the accumulation of facts of . . . a world falsified and impoverished by naturalistic tendencies of our time. . . . The world overflows nature, recapturing all the contours and richness that it has in our concrete life; it is a world of things interesting and boring, useful and useless, beautiful and ugly . . . ridiculous and anguishing."

In "The Hermitage at the Center" (*Collected Poetry and Prose* 430), Stevens dramatizes this movement from the world of "our time" toward that "center." Our time's "naturalistic tendencies" are instantiated as "leaves on the macadam that make a noise," and the noise of our world is counterpointed by the noise of the world recovered as "hermitage" at the "center." The idealized place has at center the idealized image of beauty, the "desired," who "Reclines in the temperature of heaven . . . / Sleek in a natural nakedness." She "attends" in the tranquil pastoral scene a "tintinnabula" of birds that are "called up by more than the sun / Birds of more wit" who create an "intelligible twittering" that substitutes for "unintelligible thought" (lines 6, 9, 11–12). Like the unpromising thick-headedness of Caliban, the presumably mindless "twittering" of birds becomes "intelligible," and on the other hand the presumable intelligibility of human thought, human language, becomes "unintelligible." The accomplishment in regaining or reconstituting that world makes the "world" that, for Levinas, "overflows nature."

The generation of the music is both of the earth itself and of the mimetic human response or participation, a contrapuntal harmonic entailing an interdependence. As opposed to Plato's view of the depreciative artist echoing braying asses and crowing cocks (noises of an inferior and contrary nature, sound effects of earth's cracks and groans), the poet and the human being share the assonance and dissonance of earth in making their music. The capable human voice, characteristically the poet's – attuned to the creatures of earth whose "noises" are, for Caliban, "Sounds and sweet airs that give delight and hurt not" – echoes the sounds of earth. Ariel repeats the sound of the island – "Bow-wow. / The watch-dogs bark. . . . I hear / The strain of strutting chanticleer / Cry, Cock-a-diddle-dow" (1.2.385–90). The "sea-nymphs" announce the "sea-change" of Ferdinand's father, in Ariel's report, by "hourly" ringing "his knell" – "Ding-dong. / Hark, now I hear them – Ding, dong, bell" (1.2.404–08). This raucous harmony is appropriate to the Dionysian revelers as well as to the companions of Caliban, who create a religion of wine, of the "bottle," and, as opposed to the builders of exalted dreams, will inherit the island of strange sounds (strange because both eerily present and withdrawing). As, in the *Bacchae*, the music of Orpheus plays for the beings of nature, as Orpheus "drew to him trees . . . untamed beasts" (Kirk 563–64; "Trees round him adoring to press, and the beasts of the wilderness" [Way 562]), the women's cries of Dionysus enlist nature's response: they dance, "calling in unison" for Dionysus, "and all the mountain and its wild creatures / joined in bacchic worship" (Kirk 726–27; "the hills, the wild things all, were thrilled / With ecstasy" [Way 726–27]). This devolvement into the voice of nature amounts to a reproval to the elevating language of Pentheus, who ostentatiously identifies his superiority as evident in his lineage and name: he is "Pentheus, son of Agaue and Echion as father" (Kirk 507).

Stevens's poet is par excellence among the tribe of Plato's "pantomimic" imitators. In, for instance, "A High-Toned Old Christian Woman," the parade of "disaffected flagellants" bears resemblance to Dionysus's followers and might make "Such tink and tank and tunk-a-tunk-tunk" as would produce a "jovial hullabaloo among the spheres" ("A High-Toned Old Christian Woman" 47). This poetry of verbal exuberance represents in sound effects the vital engendering of a world. The world thus made is a squawky assertion of a satisfactory human existence on earth. The necessary devolution becomes, in the poetic voice, an undermining of the exalted structure of the human architecture based in logos. The title of the poem, "Of Mere Being" (476–77), establishes this necessary renewal or retrieval or generation of essence. In the imagination of the ultimate horizon of being, "beyond the last thought," the image of a palm, and in it a "gold-feathered bird" singing, appears as an answer to traditional philosophical assumptions about knowledge and happiness – which have been the placing of a heroic icon of the human at the center of the universe, heroic as foremost among beings and as paradigm of the abstracting power of the intellect. The bird's song is "without human meaning, / Without human feeling, a foreign song." The appearing of the bird on the horizon and its transcending of "thought" identify it as an essence of being human that can be revealed only in bracketing what has been understood as the essentially human, the intellect. The import of the image is that "You know then that it is not the reason / That makes us happy or unhappy." The ineluctable truth of the visual and auditory image appears finally in the poetic decisiveness of sound per se: "The bird's fire-fangled feathers dangle down." The gratification of sound effect in the voice of the poem, for the reader, realizes the noisy music that can be heard among the multitude of diverse voices expressing a mimesis of "what will suffice."

Instances of this mimesis abound in Stevens's poetry, often representing the truth that it is not "the reason / That makes us happy." One further example demonstrates the delight and non-sense of this poetry: "Snow and Stars" begins, "The grackles sing avant the spring / Most spiss – oh! Yes, most spissantly. / They sing right puissantly" (108). The poem ends in an echo of Ariel's song about Ferdinand's father. If the devil took winter away, "It would be ransom for the willow / And fill the hill and fill it full / Of ding,

ding, dong." The ultimate meaningfulness of poetry negotiates the region of earth, is ambulatory (in William James's sense of the philosopher's ambulating), and joins in the song that is not primordially a logic of language but an extravagant (and oxymoronic) "intelligible twittering."

* * *

If the human being or poet as the voice of the human joins in the music of earth, that descent to participation transforms, perhaps transmogrifies, the celebrant. The human image, then – as opposed to the noble image appropriate to exalted worlds, as presented by Pentheus's image of himself for the world, and as seen by Miranda when she first encounters the noble images of Ferdinand's father and his entourage, and declares, "O brave new world! . . ." – befittingly becomes comic, even grotesque. Opposed to the heroic image of King Pentheus, the comic figures of Cadmus and Teiresias in *The Bacchae* produce an image of the aged men dancing, paradoxically, the youthful dance of age. There is something foolish in Cadmus's avowal that he "should not tire, all night or all day long, / of striking the earth with the thyrsus. We have joyfully forgotten / that we are old" (Kirk 187–89; "I shall not weary, nor by night nor day, / Smiting on earth with the thyrsus. We forget / In joy our age" [Way 187–89]). Along with Teiresias's assurance that he too feels young enough to dance, their being costumed in the bacchantes' trappings, their linking arms in a decision to walk creates a ridiculous image, to which Pentheus responds when he arrives. He is indignant, scoffing at their appearance, Teiresias "clad . . . in dappled fawnskin" (Way 249), his grandfather "O sight for laughter! – / Tossing the reed-wand" (Way 250–51). He takes "shame / Beholding these grey hairs so sense-bereft" (Way 251–52; "I am ashamed, old man, / to see the foolish senility of the pair of you" [Kirk 251–52]). Delivering a lecture on Dionysus's truth and power, Teiresias warns Pentheus not to "Deem thyself wise," though he might "seem so to [his own] jaundiced eye" (Way 311–12): "I, then, and Cadmus, whom thou laugh'st to scorn, / Will wreathe our heads with ivy, and will dance – / A greybeard pair, yet cannot we but dance" (Way 322–24); "Therefore I and Cadmus, whom you make fun of, / shall crown ourselves with ivy and shall dance – / a gray-haired couple, but all the same dance we must" (Kirk 322–24). Attuned to the forces of earth unleashed by Dionysus, the old men not only appropriately pay tribute as if youthful celebrants, but also do it because they "must."

In *The Tempest*, the grotesqueness of this image is repeated in the beast with four legs, Caliban and Trinculo under the cloak, which Stephano encounters and by which he is amazed. Caliban is identified, of course, as the island's only "human shape" (1.2.284), and minimally so. He is referred to as a fish and monster, sometimes as both or amphibious: "half a fish and half a monster" (3.2.26). But his monstrous image is compounded by union with Trinculo. Trinculo finds him in a fright, lying down, wonders whether he is a "man or a fish" (2.2.25), and, as the storm approaches, decides to "creep under his gaberdine" (36), an action making "strange bedfellows." Stephano, drunk, finds this apparition a "monster of the isle with four legs" (61). He responds to Caliban's pleas for mercy with wonder that the monster knows his language, and tries when Trinculo speaks from the other end of the cape, making "a most delicate monster" with "four legs and two voices" (84), to heal the monster by pouring wine into its two mouths. The comic antics of these types reiterate the comic figure that Teiresias and Cadmus make together, and the depreciation, like that of the two old men, brings them close to the music of earth. It is after Caliban accepts their new society, declaring that, seeing Stephano, he will "adore" him (2.2.130) as his new master, that Ariel frightens them with a tune and Caliban reassures them about the island's "noises" and "sweet airs." Caliban's voice, attuned to the "intelligible twittering" of Stevens's birds, conjures the creatures of the island and echoes them. His voice grows mellifluous in appealing to his new confederates' senses:

> I prithee, let me bring thee where crabs grow;
> And I with my long nails will dig thee pig-nuts,
> Show thee a jay's nest, and instruct thee how
> To snare the nimble marmoset. (2.2.157–60)

His voice grows mimetic in his stuttering song about his new freedom: "'Ban, 'Ban, Ca – Caliban / Has new master, get a new man" (173–74).

This depreciated human image becomes in Stevens's poetry the positive image of, "here and there, an old sailor, / Drunk and asleep in his boots," who "Catches tigers / In red weather" ("Disillusionment of Ten O'Clock," 52–53).[11] The grotesque figure becomes an instance of the empowered imagination, the poet stripped of the illusory ennoblements of the past for the sake of discovering a new nobility in things as they are, in discovering or making what will suffice. In "The Man with the Blue Guitar" (xv), Stevens proposes Picasso's picture of the guitarist, whose anatomy in the painting is cramped to fit the frame, as an image of ourselves. The man with the blue guitar is "'a hoard of destructions,' a picture of ourselves," an image by which the I might read itself: "deformed, a naked egg" (141). The mimetic relation of "the monster" to "myself" produces an interpenetration, "two things, the two together as one," and the self as musician becomes "more than a monstrous player of / One of its monstrous lutes" (143). The poet/musician will "patch it as I can" (135), "singing" the heroic image, "a hero's head, large eye / And bearded bronze," but that image is "not a man" and the singer must "reach through him almost to man." The consequent music will, then, be the music of earth: "To bang it from a savage blue, / Jangling the metal of the strings" (137). The poetry of earth is "made" in an Aristotelean mimesis by the commonplace poetic voice and participates in the music that is both made and experienced. It is an expression of this embeddedness in the place and time of our being. The essential interdependence of that making with the objects of experience entails, in Stevens's poetry, a loss of clear distinction and firm opposition between subject and object. The intentional subject and object of experience are united in the activity.

In "Of Modern Poetry" too, the musician is a contortionist. Since the past, in which "the scene was set" and poetry "Repeated what / Was in the script," is now only a "souvenir," "Modern" poetry has "to learn the speech of the place" and face the people of the place, to "think about war," and to "find what will suffice" (218–19). Poetry is likened to an actor who has to "construct a new stage" and at the same time "has to be on that stage" performing. The making of a world, a theater, is a phenomenological construction of what is "found" rather than what is already presumed to be. The intricate elicitings of language and being find the actor speaking "with meditation . . . words" that "in the delicatest ear of the mind, repeat / Exactly, that which it wants to hear," and the "invisible audience listens, / Not to the play, but to itself." The audience is neither precisely subjective nor objective, but perhaps both, or perhaps a new "objectivity" that opens within a subjectivity. This "finding of a satisfaction" by "modern poetry" entails us in an uncertainty of internality and externality. Its "making" of a world transcends, as Aristotle would argue, the facticity of the historian's recording, becoming a making of a world that takes the objective to itself in its new reality. Evidently the musician/actor is, like the man with the blue guitar, a kind of contortionist. The visual representation derived from Picasso's painting becomes an intellectual gymnastics in "Of Modern Poetry."

Even more grotesque in Stevens's poetry are the images of worms that are poets of the earth, creating and affirming the beauty of earth, both, as for Caliban, its "noises" and its "sweet airs." The worm is an intrinsic image of the contorted or contorting, the "eccentric," the elemental participant in the decomposition and consequent fertility of earth. Yet the worm, as a grotesque image of the poet, has the capacity to appreciate beauty and to create order. The worm companion of "Metamorphosis," a "pretty quirk" (238–39), is an accomplice in the poetic voice's expression of metamorphosis as a declination of the seasons toward winter. As that voice echoes the disintegration in a disintegration of language – September has become fragmented into three parts, "Sep – tem – ber" – a literal falling apart, fragmentation, or dissolution of components, the worm is asked to reshape itself, transform its "quirkiness" and "make o, make o, make o" (perhaps in order to begin the orthographic sequence of what remains of October with an "O": "Oto – otu – bre"). The iterable "O" remains, of course, the idea of perfection as established in Ptolemaic cosmology and in Plato's ideal One. The "old worm's" versatility ranges from his natural earthy quirkiness to his ability to make ideal form.

The worm speaking in "Bantams in Pine-Woods" has a quarrel with "Chieftain Iffucan of Azcan," the poet who is a "Damned universal cock," a bantam chicken who has constructed an illusion of himself as a "ten-foot poet among inchlings" (60). This pretentious poet enlarges and exalts himself, creating an illusion of himself consistent with the exalted illusions of world and human being of the past. The speaker, an "inchling" (inch-worm), cries his defiance to the grandiose poet, affirming that he, the inchling poet, is "the personal," that he *is* his "world," a philosophy of being that places him on the earth and aligns him with the poet of "Of Modern Poetry," who makes his world as he performs in it. The inchworm echoes the pompous utterance of the ten-foot poet, his "hoos," in declaring that he "fears not portly Azcan nor his hoos." The comic, diminutive worm-poet can, like the worm of "Metamorphosis," transform his shape, in Protean change, in that he "bristle[s] in these pines" and "points their Appalachian tangs." Giving point to or imitating the pointed pine needles, he expresses his affinity to and capability in performing the essential beings in his world. He is indeed his world in the phenomenological sense that "he" and "it" are inseparable as performance, activity; he might be thought, as well, to be a consistent representative of Heidegger's conception of the *dasein*. "Being there" is to exist in and of, not only in relationship to but also by means of, the "there-being."

Such versatility of the poet in Stevens's poetry, in bodily metamorphosis and in producing mimetic sounds, shifts the quality of the comic further toward the grotesque in "The Worms at Heaven's Gate" (40). The worms declare their mission to be bringing "Badroulbadour," an idea of beauty now dead, "out of the tomb" to heaven. Their presentation is of a decomposed object being reconstructed as beauty. Putting her back together is a feat of making or reconstructing what, like the year in "Metamorphosis," has decayed. As with the making in "Of Modern Poetry," these poets chant their performance and construct the renewed reality at the same time. As makers, they perform a presentation, chanting images of ideal-form distortion while composing the object: "Here are, one by one, / The lashes of that eye and its white lid. . . . / Here are the lips, / The bundle of the body and the feet." The deformations and noises of earth are a poetry made by the "modern" poet, echoing but distorting what has been the standard of form, the Platonic idea of reality. The Ptolemaic circle becomes eccentricity, deformed images, roundabout motions, and squawky sounds, just as in "A High-Toned Old Christian Woman" the palms of the poet's religion, as opposed to the old woman's, are "squiggling like saxophones" (47).

The Ideal appears in such devolutions, however, in that the essences of being are retrieved, or discovered, made or found, in the world as detritus. The energized poetic activity is an expression of the need to build a habitable world, and the reconstruction appears as an awareness of the need. What Stevens's comic or bizarre poet achieves is a recovery of the idea of a fruitful earth, a world of belonging and possibility, the possibility and belonging of Heidegger's *da-sein*, in which being is realized. Stevens's poet shares this impulse with the Dionysian celebrants of a "new" religion, and with the natives of Prospero's island, Ariel and Caliban, whom Prospero found there and who preserve a recalcitrance to his upliftings. Like Pentheus's Apollonian architecture, Prospero's bondage, his training Caliban, his teaching the earthly creature Prospero's own language, causes rebellion against this "civilizing." Prospero and Pentheus discover that recalcitrance of the earth, and Prospero in abdicating his authority admits that Caliban is his own: "This thing of darkness I / Acknowledge mine" (5.1.278–79).

In essence, then, the idealization of a world, proclaimed in the "blessed rage to order" of Stevens's poet ("Idea of Order at Key West" 106), and identifiable as a residual impulse toward the transcendent, as in Prospero's turning to his books and Pentheus's imposing of his high-minded values, provokes or balances against a paradoxical acceptance of earth: the advice of the Chorus in the *Bacchae* to accept the conditions of earth, Prospero's ceding of the island to Caliban, Stevens's paradoxical recovery or discovery of the immanent image of nobility within the world of detritus. In this impulse there might be felt, as in the Dionysian, a plunge of the divine to earth, which is the essential design in such works as Ovid's *Metamorphoses*, if the transformation is to occur, if the flower is to take the place of Narcissus's corpse. The forces of earth, the danger and destruction, withdrawal into the darkness, along with the devolutions of nobility, the comic and grotesque, are concomitant to this regeneration in the loosing of temporal power and beauty. The poet/musician of Stevens praises, in his "Metamorphosis," the anticipation of renewal, of refreshing, at the end of the year's darkness. Spring has not come, is not a "presence" in the poem, but the pattern of devolution entails anticipation of renewal. In this praise in "Of Modern Poetry," the musician performs the music of Caliban on his "twang[l]ing instrument."

* * *

As an afterword, I would suggest that this descent to the music of earth can be heard variously echoing across time. The pervasiveness of the paradoxical motive may become philosophically grounded in certain thought in certain times, for instance in the Romanticism of William Wordsworth. In the "Ode: Intimations of Immortality from Recollections of Early Childhood," for example, the poet records the sense of dismay when the adult senses the loss of childhood's harmony with nature and the "glory" that has "passed away . . . from the earth," the memory of a lost divine world.[12] The sense of alienation, of abandonment, he feels in having grown habituated to the "dialogues of business, love, or strife" to which he has "fit his tongue," is assuaged by recognition that his expression of solitude and "thought of grief" is not his alone but is shared with voices of nature. A solitary tree, the "single Field," and a "Pansy at my feet," all "the same tale repeat . . . of something that is gone." The blessedness that comes of this knowledge of belonging fully on the earth, its delights and its darkness, entails the poet in the affirmation of the metamorphosis that transcends the devolution. The "perpetual benediction" he can sing is not merely for the "delight and liberty" possible on earth; his "song of thanks and praise" is for the achieved awareness of the "obstinate questionings / Of sense and outward things, / Fallings from us, vanishings." In Wordsworth's apprehension of divine purpose, the potential ascent necessitates an acceptance and identification with "the meanest flower that blows."

For a final instance, we might trace the turn to the earth as an impulse in another modern poet, as the turn toward, the pressing down upon, the earth in Robert Frost's "To Earthward."[13] In Frost's poetry the search for transcendence manifests a pull in the upward direction which, as for Stevens, is a suspect or unattainable impulse. The Romantic desire to apprehend cosmic design may be sought in "Spring Pools," which "still reflect / The total sky almost without defect," yet which predominantly reveal the ephemerality of the pools and the flowers (245). The *ubi sunt* motif of Wordsworth's poem is echoed here in that the pools, brief as they are, are endowed by what is already gone, the "snow that melted only yesterday." This contingency of things in nature is not to be lamented in "To Earthward," but rather is celebrated. The insistent urge in this poem is to be wholly earth-marked, to feel the "weight and strength" of "the earth as rough / To all my length." The essence of being, in the progress of this poem, is from the pleasures of youth, the sweetness and beauty that were gratifying because they satisfied the need for "ache" and they "stung," to the mature appreciation of only the most powerful pleasure/pain of the senses, the satisfaction of a craving for "the stain / Of tears."

The poet contorts himself in straining to see beyond natural presence, and he is in "For Once, Then, Something," "taunted" for this impulse, for "having knelt at well-curbs / Always wrong to the light" (225). He is rewarded for this transcendent impulse by seeing only his own image in the surface of the water "godlike, / Looking out of a wreath of fern and cloud puffs," and his accomplishment climaxes when a drop of water falls from a fern and troubles the image he thinks he sees, "something white, uncertain," that might be "Truth" or a "pebble of quartz." The ineluctable fact of a drop of water tests his exalted aspirations and reminds the poet that he always has a comic aspect. An ironic, mocking tone pervades the poem "The Aim Was Song," which is about the musician's improvement on, or correcting of, nature in imitating the wind's blowing "loudest . . . In any rough place where it caught" (223–24). The wind's noise is brought into "measure" and converted, and the wind's appreciation of the human voice is wry. For both poet and wind, "the aim was song," and the resultant "word and note" of the poet becomes "The wind the wind had meant to be." Again the echoing sound, the poet's whistling, recalls the noises Plato's imitator likes to make. As opposed to the natural force of the wind, the poet's song might be regularizing and humanizing; but the image of the poet thus produced approaches the comic posture of the poet in Stevens's poetry, the old men in Euripides's play as they assume the roles of Bacchantes, the creatures of earth like Caliban and Stevens's birds who, rather than being tamed, regularized, made into Apollonian "measure," continue to make sweet noises of earth.

Florida Atlantic University
Boca Raton, Florida

NOTES

1 I cite the *Bacchae* from translations, comparing several types. For an older, semantically conservative version, I use the Loeb Classical Library, translated by Arthur S. Way (Cambridge, MA: Harvard UP; London: Heinemann, 1912) (cited as "Way"). It conveniently sets up Greek text and English translation on opposite pages. For a scholarly text with substantial explanatory notes I use the Prentice-Hall Greek Drama Series, translated by Geoffrey S. Kirk (Englewood Cliffs, NJ: Prentice-Hall, 1970) (cited as "Kirk"). I occasionally, when significant, cite comparisons of these with a prose translation downloaded from The Internet Classics Archive <http://classics.mit.edu/Euripides/bacchan.html> (cited as "Internet").

2 Wallace Stevens, *Collected Poetry and Prose* (New York: Library of America, 1997), p. 643. Subsequent citations of Stevens's works are from this text.

3 I have, in developing ideas about dramatic literature, pursued the *theatrum mundi* metaphor as essentially a playing between worlds, in the first place the world of the stage and the world of the audience. In my book *Human Shadows Bright as Glass: Drama as Speculation and Transformation* (Lewisburg: Bucknell UP, 1997) I developed implications from Edmund Husserl, Martin Heidegger, and Hans-Georg Gadamer in trying to effect a phenomenological analysis of this idea. Since this book, I have continued to pursue thoughts that, in general terms, would characterize the theater-dream metaphor as a version of the idea of possible worlds. A recent book by Mihai I. Spariosu, *The Wreath of Wild Olive: Play, Liminality, and the Study of Literature* (Albany: State U of New York P, 1997), is an illuminating discussion of "liminality," which is a matter of worlds in play and in relationships with others. Spariosu argues that in the final analysis, the poem is liminal. Although he mentions the stage only metaphorically, I would agree and argue that the stage is a version of the poem that stands ludically and irenically between worlds in opposition and in strife.

4 Emmanuel Levinas, *Discovering Existence with Husserl*, trans. Richard A. Cohen and Michael B. Smith (Evanston, IL: Northwestern UP, 1998), p. 57.

5 J. S. Leonard and C. E. Wharton, *The Fluent Mundo: Wallace Stevens and the Structure of Reality* (Athens, GA: U of Georgia P, 1988) respond (pp. 83–102) to Thomas J. Hines's reading of Stevens's poetry as developing, early to late, from Husserlian toward Heideggerian insights: *The Later Poetry of Wallace Stevens: Phenomenological Parallels with Husserl and Heidegger* (Lewisburg, PA: Bucknell UP, 1976). American Pragmatism's influence on Husserl's phenomenology has been observed by Herbert Spiegelberg, *The Phenomenological Movement: A Historical Introduction*, 2 volumes (The Hague: Nijhoff, 1960).

6 Martin Heidegger, "What Calls for Thinking?" in *Basic Writings*, ed. David Farrell Krell (New York: Harper & Row, 1977), p. 350.

7 William Shakespeare, *The Tempest*, ed. Northrop Frye (New York: Penguin, 1970). The ideas of music I intend here range from the classical conception of the "music of the spheres" to the Dionysian music that affronts divine harmony – gives it, one might say, the raspberry. Music as idea and various aspects of the text has been explored insightfully by Russ McDonald, "Reading *The Tempest*," *Shakespeare Survey* 43 (1991): 15–28, reprinted in Donald Keesey, *Contexts for Criticism*, Third Edition (Mountain View, CA: Mayfield, 1997), pp. 108–20. "Although the structural and narrative replications are more likely to be the subject of critical interest than the aural, most listeners find themselves beguiled by the musical repetition of vowels and consonants, reduplication of words, echoing of metrical forms, and the incantatory effect of this musical design" (p. 111). Although McDonald cites another critic observing that the "enchanted island reverberates with sounds hinting at tunes that never appear fully formed" (p. 114), which

could be taken to imply that inchoate music entails or arises from the nonmusical or cacophonic, the concern is for the Keatsean idea of "unheard music," which is of course idealized and sweet. McDonald cites and observes the passage I take as epigraph, calling it Caliban's "most memorable speech"; yet his concern is not for the idea of "noise," but for the way the passage suspends gratification. "We seem to be about to receive the riches of meaning which remain forever elusive" (p. 115). It is likely, of course, that Caliban remained a potential image of the poet from early in Stevens's career. Beverly Coyle ("Remembering Holly," *Wallace Stevens Journal* 16 [1992]: 205–21) observes that Stevens's and his daughter Holly "had something of the ugly duckling in them (Stevens's called himself Caliban in an early poem)" (p. 213).

[8] The Italian city of Palmanova, south of Venice, was first designed in 1593 and "is a perfectly shaped polygonal city frozen in its tracks for all of four centuries." Spiro Kostof, *The City Shaped: Urban Patterns and Meanings through History* (Boston: Little, Brown, 1991), pp. 160–61. Italian Renaissance artists derived support from a recent edition of Vitruvius, which interpreted Vitruvius's "wind-rose" city design "literally as a radial city" and which thus supported the idea "that their radial city plans were reviving long-forgotten principles of Classical urban design" (p. 185).

[9] The noises of earth in *The Bacchae* are appropriate to the god Dionysus, who is called, among other epithets, Bromios, a name that Kirk suggests means "'the roaring one.' It may be connected with the tendency to see bulls or lions as special embodiments of his power, or with the roar of thunder as Semele was struck by lightning at his birth" (p. 65).

[10] Hans-Georg Gadamer's way of talking about this "concreteness" of life is to insist on the essential kernel that remains in growth and development, in *bildung*. *Truth and Method*, second edition, trans. Joel Weinsheimer and Donald G. Marshall (New York: Crossroad, 1989). In talking about dramatic experience as a "transformation into structure," he sees the dramatic event as realizing a structure that includes the audience, by which the play "achieve[s] ideality, so that it can be intended and understood as play." The concreteness of this world made in the play is self-realization, and the persistence of the kernel of *bildung* is manifest. Insofar as the play entails the audience in this development, as "transformation," the activity can be expressed as *bildung*, "keeping oneself open to what is other – to other, more universal points of view" (17). The other is not abstract but concrete: "For in Bildung there is Bild. The idea of 'form' lacks the mysterious ambiguity of Bild, which comprehends both Nachbild (image, copy) and Vorbild (model)" (11).

[11] In response to this figure of the grotesque poet, I have already directed attention to the Nietzschean interpretation of the Archilochean, satyric poet as Dionysian (World Phenomenology Institute Conference on Phenomenology and Literature, Cambridge MA, 2000).

[12] William Wordsworth, "Ode: Intimations of Immortality from Recollections of Early Childhood," *Poetical Works* (London: Oxford UP, 1958), Vol. 4, edited by Ernest de Selincourt and Helen Darbishire, l, 18.

[13] Robert Frost, *Poetry*, edited by Edward Connery Lathem (New York: Holt Rinehart, 1969), pp. 226–27.

CHER KRAUSE KNIGHT

WHAT TIME IS IT?: SUBVERTING AND SUPPRESSING, CONFLATING AND COMPRESSING TIME IN COMMODIFIED SPACE AND ARCHITECTURE

Critic Fredric Jameson proposed that in the late capitalist age temporal references and understandings have been displaced by spatial ones, resulting in hyperreal experiences. The hyperreal induces consumption through spectacle, the mastery of which relies on the subversion or confusion of temporality. At shopping malls, Las Vegas casinos, and the Disney theme parks, time becomes the quintessential commodity: to be romanticized, recontextualized or eradicated altogether at the will of designers, and to the delight of consumers. The representation of time at these places designed specifically – although not exclusively – for consumer activities, infers a kind of finite infinity.

The postmodern era has reconfigured our conceptions of time-space, particularly evident in the architecture and authored places designed to stimulate consumer participation. By enacting a shared matrix of time subversion techniques, elimination of seasonal shifts; inability to gauge time of day, or day of the week, and collaging of the past, present and future, these places create a sense of timelessness that is ultimately revealing of our own age. Since a visitor cannot pin down referents to or enframements of time, time is neutralized as a marker. The patron often loses track of time altogether, and gets metaphorically lost in a seemingly endless chain of consumption opportunities, which is a primary goal of the designers. While purporting to be democratic, these places actually blur the distinctions between public and private, inside and out. They are open to the public-at-large, but only those with the money to sustain continued participation in the consumer spectacle remain within. The inside of these spaces is clearly demarcated from the outside; success depends on an image of accessible exclusivity, and the ability to block out the external world and its temporal allusions. Via freeze-dried representations of time, these consumer spaces suggest specific historical periods and places, in which the ephermeral is rendered as eternal, enduring and unidentifiable. The underlying irony of this condition belies that of our era itself: only through the timely do we arrive at the timeless.

America is a society of consumers. The Cold War ushered in an era of prosperity for Americans, who had more leisure time and disposable income than their predecessors. As city centers deteriorated, these people flocked to

A.-T. Tymieniecka (ed.), Analecta Husserliana LXXVIII, 325–336.

the suburbs with their newly acquired wealth, and patronized the shopping centers that began to proliferate there. The experience of bartering at open-air bazaars or meandering along city boulevards was replaced by a visit to the shopping mall, where the fruits of capitalism were proffered.[1] In 1956, the first fully enclosed, air-conditioned suburban mall, Southdale Shopping Center designed by Austrian-born architect Victor Gruen, opened in Edina, Minnesota. By the early 1980s there were already over 28,500 shopping malls in North America.[2] In *The Consumer Society*, Jean Baudrillard argues that gratuitous consumption supports our culture and makes us feel more alive. Time is a commodity itself, something to be consumed or wasted, or in the case of leisure time, something we are obligated to *spend*. The goods we buy are purposefully anachronistic so that we must constantly replace them.[3] Our culture has already witnessed the dovetailing of theme park and shopping mall in places such as Bloomington Minnesota's epic Mall of America, designed by the Jerde Partnership Architects, which has a Knott's Berry Farm amusement park at its center.[4] While the evolution of the shopping mall is beyond my purpose here, the form has clearly morphed since Southdale, although that too contained a children's zoo and the "Garden Court of Perpetual Spring," attractions in their own right.

Today malls are monumental spaces, offering a variety of activities (shopping, eating, movie theaters, and amusements) and themes (even if only loose ones) within a larger unifying environment. Much like Disney World, they are introverted, climate-controlled worlds that tend to be patronized by homogenized troops of leisured pedestrians who can afford the diversions they offer. In *America*, Baudrillard proposed that the consumer success of Las Vegas also relied on its break from the outside world. The casinos are lush oases of air-conditioned, artificially lighted comfort that exile the realities of desert living.[5] A sense of time is lost, or at least confused, as many casinos have no windows or clocks to gauge the passing hours of day and night. Disarming temporal sensibilities lowers our resistance to consumerism. Our spatial sensibilities are also challenged. It is easy to enter both the casino and the mall, but their space is often disconcerting, what Jameson called "hyperspace," ensnaring us in a consumer spectacle of accumulation and profusion. Although Disney's spaces are easily navigated, Walt Disney endeavored to eradicate temporal markers, observing of Disneyland, "Clocks and watches will lose all meaning, for there is no present. There is only yesterday, tomorrow and the timeless land of fantasy."[6]

Sociologist John Hannigan notes that our contemporary culture is witnessing the rise of "shoppertainment, eatertainment and edutainment," in which shopping, dining, entertainment, commerce, technology, and education overlap. As mail order, cable shopping networks, and specialty or discount stores lure customers away, malls have retaliated with exciting environments. Shopping has become "a non-utilitarian, visual experience . . . and adventure," and the shopper, a contemporary *flâneur*.[7] Both Mark Taylor and Peter Gibian attribute the contemporary mall's form to a combination of nineteenth-century shopping arcades, particularly those of Paris, and department stores. These earlier prototypes were self-contained environments that split the space of production from that of consumption, and problematized the binary oppositions of outside and inside, public and private. Likewise, Disney World, the mall and Vegas' themed casinos offer escapist worlds in miniature, where the poverty, social tensions and service functions of daily life are banished from view, and visual fantasy combats external realities.[8] (As Neil Harris observed in *Cultural Excursions*, today's malls reserve meticulous theming for their interior space, while the exterior architecture often looks "like a pile of blocks.")[9] Cut off physically and psychologically from the outside world, these efficient urban models are so dazzling that many patrons pay little attention to their constraining enclosure, which is even less noticeable in an outdoor setting. At these places we trek through a vast stage set of dynamic moving parts including elevators, escalators and fountains that emphasize our own motion. In particular, Gibian relates the mall to Walter Benjamin's "city shock," in which crowds bump and collide against each other in "snap views of passing time."[10] In *Consumer Culture and Postmodernism*, Mike Featherstone also noted that the stores and arcades of Paris were "dream worlds" of novelty and constant change for Benjamin, proffering what seemed like endless consumer goods. The role of the *flâneur* has become increasingly playful with the rise of postmodernism, as we have even more goods and experiences to sample, and are cogent of the fantastic simulacra used to sell them to us.[11]

In "I've Seen the Future and It's Fake," Margaret Crawford underscores the interrelationship of television, theme park and shopping mall as realms of experience separated from the outside world through manipulations of time and space. These places offer a coherence of vision not found in daily life, despite their radical compression of themes remote in space and time, which challenges our traditional perceptions of the world. These abrupt shifts between fantasy and fact are also found as we surf the television (and now the World Wide Web). Crawford suggests that the underlying premise of these places, and television, is that "controlled fiction" is more interesting than "messy reality."[12] These observations are similar to those made by William

Severini Kowinski in *The Malling of America*. The author acknowledges that mall-industry insiders have high praise for the Disney parks, and it seems more than coincidental that Disneyland and the first enclosed malls were built in the mid-1950s. Mall scale and design often allude to a quaint Main Street where people are friendly, the businesses prosperous, and the environs pristine. Kowinski claims that consumers willingly partake in the mall's fantasy because we have been conditioned by television, which grooms our perceptual habits. A few simplified, artificial sets represent "complete reality" scaled-down; the ads' quick communication and repetitive imagery are common; incongruities between different programs do not give pause; and the viewer is simultaneously stimulated and lulled into complacency. The mall is like a television we walk around in, constantly playing its "Retail Drama."[13]

Political scientist Benjamin Barber laments the vast malls that supplant America's Main Streets, where shopping was integrated with other public activities. The mall has now become a tourist stop and a destination point, a "neighborhood" of enclosed public squares that is the only community space in many suburbs.[14] Architecture critic Ada Louise Huxtable also dislikes the current mall culture, which she views as a "cocoon" that protects us from "assorted discomforts and also diversity."[15] Spiro Kostof, an architectural historian, observes that theme parks and malls are creating a spatially discontinuous landscape. The privatization of public spaces raises pertinent socio-political issues. Although these places are owned and operated by private corporations, they are offered for public consumption, which complicates, for example, the rights of free speech on their property.[16]

San Diego's Horton Plaza, designed by Jon Jerde, exemplifies the contemporary mall format, offering a variety of themed environments as disparate as an Italian hill town and a 1950s-styled diner. As John Sears remarked in his study of American tourist attractions, the Plaza constitutes "a manufactured culture (that) becomes one of the commodities offered for consumption." Like Disney World, Horton Plaza is *inward*-looking space, literally blocking out its downtown site. While this effect hampers urban integration, Sears notes its heightening of tourist appeal: "Entering Horton Plaza . . . is to enter a different world, a world of fantasy that transcends the ordinary, everyday world of the city or the suburb where people live."[17] This transcendence also fuels Caesars Forum Shops in Las Vegas, built in 1992, a collection of upscale stores and eateries set in the guise of a quaint Roman town among ruins. Its immense space of 250,000 square feet encloses many notable elements, including an illusionistic curved ceiling painted to look like the sky that changes from dawn to dusk and back again, in the span of a half-hour. As Gibian asserts, this effect exists not to make visitors feel they are outside, but rather to celebrate the designer's technological mastery and the mall's insulation from the unpredictable weather of the "real" world.[18] Terry Dougall, of Dougall Design Associates who planned the Shops, asserts, "This isn't great architecture, it's great theater."[19] The Forum is an incredibly successful retail outpost, outperforming many other American shopping centers. Vacation shoppers are willing to immerse themselves in indulgent fantasy while trying on a new pair of chic shoes. The Shops include expensive boutiques such as Versace and Ralph Lauren, but there are also moderately priced offerings for the 50,000 people who trek through the Forum every day. Obviously, the discrepancy between ancient Roman "architecture" and contemporary goods is not an affront to these shoppers' senses;[20] it is a pleasant juxtaposition of fantasy with function, which they recognize and appreciate.

Kowinski identifies an analogy between the layout of Disney's Main Street, U.S.A., a shopping promenade leading to Cinderella's Castle, and the contemporary mall where small shops give access to large department stores, "the consumer equivalent of a citadel of wonders."[21] But despite Walt's assertion that "Main Street is everyone's home town – the heartline of America," its target audience is less inclusive. As Michael Real proposed in his study of Disney World as a morality play, Main Street specifically celebrates the small Midwestern town at the turn of the century and its middle class folk.[22] Here, where a "marinade of nostalgia and sentiment is meant to ostensibly bespeak a kinder, gentler world," the visitor consumes goods and idealized visions of America's past. Consequently, Main Street has become a model for mall designers who prompt visitors to leave their cars – and the world – outside to get lost in the consumer spectacle.[23] As the only entrance and exit to the park, Main Street's veiled reality as a mall (the buildings seem separate but are actually interconnected for unfettered access), inscribes shopping as the first and last activity in the Magic Kingdom.[24] As sociologist George Ritzer observed, such coerced exposure to consumption opportunities is unsettling because those visiting the park are often unaware of the carefully structured though subtle constraint. Ritzer reminds us: "But, as Foucault showed in his work on the Panopticon, gentle and subtle control can be far more troubling than the blatant and brutal forms. People do not know how they are being controlled. . . . Without such knowledge, it is difficult, if not impossible, to question and rebel against the control."[25]

Main Street's down-scaling, which produces a sense of romanticized intimacy through forced perspective, also manipulates visitors to feel a sense of mastery, as though they were playing with dolls or a train set.[26] Louis Marin attributes Main Street's success to its conflation of adult reality (the consumption of contemporary goods) and childish fantasy (purchased in a Victorian setting), but laments that it conceives of life as "an endless exchange and a constant consumption."[27] In "Travels in Hyperreality," Umberto Eco echoes this argument, suggesting that the fanciful veneer of Main Street induces consumer participation that feels like nostalgic play. While he maintains that the goods purchased here are very real, "What is falsified is our will to buy, which we take as real, and in this sense Disneyland is really the quintessence of consumer ideology."[28] Thus, Main Street disarms our pretensions and worries so that we feel part of a community – if only a temporary one – in which we easily interact and demonstrate our membership through consumption.

The distinction between reality and illusion is further blurred by the details of Main Street's atmosphere. Entering the park under the railroad station, visitors are brought to the first "visual node," a meticulous public square enhanced by Victorian-style furnishings. Names on the windows imply that real shopkeepers work here, but in actuality these are tributes to Disney's designers, the Imagineers, and Walt's father. Curios from the turn of the century are integrated with contemporary props.[29] Karal Ann Marling, a historian of art and popular culture, identifies an ironic turn on the antique shop here: rather than a contemporary place that sells old goods, Main Street is an old-fashioned-style place that sells new ones: "Period décor legitimated consumption by equating the business of Main Street USA with the historical fiber of the nation."[30] David Joselit concurs that Main Street is "engineered to make consumption seem wholesome and patriotic – even *historic*. . . . Advanced global capital is domesticated, made to seem as friendly and innocuous as small-town life, which in the late twentieth century has all but disappeared."[31] But according to Imagineer John Hench, Disney's Main Street merely edits out the cynicism and contradictions – "dreary buildings" and "gloomy colors" – that pockmarked actual main streets. Instead, Disney focuses on the "collective dream" of Victorian America in which "progress was a good word, and where there was an intense optimism about what we were doing with our lives, and where we were going."[32] Hench concludes, "Ours is a kind of universally true Main Street – it's better than the real Main Streets of the turn of the century ever could be."[33]

All of the elements of Disney's Main Street were carefully selected to capture small-town America not exactly as it was, but perhaps as it should have been. Cultural geographer Richard Francaviglia's important work on

Main Streets as physical and psychological loci of meaning in American culture pays special attention to Disney's Main Street as an improved version of his childhood reality. Francaviglia traced the source for Disney's Main Street back to Marceline, Missouri, where the young Walt spent some of his early childhood.[34] Even though the Disney family was in Marceline only for a few years, it seemed to be developmentally important for Walt, who idealized those days as happy ones, especially compared to the hard life his family would face later in Kansas City and Chicago. For Walt, Marceline was a symbol – both generally and personally – of an innocent American past and a cheerful childhood. On his Main Street, only pleasantries are encountered: an ice cream shop, a bakery, a general-store-like Emporium, barbershop quartets, and the like. Walt said that he wanted Main Street to be a Norman Rockwell painting come to life. And it is, both in its nostalgia for a time past, and in its tendency to ignore the hard facts of history.[35] Here Walt falls victim to his own jingoism, blind to America's adversities. But what else could be expected from a man who said of his own patriotism, "Actually, if you could see close in my eyes, the American flag is waving in both of them and up my spine is growing this red, white and blue stripe."[36] Disney may have been guilty of near-sighted sappiness, but not of insincerity.

Stepping onto Main Street must have been like a step back in time for Walt. Shortly after the turn of the century, Marceline's Main Street, with its quaint shops, architectural detail and bustling business, still had a Victorian quality. This became a basic pattern for the Disney Main Street, which Imagineers reinterpreted to produce a deliberately themed environment. Instead of copying a building detail for detail, the Imagineers synthesized elements from many. Architectural historians Colin Rowe and Fred Koetter have stated that Disney's Main Street was less a project of idealization than "a filtering and packaging operation" that eliminated the "unpleasantness of tragedy, of time and of blemish."[37] As historian Michael Wallace notes: "The Disney people don't consider this retrospective tidying up an abuse of the past; they freely and disarmingly admit its falsification, pointing out that this is, after all, just entertainment. But they also insist they are bringing out deeper truths." These "truths" present the Victorian era as optimistic, prosperous and progressive, an image that is ultimately more reassuring than that of immigrant poverty, political protests, and lynchings, among other unpleasant realities. To Disney this de-racialized and de-politicized past symbolized "the real America."[38] For Wallace, this approach is representative of 1950s Hollywood, which employed "selective amnesia" to manufacture a "happy past" corresponding to America's "contented present." It is not that

Walt paints an untrue picture of the past, but that it is only partial.[39] The era conjured up on Main Street is purposefully unspecified. Francaviglia suggests that it is evocative of circa 1890 to 1910, a transitional time when electricity and the automobile would forever change the historic and physical landscape of America.[40] This period has been highly romanticized in popular culture, and by Walt, who selectively edited out its more unhappy or unsavory elements.

Francaviglia attests to Main Street's importance for urban planners, who admire how cleverly Walt and his Imagineers employed terminal landmarks, architectural detail, and playful scale, and managed pedestrian flow. Main Street is a primary example of well-orchestrated social engineering, which induces consumption and makes us receptive to Disney's ideology and the park as a whole. He declares, "That Disney was 'one of us' is underscored by the mass appeal of his Main Street. We feel comfortable there as we fill the role of consumers of both products and place."[41]

Francaviglia asks important questions in his study: How accurately had Walt remembered Marceline? Was it truly representative of small-town America?[42] How much has history been altered to fit the Disney mold? But he concludes that the attacks of most critics are backlash against Walt's successful recipe for social engineering, which cleverly confuses the abstracted small-town experience with American historical reality and produces a universally appealing place. Thus, Francaviglia asserts, "In the late twentieth century, it has become apparent that Disney either knew more about image building than given credit for, or that he was a pioneer in shaping values: Disney intuitively knew that images were as important as substance."[43]

In some respects Walt was a postmodern capitalist, who made images into a commodity, blurring distinctions between the real and the imaginary. At his Main Street, the icon has become so powerful that it now affects the way we view Main Streets out in the world. Francaviglia maintains that while Disney's Main Street asks us to suspend disbelief, it is also an optimistic exercise bringing an appreciation of our past to a larger public, suggesting how its values might foster civic responsibility and moral behavior in the future. On Main Street, Walt celebrated the qualities that he believed made America a great nation, embodied in the small town. To some critics, this seemed like a celebration of insularity and bigotry, but for many visitors, it was a delicately reworked presentation of the past that reassured and pleased them.[44]

Architectural critic Dolores Hayden questions whether "an American sense of place" can be created through architecture, design and urban planning. Hayden believes that "average" Americans seek a "common language" seemingly found at places like malls, casinos, and the Disney theme parks, but cautions that space is ultimately a commodity. Land developers, designers and architects who want to make us feel comfortable – and loosen our purse strings – use the vernacular to present their visions as familiar and easily understood. These appropriations may be based on specific places or times but are seen as representative of a larger whole. A sense of place, she contends, must evolve naturally at its own particular pace over time, and cannot be willed through a regurgitation of styles.[45] Yet the carefully constructed imagery and illusion of Disney's Main Street have produced a meaningful place for its visitors, even when they recognize its deliberately structured conceit. As writer Richard F. Snow reminisced of his first trip to Disneyland in 1957, it was truly "magical" – "I wanted to stay in this place forever.'" Snow was particularly enchanted by Main Street, remarking that he believed it was "a triumph of historical imagination," despite its obvious editorial idealization of the past. Snow concluded, "Sinclair Lewis found his Main Street foul with hypocrisy . . . and blighted aspirations. He wasn't wrong. But neither was Walt Disney."[46]

Emerson College

NOTES

[1] Dennis Alan Mann, "Ritual in Architecture: The Celebration of Life," in *Rituals and Ceremonies in Popular Culture*, ed. Ray B. Browne (Bowling Green: Bowling Green University Popular Press, 1980), pp. 73–74.

[2] Mark C. Taylor, *Hiding* (Chicago: University of Chicago Press, 1997), pp. 259–60.

[3] Jean Baudrillard, *The Consumer Society: Myths and Structures* (1970; London: Sage Press, 1998), pp. 46 and 99. The seven-day-and-night-a-week schedule many malls keep erases the challenges of clock and calendar, so everyone can find a time and day to shop. Jean Baudrillard, "Consumer Society," in *Selected Writings*, ed. Mark Poster (Stanford: Stanford University Press, 1988), pp. 34–35.

[4] Mall of America is actually four malls under one enclosure; it also features hotels, office towers and a convention center. Patrons often stay on site to take full advantage of the complex. See Margaret Crawford, "The World in a Shopping Mall," in *Variations on a Theme Park: The New American City and the End of Public Space*, ed. Michael Sorkin (New York: Noonday Press-Hill and Wang, 1992), pp. 3–30. While focused on the West Edmonton Mall in Alberta, Canada, Crawford's larger framework of hyperconsumption is useful.

[5] Jean Baudrillard, *America*, trans. Chris Turner (1986; London: Verso, 1988), pp. 127–28.

[6] Walt Disney, quoted in "Here's Your First View of Disneyland," *Look* 2 (November 1954), p. 86.

[7] John Hannigan, *Fantasy City: Pleasure and Profit in the Postmodern Metropolis* (London: Routledge, 1998), pp. 89–91. Peter Gibian, "The Art of Being Off-Center: Shopping Center

Spaces and Spectacles of Consumer Culture," in *Mass Culture and Everyday Life*, ed. Peter Gibian (New York: Routledge, 1997), p. 270.

[8] Taylor, op. cit., pp. 293–94. Gibian, op. cit., pp. 261–63, 271–73 and 279. Baudrillard suggests that the shop window confuses inside and outside, being neither private nor wholly public. The window is a stage set for goods that symbolically solicit the gaze, extending the invitation to shop in a "real" place. Baudrillard, *The Consumer Society*, op. cit., p. 166. He is also struck by the arrangements of goods, which may appear haphazard but are carefully organized in "a *calculus* of objects" against the "frenzy of purchasing and possession" that comes with the "simple profusion of commodities." Baudrillard, "Consumer Society," *Selected Writings*, op. cit., p. 31.

[9] Neil Harris, *Cultural Excursions: Marketing Appetites and Cultural Tastes in Modern America* (Chicago: University of Chicago Press, 1990), p. 282.

[10] Gibian, op. cit., pp. 238–39, 258 and 282–83.

[11] Mike Featherstone, *Consumer Culture and Postmodernism* (London: Sage Press, 1991), pp. 23–24.

[12] Margaret Crawford, "I've Seen the Future and It's Fake," *L.A. Architect* (November 1988), pp. 6–7.

[13] William Severini Kowinski, *The Malling of America: An Inside Look at the Great Consumer Paradise* (New York: William Morrow and Company, 1985), pp. 65, 67, 69–73 and 355.

[14] Benjamin R. Barber, *Jihad vs. McWorld* (New York: Times Books-Random House, 1995), pp. 97, 128 and 130–32. The author suggests that the suburbs themselves (where over half of all Americans live) will become themed, and cites Orange County, California as an example. For more on Orange County see Edward W. Soja, "Inside Exopolis: Scenes from Orange County," in *Variations on a Theme Park*, op. cit., pp. 94–122.

[15] Ada Louise Huxtable, *The Unreal America: Architecture and Illusion* (New York: New Press, 1997), p. 107.

[16] Spiro Kostof, *The City Assembled: The Elements of Urban Form through History* (Boston: Bulfinch Press – Little, Brown and Company, 1992), pp. 185–86. In 1968 the United States Supreme Court declared that no distinction could be made between malls and other public spaces like city streets in regard to First Amendment Rights, a decision that was quickly reversed in 1971. In 1980, however, freedom of speech was protected by the State Constitution of California, which identified malls as public space.

[17] John F. Sears, *Sacred Places: American Tourist Attractions in the Nineteenth Century* (New York: Oxford University Press, 1989), pp. 213–14. Horton Plaza has a senior citizen residence and a hotel.

[18] Gibian, op. cit., p. 273.

[19] Terry Dougall, quoted in Donald Shillingburg, "Entertainment Drives Retail," *Architectural Record* 182 (August 1994), pp. 84 and 88. See pp. 82–89 for a detailed discussion about design and retail performance.

[20] Aaron Betsky, "Theme Wars Rage in Vegas," *Architectural Record* 180 (August 1992), p. 29. In 1993 the Forum Shops sold an average of nine hundred dollars' worth of merchandise per square foot of space (making it the most lucrative shopping space in the United States per square foot), whereas sales of four hundred to five hundred dollars per square foot are considered high performances at most American shopping centers.

[21] Kowinski, op. cit., p. 70. References to Disney's Main Street throughout this article are drawn from research on the Main Streets of both Disneyland and Disney World.

[22] Michael R. Real, "The Disney Universe – Morality Play," in *Mass-Mediated Culture* (Englewood Cliffs: Prentice-Hall, 1977), p. 53.

[23] Stephen M. Fjellman, *Vinyl Leaves: Walt Disney World and America* (Boulder: Westview Press, 1992), p. 394. Also see Richard V. Francaviglia, *Main Street Revisited: Time, Space, and Image Building in Small-Town America* (Iowa City: University of Iowa Press, 1996), pp. 164–66 and 175. Francaviglia, "Main Street U.S.A.: A Comparison/Contrast of Streetscapes in Disneyland and Walt Disney World," *Journal of Popular Culture* (in-depth section. "The American Amusement Park," ed. Margaret J. King) 15:1 (Summer 1981), pp. 141–56, offers detailed accounts of the specific shops and structures on both Main Streets.

[24] George Lipsitz, "Consumer Spending as State Project: Yesterday's Solutions and Today's Problems," in *Getting and Spending: European and American Consumer Societies in the Twentieth Century*, eds. Susan Strasser, Charles McGovern, and Matthias Judt (Cambridge: Cambridge University Press, 1998), p. 138. Disney makes it easy to consume; packages are held at entrances to be picked up at the end of the day, or delivered to any of the Disney resorts. One can gather up as many goodies as one likes and not have to lug them around.

[25] George Ritzer, *The McDonaldization Thesis: Explorations and Extensions* (London: Sage Press, 1998), pp. 144–45.

[26] At Disneyland, Main Street's first floor is built to nine-eighths scale, the second to seven-eights, and the third to five-eighths. Most upper stories are false fronts with an enclosed catwalk for maintenance and security workers. The Mark Twain boats, railroad and all Main Street vehicles are also five-eighths scale, and the castle's imposing stones get increasingly smaller toward the top. Each brick, shingle and gas lamp on Main Street is also five-eighths scale, and the trees are trimmed accordingly. At Disney World, Main Street is more ornate (a "hyper-designed" Victorian atmosphere) and its buildings have full-sized upper stories. Francaviglia observes; "Disney World's Main Street USA is a caricature of Disneyland's Main Street USA, which in turn is a caricature of the Main Streets of places like Marceline and many other towns." Francaviglia, *Main Street Revisited*, op. cit., p. 162.

[27] Louis Marin, "Disneyland: A Degenerate Utopia," *Glyph*, Johns Hopkins Textual Studies 1 (Baltimore: Johns Hopkins University Press, 1977), pp. 62–63.

[28] Umberto Eco, "Travels in Hyperreality" (1975), in *Travels in Hyperreality: Essays*, trans. William Weaver (1967; San Diego: Harcourt Brace Jovanovich, 1986), pp. 43 and 48.

[29] Francaviglia, *Main Street Revisited*, op. cit., p. 162. David Koenig, *Mouse Tales: A Behind-the-Ears Look at Disneyland* (Irvine: Bonaventure Press, 1994), pp. 42–44. Sometimes the juxtaposition of the old-fashioned and the new can be jarring. When Ira Wolfert visited Disneyland and pointed out the modern ceilings and lighting in Main Street stores, Walt replied; "I'm sorry you noticed that. We had to change the gaslights here – people complained that they made the merchandise look too gloomy." Ira Wolfert, "Walt Disney's Magic Kingdom," *Reader's Digest* 67 (April 1960), pp. 144–45.

[30] Karal Ann Marling, *As Seen on TV: The Visual Culture of Everyday Life in the 1950s* (Cambridge: Harvard University Press, 1994), p. 116.

[31] David Joselit, "Architecture: Moguls on Main Street," *Art in America* 87:1 (January 1999), p. 53.

[32] John Hench, quoted in Anthony Haden-Guest, *The Paradise Program: Travels through Muzak, Hilton, Coca-Cola, Texaco, Walt Disney and Other World Empires* (New York: William Morrow and Company, 1973), p. 237.

[33] John Hench, quoted in Paul Goldberger, "Mickey Mouse Teaches the Architects," *New York Times Magazine* (22 October 1972), p. 95.

[34] The following discussion of Disney's Main Streets is based on Francaviglia, *Main Street Revisited*, op. cit., pp. 145–84. These arguments were first put forth in Francaviglia, "Main Street

U.S.A.: A Comparison/Contrast of Streetscapes in Disneyland and Walt Disney World," op. cit., pp. 141–56.

35 Walt told Ken Anderson, an animation storyman, "You can make some paintings like Norman Rockwell's, and I'll build models of them." Walt Disney, quoted in Randy Bright, *Disneyland: Inside Story* (New York: Abrams, 1987), p. 39. Both Disney and Rockwell were highly successful, business-minded artists who appealed to the middle class and considered themselves champions of the American way of life (as they viewed it, especially for white Anglo-Saxons). They enjoyed popular success, but often received harsh criticism as well. Their sentimental idealizations of small-town life seem like attempts to correct unhappy childhoods, and come to grips with the modern world. For more on Rockwell see Michele H. Bogart, *Advertising, Artists, and the Borders of Art* (Chicago: University of Chicago Press, 1995), pp. 72–78.

36 Walt Disney, quoted in Dave Smith, ed., *Walt Disney: Famous Quotes* (Lake Buena Vista: Walt Disney Company, 1994), p. 80.

37 Colin Rowe and Fred Koetter, *Collage City* (Cambridge: MIT Press, 1978), p. 46.

38 Chris Rojek, "Disney Culture," *Leisure Studies* 12 (1993), p. 127.

39 Michael Wallace, "Mickey Mouse History: Portraying the Past at Disney World," in *History Museums in the United States: A Critical Assessment*, eds. Warren Leon and Roy Rosenzweig (Urbana: University of Illinois Press, 1989), pp. 161–63.

40 Only at Disney could horse carriages and their replacements, such as the omnibus, coexist. Francaviglia, *Main Street Revisited*, op. cit., p. 156.

41 Francaviglia, *Main Street Revisited*, op. cit., p. 162.

42 For example, at night the Main Street buildings are outlined in lights, a visual buzz that was rarely seen on the Main Streets of America, and is more typical of the lighting found at expositions. Francaviglia suggests that the Imagineers were influenced by the "Yesteryear's Main Street" exhibit in the Museum of Science and Industry, Chicago, sponsored by General Motors. Francaviglia, *Main Street Revisited*, op. cit., pp. 160–61.

43 Francaviglia, *Main Street Revisited*, op. cit., p. 157.

44 Gerald Kurland, *Walt Disney: The Master of Animation*, Outstanding Personalities series (Charlotteville: SamHar Press, 1971), p. 28.

45 Dolores Hayden, "An American Sense of Place (with an Afterword)," in *Critical Issues in Public Art: Content, Context, and Controversy*, eds. Harriet F. Senie and Sally Webster (New York: Icon Editions – HarperCollins, 1992), pp. 261–69.

46 Richard F. Snow, "Disney: Coast to Coast," *American Heritage* 33:2 (February-March 1987), pp. 22 and 24. Snow admits he became disenchanted with Disney as "a sort of institutionalized self-congratulatory blandness," but later realized that Walt had "a far more imaginative, less sentimental vision than I had believed."

J. C. COUCEIRO-BUENO

THE PSYCHOMETAPHOR

I

The concept of the metaphor and its theorisation have proved to be crucial elements within the framework of cross-disciplinary thought, capable of carrying out a thorough investigation into all the consequences of the epistemological transformations brought about by the so-called "linguistic shift" in our "conception of the world" (*Weltanschauung*). As far as psychology is concerned, the metaphor represents the hope that research will not be centred overwhelmingly on studies of a quantitative nature, or probabilism, which, expressed in a Husserlian manner, would entail reducing the concept of science to the science of facts, thus inevitably implying a restriction of the very concept of science.[1] In this sense it must not be forgotten that in his *Investigaciones Lógicas*,[2] E. Husserl dispensed with philosophy's distortion by the field of psychology, guided exclusively by natural science. This allowed for the recreation of a reflexive structure through the use of concepts such as the metaphor, which possesses a proven capacity for both transgressing the limits of specialised discourse and for linguistic renovation, thereby affording us a congruent perspective of reality. Consequently, the psychometaphor can be taken as any integrating model which may result from a "metaphorical transposition" within the framework of cross-disciplinary study. Its most immediate consequence is its effect on specialised discourse whose cognitive procedures have been altered: namely cognitive psychology, linguistics, anthropology, hermeneutics, etc.

In line with this train of thought, I maintain that, for a variety of reasons, there have been earlier contributions of considerable relevance to the current proposal of the psychometaphor which, in my opinion, have not been given their due consideration. These theories come from a range of trends in cognitive psychology, theories which, it must be said, have provided the strongest models used in explaining metaphors: the metaphorical models of Lakoff & Johnson, Harré, etc. The aim was to establish a concept of psychology capable of providing an answer to the continuous trans*figur*ations of our "images of the world" caused by scientific research on a global scale. Now, my secondary aim is to point out the unwaivable (or inevitable) encounter between cognitive psychology and the hermeneutic tradition in particular, and continental philosophy in general.

A.-T. Tymieniecka (ed.), Analecta Husserliana LXXVIII, 337–346.

Prior considerations by way of a declaration of principles require that the following question be posed: What contribution can philosophical hermeneutics make to psychology? The following points summarise its main contributions:

a) Philosophical hermeneutics[3] in its basic and theoretical form, should be useful to psychology by allowing it to reach a correct understanding of its nature. In other words, it raises the issue of whether the strategies and methods employed are those required by psychology. Hermeneutics considers all theories primarily as "means of observing the world" (*Weltanschauung*), and secondly as means of ordering a series of events. Our interpretative experience absorbs information as a means of shaping our experience; as a result, the events themselves will be understood as the conceptual and narrative experience of our own personal experiences.
b) Hermeneutics also calls into question the primacy of subjectivity as the fundamental basis and rule of knowledge. Our conscience inevitably finds its own meaning and significance outside itself. Such a theory constitutes a real epistemological and ontological challenge to psychology. Against the primacy of the basic and fundamental *cogito*, a postulate which, it must be said, becomes untenable in the light of the strictest neostructuralist and neuroformalist contributions, psychology will face supra-individual entities such as language, metaphor, symbols, analogy, etc., which may afford it an overall vision (*sinoptikos*, in Platonic terminology).
c) Hermeneutics offers us what could be referred to as a "hermeneutic criterion"; it supports the idea that the direct observation of behaviour offers only a small insight into its full determination. In human beings, a function (or action) can never be reduced to observable behaviour, as it inevitably involves self-narration. From this perspective, hermeneutics may constitute a suitable tool to provide us with insights into psychological research. It would therefore be responsible for the co-ordination of the fundamentals of psychology: a neural framework, an observable psychological state, and human experience taken as narrative experience (essential initial structures which provide individual phenomena with meaning and significance).
d) Hermeneutics contributes to the growing awareness in this scientific later-modern age that all epistemological postulates may be considered from a range of perspectives (crossing disciplinary boundaries), thereby enabling them to restore the "multiple unity" of psychological theories due to their heterogeneous nature. If we discard such integration of diversity, then any

psychological concept would be mutilated at birth. Combining material from a series of sources under a new perspective (physics, biology, linguistics, arts, etc.) implies going beyond the outlandish idea (predominant in certain academic circles) that it is only worth carrying out extremely detailed research within highly specific fields or areas.

The cross-disciplinary approach that we propose could enable researchers in psychology to discover unsuspected links between "events" and "ideas" which at first sight may appear completely unrelated, but which, from a structural and metaphorical perspective, do in fact share certain similarities, and may even on occasion prove to be isomorphic. Today we know that the network of interrelations which connect the human being with the rest of the *fisis* is so complex that heteroclitic elements mutually intervene to an enormous extent.

e) Hermeneutics provides psychology (or rather it could do so, if certain persistent ideas which have ideologically restricted its own development could be overcome), with a new relationship with the traditional idea of truth, which would no longer be understood exclusively on the basis of demonstrations or *protokollstze*, but rather in relation with the universal nature of the thinking mind developed by Kant in the third of his critiques. This universality would come about as the result of contingent and unlimited consensus whose main interlocutor would be the unrestrainable expansivity of the metaphor. From this perspective, emphasis could be given to the concept of sensorial experience as the experience of truth. Its linguistic translation would consist of the eminent recovery of emotions at a safe distance from those theories which considered them to be "interference" that meddled with and distorted cognitive processes.

Sensorial experience (aesthetics/aisthesis) is unusual in that its very significance resides within itself. According to hermeneutic experience, perceiving is "taking something as true," and is therefore removed from the idea that it is a mere "collector of impressions" and is closer to the ontological theory based on the Heideggerian concept of *geworfenheit* (a way of feeling, emotional capacity).

In his work *Ser y Tiempo* (1927), Heidegger claims that the Dasein, in terms of the being-in-the-world, not only already always possesses a certain understanding of all its meanings, but also always has a certain degree of disposition. Things do not only have a theoretical meaning, but they also possess an emotional significance. These emotions are not merely involved in our understanding of the world, but are also a kind of more basic pre-understanding of understanding itself. "Feeling" is an initial global way of approaching the world which creates its own sense of understanding.

f) The most important and fundamental contribution of hermeneutics is that which I refer to as the psychometaphor: a contextualised metaphor intervening as a central concept which attempts to remind us that only from research based on cognitive psychology and philosophical hermeneutics will it be ontologically possible to study metaphors in a scientific context characterised by continuous transformations and transposition that alter the univocal and stable ideas of the physical and psychic frameworks.

II

The theorisation of metaphorical models,[4] Lakoff and Johnson's theory of metaphors,[5] and R. Harré's rhetorical proposal,[6] are part of the theoretical precursors of this paradigmatic change in the heart of the field of cognitive psychology. The closest precedents to the metaphorical models are the well known studies of "cognitive maps" by Tolman and Lynch (1960). Of particular interest in this line of study is the link between the metaphor of the mind as a cartographic map and the analogous model described by Black in *Models and Metaphors* (1966), which allows us to theorise in-depth about a kind of spatial reasoning whose most outstanding feature is its dynamic (material and procedural) and hermeneutic nature.

Running parallel to the idea of cognitive maps, aesthetic ontology refers us to the implantation carried out by human beings in their linguistic acts. The aesthetic subject establishes ontologically sensitive outlines which, rather than assisting in representing a city, serve instead to monumentalise individual and collective histories which are likely to confer an identity upon the subjects ("Me" taken as "Us") and groups so that their ontologically oriented "perspectives" can enable them to be aware of themselves and others. It may therefore be claimed that the metaphor constitutes a *monument* in discourse. Its former status as an ornament is shifted to that of its cognitive presence as a monument, representing the climax of all tales (*mythos*). Its objective is to restore that *illo tempore* when mankind began to be aware of its existence and explain its origins. The constant repetition of discursive monuments constitutes the drama of the origin of mankind, of the collective desire to survive. Aesthetic monuments that generate discourse through metaphorical utterances represent the abysmal base upon which the distinguishing collective identity rests.

Within this context we must not overlook the Heideggerian metaphor (or psychometaphor, to use my own term) of *lichtung* (luminosity, clarity, clarification), which represents a key shift in his thought. It is here that Heidegger reaches that climax which guarantees survival throughout time, with the same capacity as certain metaphorical utterances to become monuments of language and "onto-orienteering maps," which the subjects take mental note of for use as a means of acknowledgement and orientation in any society, in the *urban centre* of thought. *Lichtung* therefore constitutes the most important linguistic monument in the historical memory, a form of clarification which enables us to see things as they really are: "dark light" which does not attempt to blot out the "light," but merely to filter out the excess light whose brightness prevents an integral and profound vision.

Although it is, generally speaking, in line with our aims, A. Newell's *System of Physical Symbols* (1987) moves away from antithetical thought and toward the ideas of Heidegger. In parallel with classical hermeneutics and information science, it understands interpretation to be an "application" or "execution" of the act of accepting, at an early stage, an expression which describes a process and then the execution of the act.

In *Languages and Memory*, R. Schank claims that the concept of understanding (vital, as is well known, in hermeneutics) includes both the memory of the experience and the expectations generated by that memory. This definition includes clear evidence of the influence of hermeneutics, especially in its reference to "previous experiences," reactivating the concept of pre-understanding, namely the prior mechanism which enables us to interpret on the basis of familiarity and which consequently leads to understanding. According to Schank, understanding consists of locating the highest level structure available to explain input, in accordance with the existing model with which it shares certain characteristics within this high level structure. Understanding is therefore a process based on the recollection of closely linked experiences which are available through the memory, and that may be expressed in an analogous manner.[7] Worthy of note is the fact that this author continually insists that understanding is expressed through analogy, or in other words, through metaphors. (We must also recall that Aristotle considered analogy to be the fourth class of metaphor.)

Among the contributions made by psychology to the explanation of the metaphor, that of G. Lakoff and M. Johnson in their work *Metaphors We Live By* (1986) is of particular importance. It represents a magnificent contribution to the psychological theory of metaphors. Their work is the result of research into the human conceptual system which led them to claim that the central issue in cognitive science is the nature of conceptuality, which, in their opinion, is of an unquestionably metaphorical nature.

In the opinion of these authors, metaphorical concepts are not only understood and structured in their own terms, but also require further terms from other concepts, which involves conceptualising a class of objects or experiences in terms of another class of objects and experiences. Furthermore, they remind us that the famous machine metaphor, "the mind is a machine" and the hermeneutic metaphor, "understanding is seeing," both have an onto-structural character. As far as the idea of concept and metaphor is concerned, they state that abstract terms are defined metaphorically in terms of more than one specific concept, due to the fact that each metaphor can only describe certain aspects of an abstract concept, and therefore requires many metaphorical definitions, each of which would reflect part of the concept.

It is important to note their idea of the metaphor as complex experiential gestalts creating a kind of multi-dimensional structure which is built up from natural experiences, shaping our experience and providing it with a sense of coherence. Metaphorical concepts are based on the fact that the "B from A" gestalt forms a complex gestalt. Consequently, they claim that an "A is B" metaphor is an abbreviation of the projection of the structure of concept B on concept A.

Also worthy of mention within this theory, are the specific implications for the human conceptual system in considering that concepts are not only the result of our experiences, but are also structured upon dominant cultural metaphors, which implies that the origin of our concepts is based on both experience and the dominant metaphors in each culture.

Taking into account all of the above, it must be stated that the generally upheld theory holds that both experiential and cultural concepts (dominant metaphors) create experiential *gestalt*. They come into clear contact with hermeneutics by stating that what remains hidden are huge areas of human understanding. The hermeneutic importance of this statement is obvious, as it reminds us that the metaphor draws attention to certain aspects but shields others (precisely those which determine explicit features or characteristics).

The cognitive psychologist and scientific philosopher R. Harré is particularly worthy of mention here due to the contributions he has made to what I refer to as the *psychometaphor*. He takes the field of rhetoric as his starting point (a discipline used by Aristotle in his theories on the metaphor). His theories are developed in works such as *El ser social: una teoría para la psicología social* (1982) and *Sintaxis y estructura de la experiencia: la gramática de sí mismo* (1986). He appeals in the first instance to Austin's "speech acts,"[8] to put forward an ontology of conversation in which the *places* are human beings and acts of conversation, the *entities* the speech acts, and the *links* the rules, scripts and story lines.

Based on these ideas, Harré then proceeds to attempt a definition of the concept of "person." He feels that the human being possesses enough skill to offer two particular singularities:

a) Social identity: public manifestations based on individual reflexive beliefs and collective beliefs;
b) Personal identity: those corresponding to everyday discursive (metaphorical) practices. All individuals have a transpositionally singular discourse, due to their speech structure, which is updated through the use of pronouns and personal verb inflexions (1986, 24).

According to Harré, the first person (I) and the second person (you) are responsible for incorporating two crucial facts into speech acts (which in grammar is referred to as indexicality):

- where the person speaking in space/time is.
- Which place the person speaking occupies, in the order relevant to the speech act in question.

The use of the third person (he) represents an alternative to the name, and in grammar is referred to as anaphora. All these linguistic transpositions lead Harré to claim that in the discursive presentation of a human being, "three persons" coexist in a figurative sense in the same human body (1982, 98).

As can be deduced from the above statements, Harré reaches the following conclusion: thought is made up of a series of metaphors (linguistic transpositions). And the result of this is that the social world and its interpersonal relations are understood *in terms of* a theatre. This implies that all social events are seen through the filter of a performance, with its settings, scenery and characters. As occurs in the theatre, in our daily life we also have to learn our lines, since in our public life we behave mostly in a stylised (stereotyped) and repetitive manner. All of this implies coordinated style and action, just like in a play. If we took up Harré's suggestion to view our lives in terms of a theatre, we would be surprised to discover the multiple "roles" we adopt during the course of our daily lives. Each role has its own appropriate style: brusque, tyrannical, friendly, gentle, talkative, etc.

It is clear that Harré openly questions the belief that a person's identity is "fixed" and is a permanent possession. He attributes this mistaken belief to the attraction exerted by logical concepts, which explains their involuntary transposition in order to understand the range of personalities displayed. However, beneath this theory lies the heavy dependency of the personality (the represented role) on the person (the individual actor), the setting and the fellow players or deuteragonists.

Individuals experience an enormous range of situations and settings, and it becomes practically impossible to distinguish the "person" from the "character," in the sense that an individual shows antithetical traits depending on the role and setting he or she is in at the time, despite being the same "person." Essentially, it is a matter of accepting the analogous existence of various "persons" within a single body.

In our daily life there is a natural tendency to believe that when we are speaking about the personality of a specific individual we are unequivocally referring to a permanent (or substantial) feature of that person, without being aware of the falsity of such a belief. When we come across the same person time and time again, we feel that we are also coming into contact with the same character that they consistently display. However, according to Harré, the truth is that it is not the personality that becomes stabilised, but the "role" that is being played at any one particular time (1986, 27).

III

Following these reflections which are obviously intended to stimulate an in-depth debate into the onto-epistemological situation of the various "forms of psychology," I would like to state the following: in my opinion, in the future psychology will have an inevitably cross-discipline nature whose most immediate consequence will involve the introduction of key hermeneutic concepts, in order to shed a little light on the current state of confusion surrounding the social sciences, and psychology in particular.

The specific contributions of cognitive psychology (the metaphorical metaphors, Lakoff & Johnson, Harré, etc.) serve to remind us of the tentative first and undoubtedly unpremeditated steps towards a new conceptual order which may act as a launching pad for psychology as a whole into the new millennium, and for a new *era* whose leading player should be the psychological metaphor (the psychometaphor), as it represents the "intersection" where scientific tradition (psychology) and humanistic tradition (hermeneutics) merge into each other. The metaphor is, by its own structure and definition, the "transposition of the name of one thing used to describe another," a crossroads for various disciplines, and the ideal place from which to rethink the methods and strategies of psychology. It is hybrid in nature, due to the fact that its principle aim is to concepts. When considered "individually," the metaphor occupies a leading role; when considered "supraindividually," the most noteworthy element is the "hermeneutic praxis," whose mission is to link scientific experience (psychology) with the sensorial experience of everyday life (*Lebenswelt*) through language, which is precisely the place in which the synthesis between the individual and nature occurs.

Therefore, the metaphor represents a hermeneutic "vision of the whole" which provides psychology with the opportunity to effect a metaparadigmatic reorientation. In other words, the concept of a paradigm would be seen in this context as a counterphatic generalistic ideal, and as a set of transpositional forms and structures.

The metaphor would also provide a renewed concept of truth. It is evident that the recurring crises are not limited exclusively to the field of psychology, but are merely a symptom of the true crisis, namely the "concept of truth" itself, leading to paradigmatic instability in the area of social sciences, which are those with which we are concerned and also the most sensitive to transmutations in the various spheres of social knowledge.

The type of truth contained in the metaphor does not form part of an exhaustive apodeictic logic, which is as irrefutable as it is unnecessary, but is rather part of an aesthetic expression, which, despite its subjectivity, also contains a sense of worth and universality that can be transmitted through the use of metaphors and transpositions with clearly visible and contrastable structures.

The metaphor, as "the most consonant dissonance," is the anomaly behind every utterance that breaks the paradigmatic unity with the precise intention of creating a "superior normality." This enables the most unexpected event to be included in a cross-discipline series which is hermeneutically aware of the fact that any change in the basic "external" conceptual constellation implies a simultaneous "internal" change in us. In my opinion, this leads us into a new dimension of psychology, which, from the explanatory and comprehensive nucleus of the metaphor, moves us into a sphere of knowledge that will immediately wipe out the blurred boundaries between disciplines. If this proves to be the case, then the future of psychology as a whole will be hermeneutic, cross-disciplinary and metaphorical. Or perhaps it won't.

University of La Coruña, Spain

NOTES

[1] Cf. Husserl, E. (1934), *Die Krisis der europäuschen Wissenschaften un die transzendentale Phänomenologie.*

[2] Cf. Husserl, E. (1953), *Philosophischen Untersuchungen.*

[3] Gadamer, H. G. (1960) *Warheit und Methode. Grundzüge einer phisopischen Hermeneutik.*

[4] I am referring to the following authors and their works: Ortony (1979) *Metaphor in Science*. Cambridge University Press / Tourangeau, R. and Stenberg, R. J. (1982) Understanding and Appreciating Metaphors. *Cognition*, 2, 203–204.

[5] Lakoff, G. and Johnson, M. (1986) *Metaphors We Live By* Madrid: Cátedra.

[6] Cf. Harré R. (1973) *Introducción a la lógica de las ciencias sociales*. Madrid: Labor/ Harré R. (1982) El Ser social: una teoría para la psicología social. Madrid: Alianza. Harré R. & Clarke, R. (1984) *Motivos y mecanismos. Introducción a la psicología de la acción*. Barcelona: Paidós. / Harré R. (1986) Sintaxis y estructura de la experiencia: la gramática y el sí mismo. *Revista de Occidente*, 56.

[7] Of particular interest is G. Bueno's conceptualisation of analogy (1970). He considers it to be a particular and simplified case of homomorphism which leads him to speak of homomorphic metaphors present at the centre of the Cartesian method where the need to represent even odours in spatial terms is necessary (Cf. Trías, E. *Metodología del pensamiento mágico* (Introduction by G. Bueno. Ed Edhass, 1970).

[8] Cf. Austin J. (1962) Performatif – Constatif. *Philosophie Analitique*, 271–281.

INDEX OF NAMES

Analecta Husserliana

The Yearbook of Phenomenological Research

Editor-in-Chief

Anna-Teresa Tymieniecka

The World Institute for Advanced Phenomenological Research and Learning, Belmont, Massachusetts, U.S.A.

1. Tymieniecka, A-T. (ed.), *Volume 1 of Analecta Husserliana.* 1971 ISBN 90-277-0171-7
2. Tymieniecka, A-T. (ed.), *The Later Husserl and the Idea of Phenomenology.* Idealism ñ Realism, Historicity and Nature. 1972 ISBN 90-277-0223-3
3. Tymieniecka, A-T. (ed.), *The Phenomenological Realism of the Possible Worlds.* The ìA Priorií, Activity and Passivity of Consciousness, Phenomenology and Nature. 1974 ISBN 90-277-0426-0
4. Tymieniecka, A-T. (ed.), *Ingardeniana.* A Spectrum of Specialised Studies Establishing the Field of Research. 1976 ISBN 90-277-0628-X
5. Tymieniecka, A-T. (ed.), *The Crisis of Culture.* Steps to Reopen the Phenomenological Investigation of Man. 1976 ISBN 90-277-0632-8
6. Tymieniecka, A-T. (ed.), *The Self and the Other.* The Irreducible Element in Man, Part I. 1977 ISBN 90-277-0759-6
7. Tymieniecka, A-T. (ed.), *The Human Being in Action.* The Irreducible Element in Man, Part II. 1978 ISBN 90-277-0884-3
8. Nitta, Y. and Hirotaka Tatematsu (eds.), *Japanese Phenomenology.* Phenomenology as the Trans-cultural Philosophical Approach. 1979 ISBN 90-277-0924-6
9. Tymieniecka, A-T. (ed.), *The Teleologies in Husserlian Phenomenology.* The Irreducible Element in Man, Part III. 1979 ISBN 90-277-0981-5
10. Wojty a, K., *The Acting Person.* Translated from Polish by A. Potocki. 1979 ISBN Hb 90-277-0969-6; Pb 90-277-0985-8
11. Ales Bello, A. (ed.), *The Great Chain of Being* and *Italian Phenomenology.* 1981 ISBN 90-277-1071-6
12. Tymieniecka, A-T. (ed.), *The Philosophical Reflection of Man in Literature.* Selected Papers from Several Conferences held by the International Society for Phenomenology and Literature in Cambridge, Massachusetts. Includes the essay by A-T. Tymieniecka, *Poetica Nova.* 1982 ISBN 90-277-1312-X
13. Kaelin, E. F., *The Unhappy Consciousness.* The Poetic Plight of Samuel Beckett. An Inquiry at the Intersection of Phenomenology and literature. 1981 ISBN 90-277-1313-8
14. Tymieniecka, A-T. (ed.), *The Phenomenology of Man and of the Human Condition.* Individualisation of Nature and the Human Being. (Part I:) Plotting the Territory for Interdisciplinary Communication. 1983 *Part II* see below under Volume 21. ISBN 90-277-1447-9

Analecta Husserliana

15. Tymieniecka, A-T. and Calvin O. Schrag (eds.), *Foundations of Morality, Human Rights, and the Human Sciences.* Phenomenology in a Foundational Dialogue with Human Sciences. 1983 ISBN 90-277-1453-3
16. Tymieniecka, A-T. (ed.), *Soul and Body in Husserlian Phenomenology.* Man and Nature. 1983 ISBN 90-277-1518-1
17. Tymieniecka, A-T. (ed.), *Phenomenology of Life in a Dialogue Between Chinese and Occidental Philosophy.* 1984 ISBN 90-277-1620-X
18. Tymieniecka, A-T. (ed.), *The Existential Coordinates of the Human Condition: Poetic – Epic – Tragic.* The Literary Genre. 1984 ISBN 90-277-1702-8
19. Tymieniecka, A-T. (ed.), *Poetics of the Elements in the Human Condition.* (Part 1:) The Sea. From Elemental Stirrings to Symbolic Inspiration, Language, and Life-Signi cance in Literary Interpretation and Theory. 1985
    For Part 2 and 3 *see below* under Volumes 23 and 28. ISBN 90-277-1906-3
20. Tymieniecka, A-T. (ed.), *The Moral Sense in the Communal Significance of Life.* Investigations in Phenomenological Praxeology: Psychiatric Therapeutics, Medical Ethics and Social Praxis within the Life- and Communal World. 1986
    ISBN 90-277-2085-1
21. Tymieniecka, A-T. (ed.), *The Phenomenology of Man and of the Human Condition.* Part II: The Meeting Point Between Occidental and Oriental Philosophies. 1986 ISBN 90-277-2185-8
22. Tymieniecka, A-T. (ed.), *Morality within the Life- and Social World.* Interdisciplinary Phenomenology of the Authentic Life in the ì Moral Senseí. 1987
    Sequel to Volumes 15 and 20. ISBN 90-277-2411-3
23. Tymieniecka, A-T. (ed.), *Poetics of the Elements in the Human Condition.* Part 2: The Airy Elements in Poetic Imagination. Breath, Breeze, Wind, Tempest, Thunder, Snow, Flame, Fire, Volcano 1988 ISBN 90-277-2569-1
24. Tymieniecka, A-T., *Logos and Life*. Book I: Creative Experience and the Critique of Reason. 1988 ISBN Hb 90-277-2539-X; Pb 90-277-2540-3
25. Tymieniecka, A-T., *Logos and Life*. Book II: The Three Movements of the Soul. 1988 ISBN Hb 90-277-2556-X; Pb 90-277-2557-8
26. Kaelin, E. F. and Calvin O. Schrag (eds.), *American Phenomenology.* Origins and Developments. 1989 ISBN 90-277-2690-6
27. Tymieniecka, A-T. (ed.), *Man within his Life-World.* Contributions to Phenomenology by Scholars from East-Central Europe. 1989 ISBN 90-277-2767-8
28. Tymieniecka, A-T. (ed.), *The Elemental Passions of the Soul.* Poetics of the Elements in the Human Condition, Part 3. 1990 ISBN 0-7923-0180-3
29. Tymieniecka, A-T. (ed.), *Man's Self-Interpretation-in-Existence.* Phenomenology and Philosophy of Life. ñ Introducing the Spanish Perspective. 1990
    ISBN 0-7923-0324-5
30. Rudnick, H. H. (ed.), *Ingardeniana II.* New Studies in the Philosophy of Roman Ingarden. With a New International Ingarden Bibliography. 1990
    ISBN 0-7923-0627-9

Analecta Husserliana

31. Tymieniecka, A-T. (ed.), *The Moral Sense and Its Foundational Significance: Self, Person, Historicity, Community.* Phenomenological Praxeology and Psychiatry. 1990 ISBN 0-7923-0678-3
32. Kronegger, M. (ed.), *Phenomenology and Aesthetics.* Approaches to Comparative Literature and Other Arts. Homages to A-T. Tymieniecka. 1991 ISBN 0-7923-0738-0
33. Tymieniecka, A-T. (ed.), *Ingardeniana III.* Roman Ingardenís Aesthetics in a New Key and the Independent Approaches of Others: The Performing Arts, the Fine Arts, and Literature. 1991
Sequel to Volumes 4 and 30 ISBN 0-7923-1014-4
34. Tymieniecka, A-T. (ed.), *The Turning Points of the New Phenomenological Era.* Husserl Research ñ Drawing upon the Full Extent of His Development. 1991 ISBN 0-7923-1134-5
35. Tymieniecka, A-T. (ed.), *Husserlian Phenomenology in a New Key.* Intersubjectivity, Ethos, the Societal Sphere, Human Encounter, Pathos. 1991 ISBN 0-7923-1146-9
36. Tymieniecka, A-T. (ed.), *Husserl's Legacy in Phenomenological Philosophies.* New Approaches to Reason, Language, Hermeneutics, the Human Condition. 1991 ISBN 0-7923-1178-7
37. Tymieniecka, A-T. (ed.), *New Queries in Aesthetics and Metaphysics.* Time, Historicity, Art, Culture, Metaphysics, the Transnatural. 1991 ISBN 0-7923-1195-7
38. Tymieniecka, A-T. (ed.), *The Elemental Dialectic of Light and Darkness.* The Passions of the Soul in the Onto-Poiesis of Life. 1992 ISBN 0-7923-1601-0
39. Tymieniecka, A-T. (ed.), *Reason, Life, Culture, Part I.* Phenomenology in the Baltics. 1993 ISBN 0-7923-1902-8
40. Tymieniecka, A-T. (ed.), *Manifestations of Reason: Life, Historicity, Culture.* Reason, Life, Culture, Part II. Phenomenology in the Adriatic Countries. 1993 ISBN 0-7923-2215-0
41. Tymieniecka, A-T. (ed.), *Allegory Revisited.* Ideals of Mankind. 1994 ISBN 0-7923-2312-2
42. Kronegger, M. and Tymieniecka, A-T. (eds.), *Allegory Old and New.* In Literature, the Fine Arts, Music and Theatre, and Its Continuity in Culture. 1994 ISBN 0-7923-2348-3
43. Tymieniecka, A-T. (ed.): *From the Sacred to the Divine.* A New Phenomenological Approach. 1994 ISBN 0-7923-2690-3
44. Tymieniecka, A-T. (ed.): *The Elemental Passion for Place in the Ontopoiesis of Life.* Passions of the Soul in the *Imaginatio Creatrix.* 1995 ISBN 0-7923-2749-7
45. Zhai, Z.: *The Radical Choice and Moral Theory.* Through Communicative Argumentation to Phenomenological Subjectivity. 1994 ISBN 0-7923-2891-4
46. Tymieniecka, A-T. (ed.): *The Logic of the Living Present.* Experience, Ordering, Onto-Poiesis of Culture. 1995 ISBN 0-7923-2930-9

Analecta Husserliana

47. Tymieniecka, A-T. (ed.): *Heaven, Earth, and In-Between in the Harmony of Life.* Phenomenology in the Continuing Oriental/Occidental Dialogue. 1995 ISBN 0-7923-3373-X
48. Tymieniecka, A-T. (ed.): *Life. In the Glory of its Radiating Manifestations.* 25th Anniversary Publication. Book I. 1996 ISBN 0-7923-3825-1
49. Kronegger, M. and Tymieniecka, A-T. (eds.): *Life. The Human Quest for an Ideal.* 25th Anniversary Publication. Book II. 1996 ISBN 0-7923-3826-X
50. Tymieniecka, A-T. (ed.): *Life. Phenomenology of Life as the Starting Point of Philosophy.* 25th Anniversary Publication. Book III. 1997 ISBN 0-7923-4126-0
51. Tymieniecka, A-T. (ed.): *Passion for Place. Part II.* Between the Vital Spacing and the Creative Horizons of Ful lment. 1997 ISBN 0-7923-4146-5
52. Tymieniecka, A-T. (ed.): *Phenomenology of Life and the Human Creative Condition.* Laying Down the Cornerstones of the Field. Book I. 1997 ISBN 0-7923-4445-6
53. Tymieniecka, A-T. (ed.): *The Reincarnating Mind, or the Ontopoietic Outburst in Creative Virtualities.* Harmonisations and Attunement in Cognition, the Fine Arts, Literature. Phenomenology of Life and the Human Creative Condition. Book II. 1997 ISBN 0-7923-4461-8
54. Tymieniecka, A-T. (ed.): *Ontopoietic Expansion in Human Self-Interpretation-in-Existence.* The I and the Other in their Creative Spacing of the Societal Circuits of Life. Phenomenology of Life and the Creative Condition. Book III. 1997 ISBN 0-7923-4462-6
55. Tymieniecka, A-T. (ed.): *Creative Virtualities in Human Self-Interpretation-in-Culture.* Phenomenology of Life and the Human Creative Condition. Book IV. 1997 ISBN 0-7923-4545-2
56. Tymieniecka, A-T. (ed.): *Enjoyment.* From Laughter to Delight in Philosophy, Literature, the Fine Arts and Aesthetics. 1998 ISBN 0-7923-4677-7
57. Kronegger M. and Tymieniecka, A-T. (eds.): *Life. Differentiation and Harmony... Vegetal, Animal, Human.* 1998 ISBN 0-7923-4887-7
58. Tymieniecka, A-T. and Matsuba, S. (eds.): *Immersing in the Concrete.* Maurice Merleau-Ponty in the Japanese Perspective. 1998 ISBN 0-7923-5093-6
59. Tymieniecka, A-T. (ed.): *Life - Scientific Philosophy/Phenomenology of Life and the Sciences of Life.* Ontopoiesis of Life and the Human Creative Condition. 1998 ISBN 0-7923-5141-X
60. Tymieniecka, A-T. (eds.): *Life - The Outburst of Life in the Human Sphere.* Scienti c Philosophy / Phenomenology of Life and the Sciences of Life. Book II. 1998 ISBN 0-7923-5142-8
61. Tymieniecka, A-T. (ed.): *The Aesthetic Discourse of the Arts.* Breaking the Barriers. 2000 ISBN 0-7923-6006-0
62. Tymieniecka, A-T. (ed.): *Creative Mimesis of Emotion.* From Sorrow to Elation; Elegiac Virtuosity in Literature. 2000 ISBN 0-7923-6007-9

Analecta Husserliana

63. Kronegger, M. (ed).: *The Orchestration of The Arts ñ A Creative Symbiosis of Existential Powers.* The Vibrating Interplay of Sound, Color, Image, Gesture, Movement, Rhythm, Fragrance, Word, Touch. 2000 ISBN 0-7923-6008-7
64. Tymieniecka, A-T. and Z. Zalewski (eds.): *Life - The Human Being Between Life and Death.* A Dialogue Between Medicine and Philosophy, Recurrent Issues and New Approaches. 2000 ISBN 0-7923-5962-3
65. Kronegger, M. and Tymieniecka, A-T. (eds.): *The Aesthetics of Enchantment in the Fine Arts.* 2000 ISBN 0-7923-6183-0
66. Tymieniecka, A-T. (ed.): *The Origins of Life, Volume I: The Primogenital Matrix of Life and Its Context.* 2000 ISBN 0-7923-6246-2; Set ISBN 0-7923-6446-5
67. Tymieniecka, A-T. (ed.): *The Origins of Life, Volume II: The Origins of the Existential Sharing-in-Life.* 2000 ISBN 0-7923-6276-4; Set ISBN 0-7923-6446-5
68. Tymieniecka, A-T. (ed.): *PAIDEIA.* Philosophy / Phenomenology of Life Inspiring Education of our Times. 2000 ISBN 0-7923-6319-1
69. Tymieniecka, A-T. (ed.): *The Poetry of Life in Literature.* 2000 ISBN 0-7923-6408-2
70. Tymieniecka, A-T. (ed.): *Impetus and Equipoise in the Life-Strategies of Reason.* Logos and Life, volume 4. 2000 ISBN 0-7923-6731-6; HB 0-7923-6730-8
71. Tymieniecka, A-T. (ed.): *Passions of the Earth in Human Existence, Creativity, and Literature.* 2001 ISBN 0-7923-6675-1
72. Tymieniecka, A-T. and E. Agazzi (eds.): *Life ñ Interpretation and the Sense of Illness within the Human Condition.* Medicine and Philosophy in a Dialogue. 2001 ISBN Hb 0-7923-6983-1; Pb 0-7923-6984-X
73. Tymieniecka, A-T. (ed.): *Life ñ The Play of Life on the Stage of the World in Fine Arts, Stage-Play, and Literature.* 2001 ISBN 0-7923-7032-5
74. Tymieniecka, A-T. (ed.): *Life-Energies, Forces and the Shaping of Life: Vital, Existential.* Book I. 2002 ISBN 1-4020-0627-6
75. Tymieniecka, A-T. (ed.): *The Visible and the Invisible in the Interplay between Philosophy, Literature and Reality.* 2002 ISBN 1-4020-0070-7
76. Tymieniecka, A-T. (ed.): *Life ñ Truth in its Various Perspectives.* Cognition, Self-Knowledge, Creativity, Scientific Research, Sharing-in-Life, Economics...... 2002 ISBN 1-4020-0071-5
77. Tymieniecka, A-T. (ed.): *The Creative Matrix of the Origins.* Dynamisms, Forces and the Shaping of Life. 2003 ISBN 1-4020-0789-2
78. Tymieniecka, A-T. (ed.): *Gardens and the Passion for the In nite .* 2003 ISBN 1-4020-0858-9
79. to be published.
80. Tymieniecka, A-T. (ed.): *Phenomenology World Wide.* Foundations - Expanding dynamics - Life-engagements. A Guide for Research and Study. 2002 ISBN 1-4020-0066-9

www.ingramcontent.com/pod-product-compliance
Ingram Content Group UK Ltd.
Pitfield, Milton Keynes, MK11 3LW, UK
UKHW020135160926
475607UK00003B/46

9781402008580